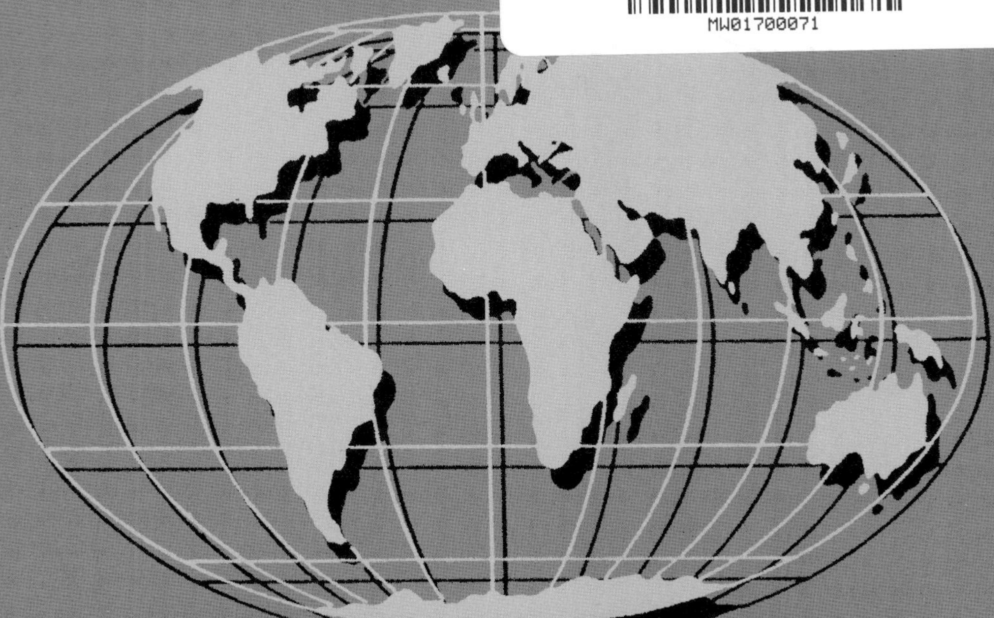

ENGINEER ...Around the World in Fifty Years

C.O. BRAWNER P.Eng.

Published by BiTech Publishers Ltd.
173 - 11860 Hammersmith Way
Richmond, British Columbia, Canada V7A 5G1

Editorial services by Jennifer and Ric Eeairsto,
Laughing Mountain Communications, Inc.,
Vancouver, BC.

Design by George McLachlan,
North Vancouver, BC.

Production by Lynn Pugh,
BiTech Publishers Ltd.,
Richmond, BC

Printing and Binding by D.W. Friesen,
Altona, Manitoba

Printed in Canada
First Printing, October 2003

Canadian Cataloguing in Publication Data

Brawner, C.O. (Carroll Oliver)
Engineer ...Around the World in Fifty Years/C.O Brawner

ISBN 0-921095-62-7

1. Engineering geology — Case studies. 2. Mining engineering — Case studies.
3. Transportation engineering — Case studies. 4. Landslides — Case studies.
5. Brawner, C.O. (Carroll Oliver) 6. Engineers — Biography. I. Title

TA705B72 2003 624.1'51 C2003-905710-0

I wish to dedicate this book first to my family, Patricia, Catherine, Randall, Suzanne and Gregory who accepted my extensive travelling with understanding and secondly to the great engineering partners at Golder Brawner and Associates — David Campbell, Glen Gilchrist, Norman McCammon, David Pentz and Robert Wilson and the many engineers, technicians and administrative staff.

The Author

C. O. Brawner, P. Eng., received a Bachelor of Science degree in Civil Engineering from the University of Manitoba in 1953 and a Master of Science degree in Civil (Geotechnical) Engineering in 1958 from the Nova Scotia Technical University, studying under Dr. G. G. Meyerhof.

Brawner spent ten years as a Soil Engineer and Senior Materials Engineer with the British Columbia Department of Highways. In 1963 he joined Golder Associates and became President of Golder Brawner and Associates. He developed that consulting group into one of the world leaders in geotechnical engineering for transportation and mining, involved in major local, national and international projects.

Later he taught Geomechanics in Mining, Geological and Civil Engineering at the University of British Columbia. At the same time he provided specialist Review

Consulting on challenging projects around the world under C. O. Brawner Engineering Ltd. He was a member of numerous Project Review Boards.

He has been a member of many Canadian and U.S. geotechnical committees on muskeg engineering, soil mechanics, mine waste structures and rock mechanics. He has chaired nine international conferences on Geotechnology in Mining and edited eleven conference proceedings. Brawner has been a keynote speaker at eight international conferences.

Brawner has also been active in geotechnical research. He developed the use of preconsolidation of muskeg for major freeway construction, the first use of roller compacted soil cement in Canada for slope protection, slope monitoring systems in surface mining, vacuum horizontal drains and drainage tunnels for stabilization of large soil and rock slides, and the drop raise procedure to fill large underground glory holes to allow safe open pit mining. In addition, he was co-developer of the dredged sand island procedure to provide drill platforms for oil and gas drilling in the Arctic Ocean.

The author of over eighty technical papers, Brawner has been an invited lecturer at over ninety universities and institutes internationally, including the United States, China, Japan and Russia.

He has received numerous awards. They include the highest award of the American Society of Mining Engineers, the Daniel C. Jackling Award, and, of the Canadian Geotechnical Society, the R. F. Legget Award. Brawner also received the President's Medal from the Canadian Good Roads Association, the Award of Merit from the Association of Consulting Engineers of Canada, and the Meritorious Achievement Award from the Professional Engineers Association of B.C. He is a Distinguished Member of the American Society of Mining Engineers, and received the McParland Memorial Medal from the Canadian Institute of Mining and Metallurgy.

Brawner has been elected a Fellow of the Engineering Institute of Canada, the Canadian Institute of Mining and Metallurgy, and the Canadian Academy of Engineers. He also received the Walter Gage Teaching Award in 1981 from the University of British Columbia.

Brawner has provided consulting engineering in over forty countries around the world, on all continents including Antarctica.

Contents

Preface

This book is about Engineering – Geotechnical Engineering. But it is different than other engineering books. It is not full of formulas, drawings, graphs, designs and specifications. Rather, it contains everyday experiences of a practicing engineer who has had the great opportunity to have worked on a wide range of challenging projects in many countries around the world. This has included working with many great engineers and a wide variety of organizations.

Experiences are described in storybook style with simplified engineering description to be understandable by any moderately technical person.

Major emphasis of the seventy-one case examples has been on landslides, rock slides, transportation geotechnology and geotechnical engineering in surface mining. Each example illustrates at least one important lesson.

The book is not chronological. Rather, projects and experience are described geographically by state, country or region. The specific approach to engineering, business ethics, moral codes, customs, labour conditions and costs, and professional experience, *et cetera*, differs widely around the globe. International consulting engineers, to be successful, must come to grips with all these considerations. Most situations are highly professional, but some are humorous, some are frustrating (the tomorrow syndrome) and a few can be frightening.

Engineering is more than a profession. It can be a lifelong trail of challenges, and for the most part, one of satisfaction. In geotechnical engineering, every major project has its unique problems. There is seldom boredom. There are often surprises.

The successful engineer must develop a wide range of skills. They include honesty, integrity, technical ability, communication capability, financial responsibility, managerial prowess, compassion and flexibility. Where travel is frequent and extensive, considerable stress can be placed on wife, family and health.

University education to learn the fundamentals of engineering is essential. Post secondary engineering programs are comprehensive and difficult. Unfortunately, the typical four-year bachelor's degree program does not include training in business and financial management, human resources, environmental awareness, engineering law and, most importantly, project supervision. Comprehensively successful engineering involves, "Management of Technology." This includes utilizing past experience, the requirements of today and, often not adequately accounted for, consideration and design for the future.

Engineering has made tremendous contributions to the development on our planet. Much of this is taken for granted. Much has received little appreciation. The reasons for this book are to illustrate the challenge and satisfaction of engineering as a career, and to challenge engineers in all related fields to tell their story. Engineering has had many successes. They deserve to be told.

1

To Be an Engineer

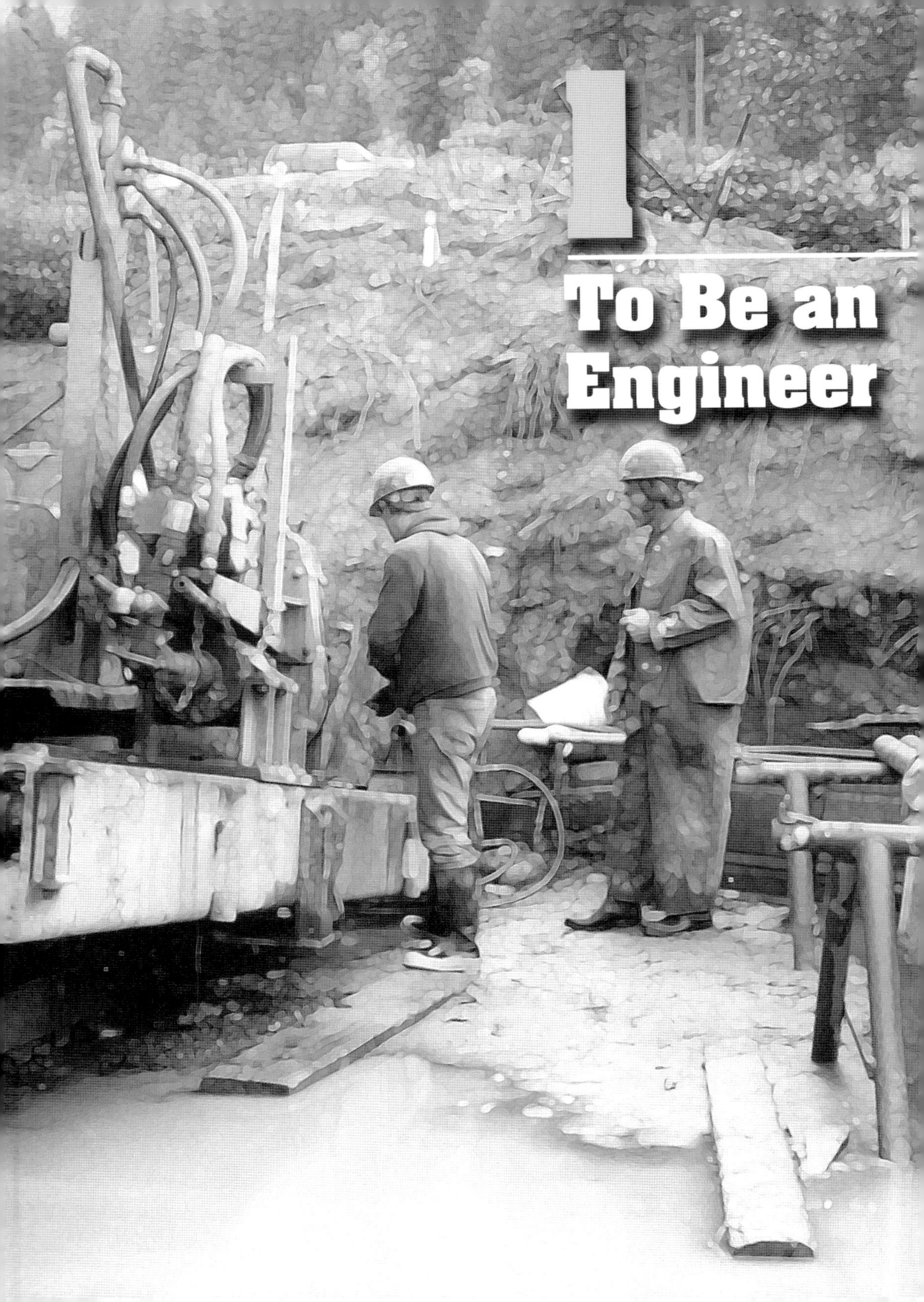

To Be An Engineer
(up to 1953)

In this introductory chapter, the author describes his early years growing up on an Okanagan fruit farm and attending the local high school. When his early interest in architecture was scuttled due to colour blindness, the Dean of the School of Architecture at the University of Manitoba recommended that he pursue engineering, in retrospect highly fortuitous advice. This was life-break number one.

When the author was involved in various hijinks at university, including a meticulously planned and executed raid on the womens' residence, he recognized the need for discipline when facing the pressure of studies. The observation of a few simple strategies is suggested. One of the most important is to avoid falling behind in completing assignments. University experience can be some of the best years in one's life. Enjoy it in full.

It is also important to develop communication skills – become "a people person." It is recommended that the prospective engineer join the Toastmasters organization or take a public speaking course.

In his last year of university, Mr. Brawner joined his classmates in seeking his first job, planning carefully for employment interviews. He accepted a job with the BC government Water Rights Branch, and with two summers of experience, was promised the position of Assistant District Engineer.

Chapter 1
To Be an Engineer

I was raised on a fruit farm in the beautiful Okanagan Valley area of British Columbia, Canada, near the community of Summerland. Our farm was three miles north of town in Garnet Valley. To the east was Rattlesnake Mountain, a picturesque peak with a steep rock cliff at the crest and pine trees, sagebrush and cactus on the slopes below. And yes, just as the name suggests, there were also rattlesnakes.

While money was scarce on our farm, there were plentiful opportunities for learning life's ways. Living off the land was an essential. Our orchard provided a great range of fruit — cherries, apricots, peaches, plums, prunes, pears and several varieties of apples. Our two-acre garden was a variety store — peas, beans, carrots, cabbage, lettuce, radishes, cauliflower, cucumbers, squash, tomatoes and potatoes. I contributed to all of them, planting, weeding, cultivating, thinning and harvesting. I also helped my mother with canning for the winter. We didn't have a refrigerator; instead we had an underground root cellar where vegetables like carrots, parsnips, turnips and potatoes were each stored in a large box of dry sand.

I had my own trap line, sort of. Five special traps to catch moles, animals like gophers with cheek pockets to carry food. Moles burrowed under the ground and ate every root they encountered, including those of our fruit trees. Some trees became so badly derooted that they fell over during a strong wind. It was always very disheartening to see a forty-year-old tree, which the year before had yielded bushels of apples, lying on the ground, never to produce again.

Rattlesnake Mountain directly overlooking our house in Garnet Valley, Summerland, B.C. Photo taken in February.

Looking northeast over the outskirts of Summerland, Okanagan Lake and the interior mountains in south-central British Columbia. A very beautiful area of Canada.

We had a team of horses that my father and I drove for plowing, developing irrigation ditches, haying and hauling to and from the orchard. We also had a cow that my younger brother Ken and I took turns milking, and once a week I churned the cream to be separated for butter.

A creek flowed through our property, and I was the family fisherman. With a willow rod, ten feet of fishing line and a hook, I often caught several eight- to twelve-inch trout for breakfast or dinner. I also hunted deer with my father in the local mountains to provide winter meat. Every year we raised one or two pigs, which I helped to feed and finally butcher.

Each spring my mother ordered fifty unsexed baby chicks. When they were fully grown the hens were kept to supply eggs and the roosters for Sunday dinner. I would pick one of the larger roosters and train him to fight. He would chase strangers, cats and dogs. My own special pet.

Mother was chief shopper and barterer. Every Saturday in season, we would take an assortment of vegetables, eggs, butter and fruit to the store where she would barter for household staples. Ken and I were each given ten cents to attend the movie matinee in the town theatre.

We took the school bus every weekday during school term. Seating arrangements were still traditional. Boys sat on one side of the bus and girls on the other.

Farm life taught responsibility, frugality, and respect for nature and the environment. Each year my dad and I felled two large pine trees on our property, then cut them up for firewood. Many a Saturday I pulled on one end of a six-foot cross cut saw, with my father at the far end.

My mother played a major role in my life. She was an excellent cook, kept our clothing clean and in repair, tended to our cuts and bruises, and saw that we did our school homework. In spite of suffering from asthma, she always had a ready smile or a pat on the back. She taught us how to use the sewing machine, iron our shirts, churn cream for butter and make our lunches for school.

Despite all the farm bounty, and despite the fact that my father worked hard from sunup to sundown in the orchard, our ten acres did not provide a living. As a result, Dad periodically worked as a carpenter's helper for local contractors and the municipality of Summerland, or for other orchardists with our horses.

Dad's education was limited — up to grade eight. In his late teens he had broken horses to ride or pull farm wagons and equipment. He impressed on me the virtues of honesty and hard work, often saying to my brother and me, "Always finish the job, keep your promises and never be late." While my parents were always pressed for money, we never went hungry.

At school I was slightly above average academically, generally a B-student. I enjoyed math in which I earned A's, but had difficulty with English literature, particularly Shakespeare, where I struggled to obtain a C. I also got an A in Industrial Arts where I made a number of furniture items and excelled at drafting. So much so that Mr. Feltham, the instructor, encouraged me to take up Architecture. This became my goal from grade eight through high school.

I got along well with my classmates, and enjoyed sports. I played Little League, Midget and Junior baseball — one year we won the provincial title. I wasn't big enough to hit home runs but I developed a secret weapon. I had the lightest and shortest bat on the team. As a result I had a fast swing and hit a lot of line drive singles and doubles. I also played hockey, which I originally learned when we were living in Saskatchewan and Alberta, prior to moving to B.C.

I had a semester as student president in grade twelve, was co-editor of the high school paper, *The Campus Chaff*, was class valedictorian for graduation, and a councillor with our "Teen Town" group. Life was good at school except for two episodes, both of which nearly cost me my grade twelve.

We had one study class each day when we were supposed to do homework or prepare for exams. A teacher was in the room to maintain discipline. On one such occasion we had Miss Walker, a new first-year teacher in her early twenties. She was blonde and beautiful, making it hard to stay focused on studying.

I sat in the back of the room with three of my buddies, Ray Moore, Norman Verrier and Jim Hack. We were whispering about Miss Walker, and Jim Hack said, "I'd sure like to take her on a date. I wonder if she'd go to a show with me?" Hack was good looking, tall, athletic, clean-cut and eighteen. I bet Moore and Verrier ten cents each that she would. Ten cents would be three dollars today. We teased Hack until he decided to go for it and raised his hand.

Miss Walker said, "Yes, Hack?" I think she expected he was going to ask her to let him go to the washroom. Instead he said, "I'm having trouble with some words. Could you help me?" She looked a bit startled, but got up slowly and came down the aisle. Hack asked her if she could explain a sentence in a poem he had to study. She read the sentence and gave him her interpretation. He said, "Thank you very much." Then he stood up and asked, "Would you go with me to the movie in town tomorrow night?"

She stared at him. Moore, Verrier and I all prompted her, "Go with him!" We should have kept our mouths shut. She obviously picked up on our teenage conspiracy. Furious, she ordered firmly, "All four of you — see me in my room immediately after classes today!"

After classes, Miss Walker assigned us a chapter in our Social Studies text, about twelve pages, to write out in its entirety and hand in to her on Monday.

I laboured over the weekend and arrived at school Monday morning with the completed assignment, but after the other three offenders. I saw Hack first. He said he had handed his in. I then saw Moore and Verrier, who said they had not. What should I do? My teenage ego and reputation were at stake. Moore was my closest buddy; I felt I couldn't let him down. I advised Miss Walker I hadn't done the assignment. She immediately reported the three of us to the principal, A.K. MacLeod.

He called us into his office, described what Miss Walker had said we had and had not done, then told us in no uncertain terms that we had disgraced ourselves by acting so stupidly and embarrassing Miss Walker. He had brought another teacher, Clark Wilkins, with him as an observer. We soon learned why. The principal pulled out a thick leather strap and said, "Moore, you're first. Hold out your hand." Moore did so. MacLeod swung the strap with a full swing. Moore winced. MacLeod swung three more times, followed by four on the other hand. Beads of sweat broke out on Moore's forehead. But he didn't make a sound.

Next it was my turn. Strike one! A sharp pain shot from my hand all the way up my arm. My hand became numb. I hardly felt the other three whacks. The other hand suffered the same sharp pain followed by numbness. I flinched but didn't yell. Verrier didn't do too well. He flinched and yelled several times and realized later he had wet himself. Afterwards, MacLeod reprimanded us sternly, "Do anything like this again and you're gone." We presumed he meant expelled from school.

My hand was swollen for several days. I never told my dad. He had warned many times, "If you ever get the strap in school, you will get it again at home," and I sure didn't want it a second time. My ego wasn't that big.

Another episode occurred the following semester. I was co-editor of *The Campus Chaff* school paper with Valerie Tingley. She was an attractive, intelligent and friendly girl to work with. We shared many laughs putting the student personals together. Frequently, when we were preparing copy, we would end up with about a quarter of a page or less unfilled. We referred to past issues by previous classes and observed that short jokes were used to fill the page. An obvious source was the *Reader's Digest,* which used the same procedure. We weren't aware of copyright laws so took total poetic license.

While working on one issue I needed about eight to ten lines to fill one final page. I decided on the following joke, which I thought was cute:

It was in October and Johnny's mother was looking ahead to Christmas. She asked Johnny, her six-year-old son, "What would you like for Christmas, Johnny?"

He thought for a few moments, then replied, "I would like a baby brother!"

She was surprised but answered, "I'm sorry; that doesn't give us enough time!"

Without hesitating Johnny responded, "Mom, put some more men on the job!"

The school paper was available midmorning next day. Three piles were placed along the main hallway for the students to pick up.

At the end of the next class Mr. MacLeod intercepted me and called me into his office. Red-faced with anger, he shoved the paper in front of my face and questioned, "What is the meaning of this? There is no place for such sexy trash in a high school paper. This is grounds for expulsion! What do you have to say for yourself?"

I was flabbergasted! I stuttered, "It was in *Reader's Digest*; it can't be that bad."

He appeared surprised at my answer. He glared at me and retorted, "From now on I want to see the copy before it's printed." He did get to see the next issue. I believe that had I not answered responsibly, or had I made light of it, he would have been more reactive. What a difference from today.

While in high school I had made up my mind that I would get to university. My parents could not help me financially and never broached the subject. I was on my own.

Chuck on high school graduation day at Summerland High School. 1947.

Summerland High School went to grade twelve. Penticton High School, ten miles away, included grade thirteen, equivalent to first-year Arts and Science, a prerequisite for Architecture. I could afford to go there if some arrangement could be made for the travel. There was no tuition charge and I could live at home.

I spoke to several other students and found that three others in grade twelve also wanted to attend Penticton High for the additional year. I approached MacLeod, indicating that the four of us would like to pursue this opportunity. Was there anything he could do to help us regarding the travel? He listened carefully and then responded, "This has never been thought of before, but it's a great idea. I'll discuss this with the school board, who control the school budget."

Two days later he asked me if the four of us would come to his office after the last morning class. When we assembled there he advised us, "I have good news. The school board has agreed to pay for a taxi to take you to Penticton each morning and pick you up each afternoon next year. They want a letter from your parents with a request for the transportation." We all had the letters the next day.

On my first day at Penticton High I asked Mr. Pritchard, the principal, if he would advise me as to which courses to take in order that I could enter Architecture the following year at the University of British Columbia (UBC) in Vancouver. Later that day he gave me the list. In addition to Math, French and English, two science courses were included — Physics 90 and Chemistry 90. I took the year at Penticton, enjoyed it, wrote the final exams and passed with a B average, in those days a high enough grade to gain entry into the University of British Columbia.

That summer I worked at the local fruit-packing house, and with money I had saved working after school and on weekends during high school, as well as playing senior baseball, I had saved $900. I had determined that this would be enough for first-year Architecture at UBC. Tuition was only $200 at that time.

I forwarded my marks from Penticton to UBC with a letter of application. However, I received a letter back advising that I had taken the wrong courses! I should have taken Physics 100 and Chemistry 100, more advanced courses than taught at Penticton. The principal had not kept up to date with UBC requirements. Thirteen students were caught in this unfortunate mistake. We never even received an apology. Today there would have been lawsuits.

I reacted with typical teenage aplomb. I decided, "I'll punish them; I won't go!" Instead, I went to work for the municipality, helping to build irrigation flumes from pre-cast concrete sections. It was hard manual work, and six months later my attitude had definitely changed! I needed a university education.

I sent a letter off to the University of Manitoba in Winnipeg, with my marks, asking if I could enter Architecture. I don't recall why I chose Manitoba. Within a week I received a reply. "We would be pleased to accept you in first year." Hallelujah!

The trip to Winnipeg meant two days on the train. The first day, through the Canadian Rockies, was especially beautiful. However, I travelled coach, so I didn't get much sleep and was tired when I arrived. At 9:00 in the morning I joined the registration line for Architecture. I was advised that the first step was to be interviewed by a professor. During this interview, the professor determined that I was partially colour blind. I can't distinguish blue from purple and have trouble with red, green and brown. He suggested, "We should see the dean, Professor Russell." During this meeting it was suggested that pure Architecture would not be a good choice. For that I needed a good appreciation of colour. The dean looked in a university reference manual and suggested I might consider Architectural Engineering. However, at that time, the University of Pennsylvania and Stanford University were the only

North American universities to teach that specialty. They were out of the question financially. Then he said, "Now that you're here, I suggest you register in Engineering. After first year you'll still have many options to choose from."

I followed his suggestion and registered. My plan was to complete first-year Engineering at Manitoba and then transfer to UBC, assuming of course I passed first year. After six years of planning to be an architect I had gone in at 9 A.M. to register in Architecture and come out at 10 A.M. registered in Engineering. But I have never regretted the change. This was in fact *life-break number one* for me.

After registration, I went over to the university residence to see about accommodation. There were a few rooms left and I signed up for a south-facing room on the fourth (top) floor. All rooms were doubles, each roommate with a single bed. The beds could, however, be set up as bunks to allow more space. My roommate was to be an Agriculture diploma student who wouldn't arrive for another month.

The residence building had been used for army accommodation during the war. It was a four-storey brick building with four separate wings. The east wing had rooms for 250 women and the west wing 350 men. Midway between the wings there was a cafeteria on the basement floor, a main dining room on the second floor, and a high-roofed auditorium on the third and fourth floors. The deans for men and women's residences each occupied a suite in their respective wings. There was a small medical section in the women's side with a full-time live-in nurse. An eight-woman kitchen staff lived in the basement of one wing.

Room and board (three meals a day) was provided for a total cost of $55 per month. Meal hours were fixed, but special arrangements could be made for a late meal. The cafeteria was open from 10 A.M. until 10 P.M. Students in the residence often came down from 9 to 10 P.M. for an evening break. There were clothes-washing facilities and a gymnasium in the basement of each wing. Nearby there was an ice-hockey rink, an indoor rifle range, and a new sports complex with a full-size basketball court.

After arranging for my room and walking around the campus, I returned to the dining room for my first meal — lunch. The dining room held about fifty tables, each seating eight. I proceeded through the cafeteria-style line-up, filled my tray, handed my meal card to the cashier for verification, then sat at a vacant table near the cashier, facing the meal line. While in the registration line I had spoken to several other freshmen to be, so I kept my eyes on the line-up. Soon I recognized a face. He looked my way and I nodded. He came over to my table, sat down, and we introduced ourselves. After about fifteen minutes the table was full. All freshmen engineers! Introductions all around. They were Norman Anderson, Jim McGuirk, Reg Blackman, Ron St. Mars, Bob Welin, Bud Whiticar and Frank McNair, and myself. We all came from Western Canada. We soon became fast friends and many of us often socialized together.

I joined the rifle club (my dad had taught me to shoot on the farm), the bowling league (where I met my future wife), played interfaculty hockey (goaltender), and curled in the engineering league. On Friday evening and Sunday we dated girls from the residence. Gays and lesbians were unheard of then. Most of our early dates became steadies. None of the group had cars; we bussed everywhere.

On Saturday nights, we men went out for our own fun. Typically we would go to a pool hall or bowling alley, play for about an hour and a half, then head off to our secret beer parlour, the Aberdeen, where we were allowed to sit in a back corner, provided we were well behaved. We were all under twenty-one, the legal age limit. We usually had three beers, shared experiences, jokes and each other's company. By fourth year we were all twenty-one and so graduated to a front-row area for steady customers. When we graduated I'm sure the staff missed us. Occasionally we would go to the Aragon Ballroom, frequented by many single young men and women. Dancing in the late '40s and early '50s was fun, much of it slow and cheek-to-cheek. Nothing came of it after hours, however, because we didn't have cars.

Our first-year course load was difficult. We attended thirty-five hours of lectures and labs — seven hours a day, five days a week. Today, engineering students typically have about twenty-four contact hours per week. And to make it more difficult, I was weak in Physics and Chemistry. Accordingly, we stayed out of mischief the first year, particularly during exams. We had a very rough exam schedule — eleven final exams in ten days. I had submitted all my assignments on time but had let some of my studying go. As a result, I failed Mechanics and Physics. Fortunately we were allowed supplemental exams. I wrote both in Vancouver in August and passed, so I could go on to second year. I had enjoyed the University of Manitoba, and Winnipeg, and so abandoned my plans to transfer to UBC.

First- and second-year Engineering were very basic; maths, sciences, and English courses, drafting and surveying. Only in third year did we delve into the practical side of engineering. It became much more interesting, so much so that I could hardly wait to get to some classes.

The quality of the professors varied. We could tell who had actually been involved in consulting engineering out in the real world. They were much more interesting and practical than those with no external experience, who taught strictly from the textbook. Those with practical experience usually described actual case examples to illustrate theoretical principles.

During freshmen year we kept a very low profile. Out of eighty students there was only one woman! There was no hazing or initiation at the University of Manitoba. Fraternities rushed a few students, but our group wasn't interested.

The second year was different. Because of my tight money situation I stayed in Winnipeg over the Christmas holidays, as did about twenty-five other students. Things got a bit boring so a group of eight decided we would play a prank on one of the wealthy live-in students from Boston who had a small Morris Minor, a British auto. I'll call him Joe. He had flown home for Christmas and left his Morris in a parking area outside the residence. He had a room on the fourth floor. In each half of the residence there were four stairways. Each flight was about eight feet wide and had one landing between each floor. The outer lower door was double and both doors could be opened at the same time.

The plan was to put Joe's car into his room. We figured we could carry it up the stairs. However, to fit it through his door we would have to remove the wheels and tip it on its side. This meant we would have to drain the gas, oil and water.

We chose the stairwell farthest from the dean's office so he wouldn't hear us. In order to ensure we could lift the car, we first tried to raise it outside. Because of the angle of the stairs we put five guys on the engine end and three on the back end. It was heavy, but we were able to move it. Next we did a test run on the outside stairs. We had left the wheels on so we could partially roll it, but this was still tough going. We lowered the car back down.

We then had a discussion to come up with ideas to make it easier. One suggestion was to have two more guys join the group and get a strong rope so two could be at the top of the landing to help pull the car up with the rope. We tried this and found it much easier. It was decided to go for broke the next morning. That evening we drained everything into separate containers, which we saved in order to replace the fluids later.

The exercise began after breakfast. We had seven sets of steps to go. It was slow going, three hours in all, but nothing tragic happened.

We pushed the car down the hall to Joe's room. There we jacked up each wheel in turn, removed it and placed a wood block under the axle. Next we took a mattress from a single bed and placed it on the floor beside the car. We rolled all the side windows down, then carefully tipped the car onto its side on the mattress. Still okay.

We had learned to open locked residence doors with a plastic ruler. That was the next step. Joe and his roommate had bunk beds, so this gave us more room. We slid the car on the mattress into the room, and gingerly set the car upright again, with the axles still on blocks. We then jacked up each axle and put the wheels back on, facing the car towards Joe's door. We had put a sign on the front window of the car, "Happy New Year Joe!"

We had word that Joe was to return the Sunday before classes recommenced, but we didn't know what flight he would be on or the time. We checked with the airline under the pretence of wanting to travel that Sunday from Boston to Winnipeg. The logical flights were through Minneapolis. There were two flights Sunday. We phoned the airline again and asked what flight he had reserved, but they wouldn't give us that information. It was decided that we would wait until Saturday; two of us would go out to the airport and explain that we had a major surprise reception for him and so we needed to know which flight he was on.

Joe Cooper, an architecture student, and I drove to the airport and spoke to the agent on duty, who said, "I'm not supposed to give out that information." We had put a ten-dollar bill in an envelope before leaving, each member of the group contributing a dollar. This was now placed on the counter with the comment, "This is for you." He took the envelope, turned and stepped through a door behind him. A minute later he returned and silently set the envelope back on the counter. We picked it up but were afraid to look. Instead we walked toward the airport entrance, then stopped and opened the envelope. He had written the flight number and time: 7:30 P.M., Sunday.

After dinner on Sunday, we phoned the airport to check the time of expected arrival, and were advised that weather had delayed the flight about a half hour. It would take Joe at least forty-five minutes to retrieve his bags and arrive at the residence by taxi.

We didn't know when Joe's roommate was returning either, but we left a note for him on Joe's door to immediately check with us when he did. The roommate arrived Sunday morning, and we let him in on the prank. He took it in good humour.

The group eagerly assembled in room 409, opposite Joe's. A warning spy was stationed at the top of the stairwell Joe would use when he arrived. He would be able to see Joe coming from two floors below. When he saw Joe, our spy was to move quickly and join us in the room where we were lying in wait.

Word about Joe's car hijacking had spread. By 8:30 P.M. not ten but about twenty-five guys were crammed into the "observation room." The timing was good. About 8:45 Joe strode down the hall.

We had opened the door to 409 just enough to see Joe put his key in the door and then begin to open it. As he opened his door we opened ours quickly to fully appreciate Joe's surprise. He stared for a short time and groaned, "What the hell is this?" He heard us laughing, and as we began to cheer and clap, he turned around, looked at us and then charged! "You bastards!" He hit about four of the guys before we could subdue him. His reaction had surprised us, and we had to hold him until he simmered down. We assured Joe that there was no damage to the car and that we would take it down Monday evening after classes.

Everyone was told not to mention the hijacking until Tuesday, after the car was removed, and indeed no one squealed. Within a few days Joe accepted it as just a prank. He didn't dare try to get even. There were too many of us.

Exam time was truly crunch time. There were so many courses and assignments that you didn't dare lag behind. I learned early to complete assignments before they were due. I also learned to summarize each course about every two weeks. When exam time came I had a complete set of shorthand study notes. I could review an entire course in less than two hours. Before each exam I would give my summary notes a thorough read and put a star beside each point that I understood and knew. The next read I concentrated on those points I was not clear on or hadn't remembered the first time. I again starred the information which I felt comfortable with. During each read I starred more sections. Finally, during the last hour, just before the exam, I did one quick read. By following this system I felt I was writing on about 90 to 95 percent course coverage. My average marks in second year jumped by 10 percent. It worked for me and made studying easier, more organized and less stressful. If I really wanted a quiet study time I would go to the Engineering building, only a half a block away, and find a vacant classroom.

After every exam our group of eight would play cards — bridge or hearts — for a relaxation break. During exams we never played for money. We needed to unwind as much as possible. Halfway through exams we went out for our usual Saturday night "games and beer." That too was a welcome break.

The mood in the examination rooms was usually tense. We always wrote our exams in the large engineering drafting room. Tables were set up in rows of three sets of three tables. Each set of tables sat six students side by side. A student writing a non-engineering exam sat between each engineering student. Professors and graduate students continuously walked the aisles screening the students for cheating or communicating. All exams were three hours long.

During one exam a disconcerting situation occurred. An attractive young woman sat to my right in the centre of the row. About an hour into the exam she rolled back her skirt to expose a white slip with black pattern marks on it — and a nice pair of legs. Pinned to the slip was a white crib sheet. Where she was sitting the scrutineer could not see her cheat sheet. Normally I took about two hours to finish and reread my exam paper. This time I took nearly the full three hours. Why miss the scenery?

Two other important lessons I learned were to always re-read the exam questions to ensure I didn't misread them, and to proofread and double-check all my calculations. With these techniques I am convinced I picked up an average five marks I otherwise would have lost on every exam. That could be enough to miss a scholarship or qualify for a non-refundable bursary.

In third year we pulled the prank of all pranks. In early spring a group of us decided we would raid the women's residence on the evening before April Fools' Day. A lot of planning would have to go into it.

We first reviewed the names of all of those in men's residence. Out of 350 students, we agreed on 40 whom we thought we could trust, and who would then be invited to join us on the "mission." We called it "Operation Peak-a-boo." Those selected were from all years and faculties. The reason for this was that, in the event any of us were caught, all 40 were to own up. We didn't think they would kick that many out of university.

Our core group then had to call a secret meeting of the 40 raiders. A brief outline of the proposal was written up, copies were made, put in a plain envelope with each person's name on the outside. "Confidential" was written on each envelope. Every resident had a mailbox in the Administration Building, and so we asked the postmistress to put the envelopes in the correct post boxes. The condition of secrecy was stressed.

At the meeting, each raider would be asked to select the girl he wanted to see, and for this we needed a list of girls and the rooms they occupied. There was a common room in each wing, and a bulletin board in each room. Until 10 P.M., men could visit the girls only in the common room, never in their rooms. Posted on the bulletin board was a list of all residents. One of our group pinched this list from the board.

Every raider had to choose a different girl, although two different selections could be in the same room. Extensive name trading went on before agreement resulted. Every raider was to carry a flashlight and wear the same attire — blue jeans, a white T-shirt, a baseball hat, running shoes, and a pair of sunglasses. This would make it harder to identify anyone. The rules were simple but specific:

1. Everyone was to practice opening their doors with a plastic ruler until they could do it quickly. This caused some suspicion because a lot of us were observed practising. Everyone developed his own excuse.
2. Absolutely no sex. Let the girls know this as soon as you entered the room. We didn't want the girls to have a major reason to turn anyone in.
3. Souvenirs were not to exceed one brassiere and one panties — autographed if you could get it. This would be a mark of success.
4. Everyone was to leave after fifteen minutes.

5. Every girl was to be asked not to squeal on the raider.
6. Every raider agreed to own up to the prank even if only one was caught and was to be punished.

Several special arrangements had to be made prior to the raid. There were no private telephones in any residence room. However, the deans of men and women had phones in their suites that had to be decommissioned. It was noted that both doors to their suites opened inward.

There was a service tunnel under the residence, connected to six other campus buildings. Several of the group were assigned to scout the tunnels and determine where the telephone lines and electrical boxes were. During this reconnaissance they became aware that a watchman toured the buildings at night and checked a time clock. Two raiders arranged to meet him. Their excuse was that they were writing their fourth-year thesis on the university communication and security system. They made two tours with him and noted the times he was in each building. Believe it or not, so that they wouldn't be recognized later, they wore false moustaches and wigs.

It was planned to enter the women's residence on the third floor, through a door between the auditorium and the women's side. However, this door was normally secured with a special lock. The resident nurse had a key, and, fortunately, her husband was taking engineering. We described our plan to him and he arranged to get us the right key. We rushed downtown with it, had two copies made, and returned the key to him within two hours. His wife never knew it was gone.

The plan developed as follows:

The watchman's schedule left a one-hour gap, from 2:30 to 3:30 A.M., when he would not be in the residence. Everyone was to be at the auditorium door at precisely 2:45 A.M.

One leader was chosen with four deputies, one for each floor. We had alerted a reporter and photographer from the *Winnipeg Free Press*, stipulating that no names or frontal photographs of the women would be taken or referred to.

At 2:30 A.M. the telephone lines to the deans' offices were cut. At the same time their outer doorknobs were tied by rope, across the hall to a massive banister. If the deans wakened, we felt they would not think to unscrew the doorknob or remove the door hinge pins.

Only lights in the deans' offices were shut off. To black out the entire residence would be too scary. At 2:45 the leader opened the auditorium door. Everyone went quietly but rapidly to the door of his female selection with plastic ruler at the ready. One minute was allowed for this movement. As that minute expired the leader gave a single whistle to the floor deputies, who then gave a double whistle to everyone on their floor.

Within a few seconds everyone entered the girls' rooms. Lights were turned on at the outset. The girls' reactions varied. Some were calm and took about a minute to realize what was happening. Girlfriends generally were surprised but pleased, and enjoyed some cuddling and necking. Others screamed and several ran out into the hall. One had on a shorty nightie and the photographer caught a nice pair of retreating buns. A number were belligerent and said they were going to report us to the dean, as well as the police.

Within five minutes things had quietened down and souvenir time took over. Some prizes had to be bartered for. Several thoughtful fellows traded their shorts. Some future dates were made. Some souvenirs were autographed — in several cases with lipstick. No sexual skirmishes or assaults occurred.

At 3:00 the whistle to return was sounded. The dean of women had been wakened and was heard pounding on her door. She was finally let out by some of the girls who undid the rope on the door handle. She was outraged and ran through the building to the office of the dean of men, pounded on his door and undid the rope. He hadn't heard a thing.

We were all back in our rooms and beds by then.

I went down for breakfast at 7:30 the next morning. Many students were already there, and the room was abuzz with excitement. Some of the girls were animatedly telling others what had happened to them. Most were thrilled to have been selected. Others were quiet listeners; obviously those who hadn't been raided. One had the inescapable sense that the status of those raided had been notably increased.

We came to learn that a number of women had threatened to go to the president. However, another group stood fast and said if they did, the accusers would be harassed out of university. Apparently a battle royal occurred, but in the end not one woman made a complaint. My selected girl and souvenir will forever remain a secret.

That morning there was an article on the raid in the *Winnipeg Free Press*, as well as several photos. It was also covered by Winnipeg radio. All hell broke loose in the administration.

The deans spoke to us at dinner that evening, expounding upon how outrageous our escapade was and the disgrace it had brought to the university. They threatened that they would root out the participants and remove them from residence. I doubt they would have, even if they had been able to identify them.

Later that evening Dr. C. Gilston, president of the university, called a meeting of all residence students. He stated that such behaviour would not be tolerated, indicating that many parents of the girls had phoned him personally, urging him to take action. He informed us that the university had instigated an investigation and all involved in the raid would be expelled.

But no one was ever caught, and within a week the excitement had died down. The girls turned out to be trusted participants. In fact, many of the girls quietly let it be known that they would like to see it became an annual event.

Life in residence was generally enjoyable for us. Very close camaraderie existed among our group. One could always find something to do. Cards became a major form of relaxation, many played pickup basketball, ping-pong was available (with occasional tournaments), and intramural sports found a large following. I had not curled before and found that to be lots of fun. We would go to the Maple Leaf Curling Club in downtown Winnipeg for seven o'clock games every Saturday morning. The ice was natural, so curling couldn't start until winter arrived. I remember one morning when the outside temperature was –40°F. The ice was so cold and the surface friction on the ice so high that we couldn't push the rock to the other end of the rink, not even if we put both feet in the hack and pushed the rock with both hands. We had to quit.

Every Tuesday night we bowled in a university league. Teams generally were comprised of three men and two women. I had bowled for many years in Summerland and became quite proficient, consistently averaging in the top five bowlers. After the games, a group of us would go next door for coffee and a sandwich. Low-key stuff, but enjoyable, just the same.

An attractive Home Economics student, Pat Norris, was on my team. We went together for four years and then married. Pat and I raised four children and managed to put them all through university. Cathy, the oldest, married Allan Peters, a successful architect builder in the Vancouver area, and is now a homemaker and mother of three children. Randy, our oldest son, is married with four children, and is a vice president with Paine Weber stock brokerage in Spokane, Washington. Suzanne is a graduate nurse with a master's degree, and has specialized in medical sales. She is married to David Bell, a Royal Canadian Mounted Police (RCMP) officer. They have two children. Greg is a graduate of Embry Riddle Aeronautical University and is now a jet captain with Continental Express in Cleveland. He and his wife Robin, sales manager with a business furniture supply company, have no children and two dogs.

As far as studying was concerned, I was a morning person. My roommate and I had separate study tables and reading lamps. I would wake about 6 A.M. to study or work quietly on assignments until 7:30. I then went to the common washroom — two per floor — and completed my morning duties. My roommate and I then headed for breakfast about 7:50. Our group generally sat at the same table. Classes were only two minutes away.

Two weeks before classes ended in third year I was hit with internal bleeding. It was serious enough that I was taken to hospital in an ambulance and given multiple blood transfusions. Hospital staff kept me very quiet and administered special clotting drugs. After three days the bleeding stopped, but they kept me another several days. I missed a total of eight days of classes at a critical period just before exams.

Dean MacDonald heard about my illness and called me into his office. He said, "Brawner, I've heard of your medical problem and want to help you. I can postpone several of your exams until fall to assist with your studying. What do you think?"

I was surprised but pleased at his concern. I had all my assignments finished and most of my courses summarized. "I think I'll be alright. Would it be okay if I write the first three exams and then discuss it with you again?" I asked.

He agreed, "That will be fine."

After the third exam I decided I would be okay to write them all. He complimented me, and I got through okay, passing all exams with a B average.

To some a B average doesn't sound great. In hindsight I have come to learn that the all-round student — someone slightly above average academically who participates in social activities and sports and who develops good communication skills — has a greater chance to succeed in later life than the pure academic. Have some fun at university! Learn to be "a people person."

The doctor who treated me at the hospital in Winnipeg diagnosed my problem as Meckel's Diverticulitus — a growth had developed on my intestine that had become infected and sprung a blood vessel leak. He recommended surgery. I said, "If I have the operation now I'll miss exams. Could it be left until the fall when I could come down to Winnipeg one week before exams and have it then?"

He thought for a minute and asked, "Over the summer, would you be near a doctor and hospital?"

I answered, "Yes, within one hour."

He returned, "I'll write a letter for you to give to the doctor in the event that the bleeding returns. It will describe the problem and advise him what to do. In any case, in the next two weeks do not do anything strenuous. Give yourself a chance to heal."

I worked as a surveyor and played baseball that summer. I had one recurrence, which sent me back to hospital for two days. Other than that all went well. I had the surgery during the first week in September 1952 in Winnipeg General Hospital, and it was successful.

In fourth year, the courses became very practical and even more interesting. We first took the theory of design, then worked on design assignments. Projects included designing a water and sewer system for a section of a city, a highway between two towns, a steel bridge, a retaining wall, a pipeline, foundations for buildings and bridges, plus water and flood control measures. We also did landslide analysis and stabilization design.

Most of us found summer jobs in engineering. We were able to make almost enough money in four months to pay for our next year. Over the four years I had to obtain a loan for only an additional $1,000. Part of that went for travel.

Bob Welin, a classmate, worked on bridge construction in the summers. In the Bridge Design Laboratory one afternoon, he was developing a design that used rivets to connect the bridge beams. Bob was drawing the plan for the rivet locations as the structural professor, Dr. Jack Hoogstraten, looked over his shoulder. He said to Welin, "You're developing an unusual rivet pattern. Have you worked on steel construction before?"

"Yes," Welin answered, "I worked on a bridge construction crew last summer."

Hoogstraten asked, "Did you ever think about the engineer who designed the bridge?"

Welin replied, "We had nothing to do with the deaf and dumb gang!"

The professor was startled, "I don't understand." Welin returned, "Some of their designs are impractical. They often locate rivets in places that we can't get the rivetting gun into. A designer should have to build what he designs."

Welin had learned an important lesson in basic engineering design. Make sure you can build what you design.

One big difference today is the use of computers by students. In my day we used a slide rule. Calculations, trigonometry, logarithms could be performed, but more slowly and with less accuracy than with a computer. The great benefits of a computer are the speed of calculations, the opportunity to evaluate many options, the development of computer models, and the wide variety of programs that are available or can be developed. Students now learn much more in a shorter time period than we ever did.

Each year the engineering students held a major event — the Power Prom. On campus it was considered the social event of the year, and was very special for the girls who were invited. The event was held at the Royal Alex Canadian Pacific Rail Hotel just after New Year's. A top orchestra was hired, and dress was formal. Mid-evening snacks were also available. Some special event, a singer, a speaker or a major draw, was usually scheduled. Our group rented a hotel suite for the evening in the event we wanted a break — to clean up on arrival or leave our winter coats. Some would have a drink or two.

In third year the student council planned a beauty contest. To make it a major event, six attractive university debutantes were selected. All were from well-known Winnipeg families, including the provincial premier's daughter.

Six engineering students were carefully selected to escort the girls as their dates. The contest was well advertised so publicity was widespread. Each girl was given the name of her escort, who was to pick her up at 9 P.M. Unbeknown to us, that information had been obtained by some medical students.

Many of the engineers and their dates had arrived at the dance by 8:30 that evening. Shortly after 9:00 agitated calls came into the Royal Alex Hotel from several of the engineering escorts: "Our celebrity dates were picked up before we came to get them!"

All six girls had been kidnapped. Parents of the girls also phoned in. They were very upset and called in the police.

We discovered later that medical students had phoned the girls before dinner and instructed, "The pickup time has been changed to 8:30. Would you please be ready then?" All were ready, suspected nothing, and so were picked up at that time. They were taken to a nightclub on the outskirts of the city where a special party of medical students had rented the entire club.

But we had no idea where the girls had gone. It was suspected it was a prank, but by whom we did not know. At about eleven o'clock a call came in saying that the girls were at a specific nightclub and that we could come and pick them up. The caller said, "We're finished with them." That club still exists so it will go unnamed.

The message went over the microphone in the dance hall. It was decided that six cars with the original engineer dates plus two extra engineers per car would carry out the pickup.

The kidnappers were obviously concerned that multitudes of engineers would descend on the nightclub and possibly start a fight. They alerted the police. On arrival, five cars of police at the front door cooled any thought of violence. The identities of the actual kidnappers were not known until the girls were asked. The debutantes said they were apprehensive when they realized they weren't with their intended dates, but that they were looked after well, and following the initial shock, actually enjoyed themselves.

The engineers had been scooped. What we never figured out was why the medical students? We knew of no conflict with them in the past. We also never determined who gave the medical students the beauty contest plan and girls' names.

The girls were brought back to the Royal Alex Hotel and the contest was held, belatedly. However, the Power Prom had been upstaged and much of the enjoyment that should have been was destroyed.

I believe there is room at university for some nondestructive and good-natured hijinks — of course within bounds. Therein lies the test.

During two summers between university semesters, I worked as a surveyor for the Water Rights Branch, headquartered in Kelowna, B.C. Anyone wishing to take water for irrigation or domestic use from a creek, river or lake was required to obtain a water licence. There were limits on the allowable volume, according to the acreage to be irrigated, or the number of household members. The water licence granted approval for a given volume of water on a priority basis according to the date of the licence. Maximum volumes of water were allowed from each water source. In some areas, such as the semi-desert Okanagan Valley, many streams had reached their licenced capacity.

Where land was to be irrigated, the area of the land had to be determined by surveying. This was my responsibility, along with a rodman. The area was measured by surveying the property, measuring angles and distances with a transit and stadia rod to all points on the irrigated boundary. This information was plotted on a scaled plan and the area measured with a planimeter.

One project was on Anachrist Mountain, east of Osoyoos, B.C. I had taken several readings from one location and was carrying the transit on my shoulder to another location down a rocky mountain slope, stepping from rock to rock. Just as I was stepping to another rock, I noticed a coiled rattlesnake on one end of the rock. I had, however, already moved to step onto that rock and couldn't stop my forward motion. As my left foot hit the rock the rattlesnake struck.

He hit my high leather boot about one-half inch below the top. My momentum forced me to jump to the next rock. My rodman, coming behind me, saw the reptile and slammed the stadia rod down on the snake, killing it but shattering the rod.

By this time I had stopped and looked down at my left boot. The snake venom was flowing in a narrow stream down the side. At that moment, the shock of the near snakebite hit me. I threw up my breakfast and lunch and felt like fainting.

Gathering myself together, I said to Keith, the rodman, "C'mon, I want to go to the drugstore and get a rattlesnake kit with syringe and anti-toxin. That was too close for comfort." Such kits were readily available in the south Okanagan, where rattlesnake bites were moderately common.

A second incident occurred near Grand Forks, B.C., an area where many members of the Doukhobor religious sect reside. They had come from Russia in search of religious freedom. Excellent farmers, they had established numerous farms in the area. Some were located outside the local irrigation district and had to develop special irrigation systems. These required a water licence. One farmer, a Mr. T. Posnocoff, had a small five-acre plot on which he had about an acre of vegetable garden, two acres as an apple orchard, and the rest in pasture for several cows.

His son had prepared the application for him and done all the paperwork. I arrived at the father's house and knocked on the door. He answered.

"Hello," he said in a heavy Russian accent. I realized his English was likely poor, so I spoke slowly as I advised him, "Mr. Posnocoff, I am here to survey for your water licence application. I must inspect your pump and measure the amount of land you irrigate. It is for your benefit. It will not take very long." Then I queried him, "Okay?"

He stared at me and I thought he nodded. At any rate, I smiled, turned around and climbed back into my government Land Rover. I drove to the edge of the pasture where I could see the entire property from one transit location. I set up the transit and took a distance reading on the stadia rod being held by Keith on one of the survey pins at a corner of the lot.

Keith had just begun to move to another location for a further reading, when suddenly he yelled, "Look out, he's got a gun!" I turned toward the house. Posnocoff was walking steadily toward me with a rifle pointed directly at me. He looked very intent.

I stepped away from the transit and held my hands extended sideways to show I had no weapon. He continued to approach, marching resolutely, until he stuck the rifle directly into my stomach!

He shouted loudly, "Go! Go!"

I backed away a little and uttered an, "Okay."

I picked up the transit and tripod, went to the vehicle, put the transit against the seat and climbed in. He continued to point the rifle toward me. I drove out to the main road and picked up Keith. He asked, "What do we do now?"

I answered, "We go to the RCMP and see what can be done."

We found their office and I explained to the officer what had occurred. I said, "It's obvious he didn't understand our intent. I don't want to lay any charges. We just want to survey his land for his water licence. Have you any suggestions?"

"Sure," he said, "I'll get another officer who speaks some Russian and we'll go out to his farm. You follow us but stop at his gate."

At the farm, the two officers spoke to Posnocoff, then convinced him to get into their vehicle. They drove out to the gate, stopped and said, "We'll take him with us and keep him in a cell. When you finish your surveying, let us know. Then we'll take him back home."

Posnocoff probably never knew we completed the licence survey. However, in due course he did get his water licence.

Even today, when I think of the gun in my gut, it brings a shiver.

In my final semester, many companies or agencies interviewed students for jobs after graduation. They would make a general presentation regarding their agency or company, describe details of the job(s) available, outline the type of person they were looking for, and indicate the salary range offered.

If anyone was interested, they signed an interview sheet for the next day. In 1953

Summerland Merchants Senior Baseball team, Okanagan International Baseball League. 1949. I played third base. Lower right.

there were many jobs available. I attended six interviews and was offered four jobs. I accepted a job offer from Imperial Oil Pipelines to be a pipeline designer. The job would be in Edmonton and paid $290 per month.

The previous two summers I had surveyed with the Water Rights Branch out of Kelowna, B.C., covering the entire Okanagan Valley. This allowed me to play senior baseball for Summerland in the Okanagan International Baseball League. The second year was a good year. I hit .323. My performance was good enough that a scout for the old St. Louis Browns came to our home and offered me a contract to play in a Florida league. I discussed the offer with my father and he counselled me, "You might be able to play baseball for ten years, but engineering would be for life." I turned the offer down.

About halfway through final fourth-year exams, I received a phone call from Allan Ker, District Engineer of the Water Rights Branch in Kelowna, asking if I would return that summer. He said his assistant was resigning in September and he proposed that if I worked there for the summer, he would appoint me Assistant District Engineer. Near my home, another year of baseball — and a promotion! I accepted immediately and phoned Imperial Oil advising of my other offer and decision.

During the last week of classes I met with Professor Hoogstraten to review a structural design project. I will always remember one of his comments — "We don't care if you forget all the formulae we've taught you. Our goal is to teach you how to think rationally."

Final fourth-year exams went very well and I graduated with a high B average. Because of a shortage of funds, I went to work immediately and did not return to Winnipeg for graduation ceremonies. But I received my certificate. That's what counted.

I returned to Kelowna to start my career as a civil engineer, surveying for the Water Rights Branch and preparing approvals for water licences throughout the Okanagan Valley.

2

Highways

Highways
(1953-1963)

After six months with the government Water Rights Branch, a provincial election resulted in a challenging career change with the British Columbia Department of Highways as a Soils Engineer. Life-break number two! Numerous interesting projects were performed. The investigation and stabilization of glacier-like 2.2-mile-long Drynoch Slide was developed. The slide had been moving for over fifty years. The rate of movement of the slide was directly related to the depth to the water table. It was stabilized by developing a seepage cutoff in a backfilled trench across the slide. Successful stabilization allowed construction of the Trans Canada Highway to proceed.

A foundation investigation was performed for a three-hinged steel arch bridge across the Fraser River in the Fraser Canyon. Abutments were founded on up to 60 feet of gravel which was consolidated with preload stressing cables into bedrock.

The Burnaby freeway section of the Trans Canada Highway into Vancouver was constructed on up to 35 feet of organic muskeg over soft clay. It was constructed using the pre-consolidation technique based on results from two test sections. Sawdust which settled below the water table was placed initially as a weightless spacer. This was a world first.

A major water outflow from under the Bear Glacier in northern B.C. washed out several miles of new highway construction. The author developed an overflow trench across the glacier to ensure the glacier did not "float" from its base in the future.

This ten-year period was very exciting. Premier W. A. C. Bennett and Highways Minister, P. A. Gaglardi oversaw a major highway exp ansion in British Columbia. This included the challenge of constructing the Trans Canada Highway through the Rogers Pass, Kicking Horse, Thompson and Fraser Canyons.

Chapter 2
Highways

I spent the summer of 1953 with the Water Rights Branch in Kelowna, and looked forward to replacing the retiring Bob Harris in the fall. However, it wasn't to be that simple. A provincial election in British Columbia in late August resulted in a surprise upset. A new political party under W. A. C. Bennett was elected — the Social Credit Party. Bennett, an astute businessman who didn't believe in borrowing money, selected his cabinet and reviewed the state of the budget. The provincial finances were in a negative position. His reaction? He froze all jobs, including the position of Assistant District Engineer in Kelowna. The position was never filled. I had been blindsided!

I discussed my situation with Mr. Ker, District Engineer. He said that as soon as the summer's work was over, he would propose I be transferred to the head office in Victoria.

In early October I moved to Victoria and found a family to room and board with. At head office, I was initially assigned to review historical creek and river flows on the upstream portion of the Columbia River, north and east of Revelstoke, B.C. Art Paget, Comptroller of Water Rights, advised me that, in order to provide better flow regulation of that large river, and to generate hydroelectricity in Canada, there was a chance dams would soon be built on the Columbia.

River regulation was also important to the United States. By providing an almost uniform annual water flow at the massive Grand Coulee Dam in Washington State, power generated would be increased by about 40 percent. In addition, substantial flow would be available for irrigation of the semi-arid central Washington agricultural area. British Columbia would receive power or money in return for the additional power generation. (Subsequent agreement between the two countries, called the Columbia River Treaty, resulted in development of large dams at Revelstoke, Castlegar and Mica Creek.)

After a month of crunching numbers and plotting graphs, I went to see Paget and asked what he had planned for me in the future. He advised, "You'll continue calculating the water flows of the entire Columbia River Basin upstream from Revelstoke this winter and spring, and then develop projected annual storage capacity for dams at two of the potential sites. When the snowmelt is complete at the Mica Creek dam site, you'll be a survey chief on dam site location studies. You'll work out of a field camp at Mica Creek, ninety miles north of Revelstoke."

I thought briefly and then said, "I'm planning on getting married next summer in July. Can my wife be with me there?"

He raised his eyebrows and answered, "No, unless she would agree to bull-cook for the survey crews!"

What a honeymoon that would be, cooking for ten to fifteen surveyors, seven days a week. I couldn't imagine Pat doing that. She was a Home Economics graduate and a teacher. I was sure her answer would be an emphatic no, plus I didn't want her living in a construction camp during our first year of marriage.

I went back to my small office and thought about what he had said. After about an hour of mulling over our discussion, I concluded the assignment wasn't for me. I wrote out my resignation, effective in two weeks. I explained in the letter that I was after a permanent job where my new bride and I could be together the following year.

I gathered up my courage and took my resignation in to Paget. He read it and looked at me with noticeable anger. He said sharply, "You stupid young fool. Here's a great chance for you to gain excellent field experience while having responsibility for the work of two field crews. I'll give you twenty-four hours to change your mind!" I was stung by his use of the word stupid. That ticked me off! I didn't earn an engineering degree over four years of hard work to be called stupid.

I returned to my office to have lunch. My mind kept churning over the turn of events. I achieved very little that afternoon; by three o'clock I decided I needed to go for a walk. The office I was in was across the street from the Douglas Building in downtown Victoria, which contained the Department of Highways offices. In the basement there was a cafeteria. I walked over there to have what turned out to be a fortuitous cup of coffee.

I sat by myself near a table occupied by four men, and couldn't help overhearing that they were talking about bridge foundations and soils engineering. One man said, "I need a young soils engineer to do field investigations, supervise laboratory testing and perform foundation designs. Someone who can also work on landslide problems. If you hear of anyone like this please let me know."

I was all ears. I had no practical experience, but, in fourth year at Winnipeg, I had received my highest mark in Soil Mechanics and Foundations Engineering.

I decided to give it a shot. I waited until they left the table and, at a discrete distance, I followed the man who had spoken. He entered a room only three doors from the cafeteria. After he closed the door I stopped in front of it. A nameplate read: "Materials Testing Branch."

I opened the door and stepped up to a receptionist, "Can you tell me who the gentleman was who just entered?" She look incredulously at me, "Why do you want to know? Who are you?"

I answered, "I'm Chuck Brawner. I overheard him in the cafeteria say he was looking for a soils engineer. I'd like to apply for the job. May I see him?"

She answered and smiled, taking on an air of importance, "That's Reg Thurber, the Senior Materials and Testing Engineer. I'll check if he'll see you."

She went into his office and returned shortly, "He will see you. Please go in."

Fortunately, I had gone to work that morning with shirt, tie and jacket. I was presentable for an interview. I went into his office; he stood up and stepped toward me. I introduced myself and put out my hand.

He said, "I'm Reg Thurber and I'm pleased you overheard my conversation in the cafeteria." We shook hands. I made certain I gave a firm grip and looked directly at him. He added, "Please sit down."

I said, "I'm sorry, I didn't count on these circumstances developing today and I don't have a transcript of my marks and degree with me, but I can get them for you tomorrow. I don't have any soils or foundation engineering experience, but my best mark at university was in Soils and Foundations, and I had the best mark in my class for my graduation thesis. I received a civil engineering handbook as a prize. I'd like to apply for the position you spoke of. I'm not afraid to work, and I can give you several references." I was surprised at how calm I felt, and how well I had introduced myself. I was normally not that aggressive.

He seemed interested, saying, "First let me give you some information on the Branch. We're responsible for field geotechnical investigations for highway design, retaining walls, culverts, landslides, foundations for new provincial bridges and buildings, and the location of gravel pits for construction aggregate. We have two of our own drills and a comprehensive laboratory, which I'll show you later. In addition, we perform tests on many materials used by the Department, such as guardrails, culverts, cement, concrete, road base gravel, centreline paint, *et cetera.* We cover the entire province from Victoria. There will be a moderate amount of travel, paid by expense account. You'll be supplied an automobile for these trips. Beside myself our staff comprises three engineers, five laboratory technicians, six field technicians, one draftsman, five field drillers and Miss Compton, the secretary. We're presently one engineer short. You would be the junior engineer, but you'd be involved in all phases of our geotechnical and materials work. The starting salary would be $287 per month with annual increments for good service." Being very green, I didn't know about fringe benefits and didn't ask. Later I found that they were very attractive.

He asked a few personal questions concerning high school, the work with the Water Rights Branch, sports activities, general health, and a driver's licence. I answered all questions with confidence and then added, "I definitely wish to be considered for the position, and I could start in two weeks."

He thought a minute and then stood up, held out his hand and said, "Chuck, you're hired!"

Wow. Half an hour earlier I was downcast, staring into my coffee, contemplating unemployment, and suddenly I was hired to be a soils and foundations engineer for the B.C. Highways Department! This was *life-break number two.*

The work was fascinating. My first project was to supervise a bridge foundation study for two bridges across the Okanagan River in Penticton, B.C., at the south end of Okanagan Lake. Thurber discussed the engineering study with me. He recommended I go to the Geology Branch and obtain geological maps and reports on the local soil and rock types, as well as the origin of the materials in that area. I should also meet with the Hydrology Branch to locate and summarize the soil logs of any water wells. In addition, I should determine the daily flows in the river, and obtain topographic maps and stereo air photos of the sites. (In my engineering I came to rely extensively on air photos because one can get a three-dimensional perspective of the land surface and identify soil, rock, landslides and surface water conditions.) He also advised me to meet with the bridge designer to obtain a site plan and information on the bridge type, dimensions and loadings, pier and abutment locations and elevations, as well as the foundation reports for the original bridges, if they still existed.

I spent two days obtaining the material and another day reviewing it. In the evenings I reread my university texts on site investigations and pile foundation design. My review of the conditions indicated the site was underlain with some 40 to 80 feet of soft clay followed by sand, and under that, dense glacial till.

The drills owned by the Materials Branch were power augers. They had a maximum drilling capacity of about 60 feet in soft clay soils, so I was concerned that they could not go deep enough. I discussed this with Thurber, and he agreed there was a problem. He said, "You'll have to hire a drilling company with a rotary drill to drill and obtain samples. Draft the letter, outlining the requirements, and ask for a proposal and cost estimate. But first look in the files and review several site investigation reports and requests for proposals."

Thurber was an excellent teacher. He gave me ideas where to find the information, then I was on my own. I spent a day going over reports and proposals, and then developed the letter to be sent to drilling contractors. I took it to him for review. He was very helpful — and tactful. He suggested some changes and additions and gave reasons for them. He did not rewrite anything. I developed the final letter for his signature. When he read it he said, "That's good." (I decided there and then that when I became a supervisor I would never rewrite an employee's report. It's humiliating, and that's not the way to train anyone.) I had passed my first test.

About a week later four proposals had come in. One price was much lower than the others. I discussed the proposals with Thurber and he said, "The low bidder has older equipment, but they have one very good foreman driller. If they'll guarantee he is on the project full time, we can accept their proposal. You must get that in writing before they start. If not, go for the second low bidder."

Thurber further instructed, "Determine when they can start. I want you to be on site for the entire program. You'll provide instructions for sampling, prepare drill logs, and package every sample. See that they are properly prepared for shipment and review the foreman's drill logs. You'll also survey the drill hole locations. Take some photos of the site for your report. Also, include photos of typical samples. Keep track of their hours of work. You'll be responsible for reviewing and approving the invoice."

I spent two weeks in Penticton supervising the program, although I use the word supervising loosely. I tried to give the impression that this was not my first project. The drilling and sampling turned out to be straightforward and the weather was good. I helped on the drill occasionally, something you couldn't do today because of union restrictions. The job was a good learning experience and I had a lot to learn. One evening I went out with the drillers for a beer and dinner. They had both played baseball so we didn't have to talk shop.

I stayed with my parents in nearby Summerland. They were very proud of my engineering degree and the job I had obtained. So was I.

Subsequently I arranged and evaluated the soil testing, concluded that pile foundations driven into the sand were required, and designed them. The design was quite routine.

Another early assignment was to evaluate a number of landslides that had recently occurred in the Kicking Horse Canyon, east of Golden, B.C., alongside the new Trans Canada Highway that was being constructed through very difficult mountainous terrain. The project extended about 10 miles from Golden east to the Yoho Park Boundary. The easterly 4 miles traversed a mountainside largely comprised of glacial soils. There were many gullies to cross, each of which carried water during rainy or snowmelt periods.

An old-time railway engineer, Jack Brown, was the resident in charge of the project for the Highway Department. He was a character. He smoked roll-your-own cigarettes and talked to the wild animals. There was a herd of mountain goats that frequented the cliffs on one side of the Kicking Horse River, and a herd of deer on the other side. Brown would bring vegetable and fruit scraps for the goats, and he put up a salt lick for the deer. He could almost get close enough to pet them.

Brown had developed an unusual procedure to place culverts in the gullies, under the earthfills, which ranged from about 15 to 80 feet high. The gullies were dry then, due to minimum rainfall. He had developed small earth dams upstream of the fills, and constructed a diversion ditch along the side of the gully to carry surface water to a culvert entrance about 6 feet below the road surface. The culvert then carried the water at a 1 percent gradient under the highway. Brown had the contractor dump a pile of blasted rock under the culvert outlet to control erosion, then place a circular flume at the far end to carry the water down the slope. He reasoned that this procedure reduced the length of expensive culvert pipe needed, plus the low soil load required a thinner gauge culvert. The result was a substantial material cost saving. Brown didn't, however, allow for heavy precipitation events. In view of his long-term experience, this design was a surprise.

During springmelt in 1954, an excessively heavy rainfall occurred, and the flow in the gullies was very high. Many gullies overflowed, washing out the small dams and diversion channels and filling the basins behind the fills until the fills became saturated. The fills had obviously not been adequately compacted, because seepage developed on the downstream slope, causing severe erosion and oversteepening the slopes. Failure of 16 fills occurred. With all the water behind it, the fill material developed into flow slides. Some of the soil flowed down as far as the Kicking Horse River. All the culverts and flumes were damaged; some were never recovered. The highway was closed for several days until temporary diversions and culverts could be placed.

Thurber sent me to provide advice. I took my wife along, since she was on her summer break from teaching in Victoria. We drove from Revelstoke to Golden via the old Big Bend Highway, 190 miles of crooked, dusty and washboard gravel road. We had hiccups for a day after the rough ride, and I don't think the car was ever the same.

Brown wasn't happy to see me there. He let me use the site plans and profiles, but offered no assistance, not even the use of his survey crew.

On my second day I was out at the Park Bridge where a replacement bridge was under construction. Fred Hutchison, resident Bridge Engineer, and Jack Brown were there. We were talking about construction of the east abutment when a large blue Buick crossed the bridge and stopped beside us on the road shoulder. Two gentlemen stepped out — Fred Brown (no relation to Jack), the Chief Engineer from Victoria, and someone I didn't know.

Fred looked at Jack Brown and said, "Mr. Brown, this is the Honourable Minister of Highways, Mr. Phillip Gaglardi." Gaglardi replied, "Pleased to meet you. How long have you been with the Department, Mr. Brown?"

Jack Brown looked at him, his cigarette dangling from the corner of his mouth and staining his moustache, and said almost disdainfully, "That's not the question. How long have you been with the Department?"

Gaglardi wasn't phased a bit. He answered, "I'm here to inspect the damage of the flooding." But before he had a chance to say more, Jack interrupted, "If you don't mind I have more important things to do!" Then he turned and headed toward his truck.

I looked at Gaglardi and said, "I've only been here three days, but I know the problem and the solution and I'll be pleased to show you and Mr. Brown around."

Brown accepted with a, "Thank you, Chuck."

Gaglardi then suggested, "You know the detours, why don't you drive?"

"Sure," I said, and we got into the Buick. It had every extra you could imagine, including a long-distance two-way radio, unusual for those days. "Flying Phil" Gaglardi was known for his fast cars. I was glad to drive.

We stopped at about six different locations and observed the damage. I explained that the culvert installation was faulty: "They should have been larger diameter, placed at the base of the gully, and constructed with a concrete apron and head wall extending at least two diameters above the top of each culvert so they could flow under a head, which increases capacity. A debris catcher should also have been constructed upstream to catch material that could plug the culvert."

I personally thought that Jack Brown had likely changed the design from the original plans, but said nothing. The contractor certainly wasn't complaining. He got all the extra work involved in constructing the temporary detours, installing new culverts and rebuilding the fills. He would negotiate this work as cost plus profit because it was extra to the contract. The mistake cost Highways a bundle.

Under these circumstances, one would expect management to fire an employee responsible for such a serious error in judgment. That wasn't Gaglardi's style. Instead, he usually arranged for their transfer to another department where they would have less responsibility. Later, during my five years as senior materials and testing engineer, I had three engineers from other departments transferred to my group. Brown retired when the project was completed.

The two abutments for the Park Bridge across the Kicking Horse River had been completed. Fred Hutchison, the site engineer, received word that the steel was being shipped by railway and to expect it in three days. I was in the Department office working on my report when he instructed his surveyors to do a final check of the distance between the bridge seat bolts of the two abutments.

Approximately two hours later the survey crew came back. The party chief exclaimed excitedly, "Fred, the distance is three inches less than when we checked it last! We checked the distance three times. One abutment appears to have moved."

Fred offered, "I'll go back with you. Chuck, could you come too?"

I agreed, changed into hiking boots, and we left in his panel truck.

The south abutment was constructed on and tied solidly into stable bedrock. If movement had occurred it must have been at the north abutment. That abutment had been constructed on steel piles driven to "refusal" in either dense glacial till or bedrock. The main line of the Canadian Pacific Railway passed only 20 feet in front of the abutment, at the edge of the river. I said to Fred, "I'll climb up the slope and look for cracks." I expected that a large slide had developed. The approach to the bridge involved a major cut, up to 90 feet deep, with a steeper than normal slope angle. This steep cut had increased the shear stress in the slope. Typical highway cut slopes are 1.5H:1V. This slope was 1H:1V.

About 300 feet above the abutment I encountered a crack about 8 inches wide and at least 20 feet deep. That was as far down as I could see. I followed the crack to the west where it encountered a bedrock cliff, then extended itself down the slope to the track. The crack extended to the east down the steep slope to the highway ditch. I couldn't follow it across the road due to recent construction, but did pick it up again on the natural slope down to the railway tracks. The tracks appeared undisturbed. I concluded that the toe of the slide was in the railway ditch.

I obtained six survey stakes from Hutchison, some nails, a survey tape and an axe. Then I climbed back up the slope and drove stakes on either side of the crack at three locations. A nail was driven in each stake and the distance measured across the crack, nail to nail, at each monitoring point. This was measured daily. Movement generally averaged ½ inch per day for the next week. However, the rate of movement was not increasing, so major movement was likely not imminent, unless heavy rainfall occurred.

I had learned in class that landslides were stabilized by reducing the driving forces; in this case by unloading the top of the slide to reduce the shear stress. I also suspected a high water table, which would reduce the shear strength at the base of the slide. Lowering the water table would require deep drainage. I discussed my thoughts with Hutchison, saying that I also wanted to confer with Reg Thurber. We drove back to town.

I was aware that more movement could affect the railway, so I advised the CP Rail Roadmaster in Golden, and then phoned Thurber. I had decided by then that I would recommend we employ the contractor on the project to remove some of the top of the slide. I also intended to propose that a hole be drilled near the centre of the slide to determine its depth, the depth to the water table, and the permeability of the soil for potential deep drainage.

I described the conditions to Thurber and he agreed. It was urgent that a drill be mobilized rapidly. Thurber said he would arrange that.

Some unloading at the top commenced the next day, and two days later a rotary drill arrived from Kimberly, where it had been used by Cominco for mineral exploration. The road contractor had by then developed an access road to the drill site.

Armco drainage tunnel installed below the failure of the Park Bridge Slide in the Kicking Horse Canyon, B.C. The tunnel extends in 200 feet with laterals at the end. The drainage along with some unloading stabilized the slide.

The prime concern was the location of the water table, so drilling was expedited, with limited sampling, until the soil became saturated. This occurred at 120 feet. From there, samples were taken every 5 feet. The dense glacial till continued to 180 feet. At that point gravelly sand was encountered. Suddenly water flowed out the drill pipe. The pressure was artesian! No wonder the slide had developed — an oversteepened slope and high pore water pressure.

The question was how to lower the water pressure. It was decided to drill the hole deeper and lower a screen inside to allow the artesian water to flow more freely. A plumber attached a valve and pressure gauge at the top so we could stop the flow and measure the pressure. There was 20 feet of water pressure head at the ground surface! This was not unreasonable since the mountain behind the slide rose about 3,000 feet above grade.

A joint decision was made between Thurber, the Railway and myself to install a 60-inch-diameter Armco tunnel liner from the railway grade into the lower slide zone. A tunnelling contractor was hired to do the installation, working round the clock, three 8-hour shifts daily. They progressed an average of 20 feet per day. At 100 feet some seepage was encountered. At 120 feet a major flow of about 300 gallons per minute was encountered. The water level in the drill hole started to reduce, and the movement slowed. The tunnel was advanced a further 10 feet and then two lateral drains, each 50 feet long, were developed. By the time the drainage tunnel was complete the water pressure head had dropped about 100 feet.

Once the slide was stabilized it was considered that the piled abutment would be stable. The east bridge seat was redesigned and reconstructed so the bridge steel would fit. The slide has remained stable since 1954.

At the same time as working on the slide, I prepared my report to reconstruct the highway. Essentially the plan was to redevelop the original design, except that the culverts were to be placed in the bottom of the gullies, culvert sizes would be increased and concrete headwalls and debris catchers would be installed. The new fills would all be placed in 12-inch lifts and properly compacted.

I got along well with Reg Thurber, and he taught me a lot about transportation geotechnology. He was outgoing, extroverted and a good communicator, someone who realized the importance of getting along with people. During my second year with the Materials Branch, he and several other engineers came up with the idea of starting a Toastmasters public speaking club for engineers. Thurber invited me to join. He realized that I was introverted and could benefit from such training.

Toastmasters meet every week, generally for dinner and for about two hours. The program is comprised of five-minute dedicated speeches, two-minute off-the-cuff commentaries on surprise topics, tactful criticism of the speeches and the speaker's diction, and the assignment to chair the meeting in general or selected "table topics." The annual program usually requires every member to give about twelve speeches, each with a specific theme.

The initial speech is an icebreaker — "Who are you?" This is the easiest for anyone to give, but I was terribly nervous when it was my turn. Somehow I staggered through it. By the end of the year I had developed much greater speaking assurance and confidence. To my surprise, I won the club competition, the Vancouver Island competition, and the provincial competition.

The experience has been a wonderful asset for me. Over the years I have presented over eighty technical papers at conferences, have been invited to give ten keynote presentations, and lectured at over ninety universities and institutes internationally. I also taught Geomechanics at the University of British Columbia in the Mining Engineering Department, and each of these opportunities has been greatly facilitated by my Toastmasters experience. I recommend to all young engineers that they develop communication skills early in their career. The Toastmasters or Dale Carnegie public speaking programs are excellent.

After three years with the Materials Branch I had become fascinated with geotechnical engineering. By this time the two other engineers with the Branch had moved on and I had become Thurber's assistant. I decided I would like to specialize and go on to take a master's degree. I applied for several fellowships. Some months later I received notice that I had been selected to receive the Brunner Mond Chemical Fellowship for $4,000. It had two requirements. My thesis topic was to be "The Use of Calcium Chloride to Increase the Length of the Construction Season in Canada," and the award was only tenable at Nova Scotia Technical College in Halifax. That was 5,000 miles away. It would cost me nearly half of the award to travel twice across the continent with my wife and new baby girl. I had hoped to obtain a fellowship tenable at the nearby University of Alberta and study under Dr. Bob Hardy, at the time Canada's foremost geotechnical consultant and professor.

I discussed a leave of absence with Thurber, who said he would agree, but the minister would have to make the final decision. I made an appointment with Gaglardi. At the meeting, I first discussed a leave of absence, which Gaglardi granted, provided I would agree to stay with Highways for two years after completing the Master of Science program. I agreed to that.

I then indicated I would be short of money, "Was there any possibility the Department would help me financially?" He thought for a while and then said he would discuss this with the Highway Board (the four senior engineers in the Department) and try to determine if there was a project I might work on while I was away, something that the Department needed and could pay for.

A week later I was advised that if I prepared a state-of-the-art report on "Highway Construction in Muskeg," the Department would pay me $1,000. I readily agreed. (Muskeg is highly compressible organic material, also referred to as peat).

I made inquiries through the University of B.C. and learned that McMaster University in Hamilton, Ontario, was doing extensive muskeg research under Dr. Bill Radforth. I arranged to spend a week with him and his research team. Most of their work was related to the Department of National Defence's interest in military vehicles' ability to operate in muskeg, and oil companies' interest in drill rig access in the northern tundra. Their research did, however, introduce me to the engineering properties of muskeg.

At Nova Scotia Tech I used the Engineering library to obtain other information from the U.S. on muskeg, and, surprisingly, from Ireland. I finished the report in November and sent it to Victoria. Two weeks later I received the subsidy, and it was a great help.

The difficulty in finding accommodation in Halifax surprised me. We had rented our Victoria house furnished for $90 per month, and had hoped to find furnished accommodation in Halifax for that same amount. I couldn't find anything suitable in the city, as rents were unbelievably high. Finally I located a furnished basement suite in Armdale, about three miles out of the city. It was very inconvenient, since we didn't have a car and would have to use a very infrequent bus.

I kept checking the newspaper ads. After about six weeks I noted a simple ad in the paper: "Two bedroom furnished apartment to rent to responsible couple, P.O. Box 110." There was no rental rate. I had answered many ads, and each answer had tended to become longer, going into details of our house in Victoria, my wife being a Home Economics graduate, I was taking a master's degree in Engineering at Nova Scotia Tech, and that I was a handy repairman around the house.

A message was soon left for me at Tech to call a "Mrs. MacLeod." I phoned and made an appointment to meet her and view the apartment.

I knocked; she answered and invited me in. The apartment was beautifully furnished, antique English furniture in the living room, a cherry-wood dining room suite, and many Royal Doulton figurines. Immediately I realized I couldn't handle the amount of rent that she could realistically justify. I said, "Your apartment and furnishings are beautiful. There's no way I can afford this." I guessed she would want at least $300 per month.

She said, "Come in, I'd like to meet you." We chatted for half an hour. No mention of cost came up. I learned that she had been an army nurse during World War I, and was now a widow. Her husband had been a successful jeweller. Each winter she went to Florida for six months. She was a lovely person and we enjoyed talking with each other. Finally I repeated, "Mrs. MacLeod, I simply cannot afford what the apartment is worth."

She looked me straight in the eyes, "Young man, I like you. What can you afford?" I hesitated to say anything. I didn't want to embarrass either of us. Then she completely startled me. She asked, "Can you afford $125 a month?" I stammered, "But that's not fair to you."

She repeated, "Can you afford $125 per month?"

My mind swirled. I had budgeted $90 per month for accommodation. I thought quickly. Somehow I had to dig up the extra $35 per month. I said, "I appreciate your offer. I accept, and promise we'll really look after things well."

She said, "You can move in November 1st and stay until April 30th next year."

She left everything for us — even some food in the refrigerator. We greatly enjoyed the apartment. It made our stay in Halifax much more pleasant. At the end of the first year Pat and I cleaned through the entire apartment. On her return Mrs. MacLeod wrote to us saying how pleased she was with the condition of the apartment. As a result she reduced the rent for the second winter to $100 per month!

I regret that my wife was never able to meet Mrs. MacLeod, but several years later I was again in Halifax, this time for a geotechnical conference. I phoned Mrs. MacLeod and invited her to dinner. She accepted.

I wanted to do it up right. I rented a car, but the only one available was a convertible. When I called for Mrs. Macleod I apologized for the car. It didn't seem suitable

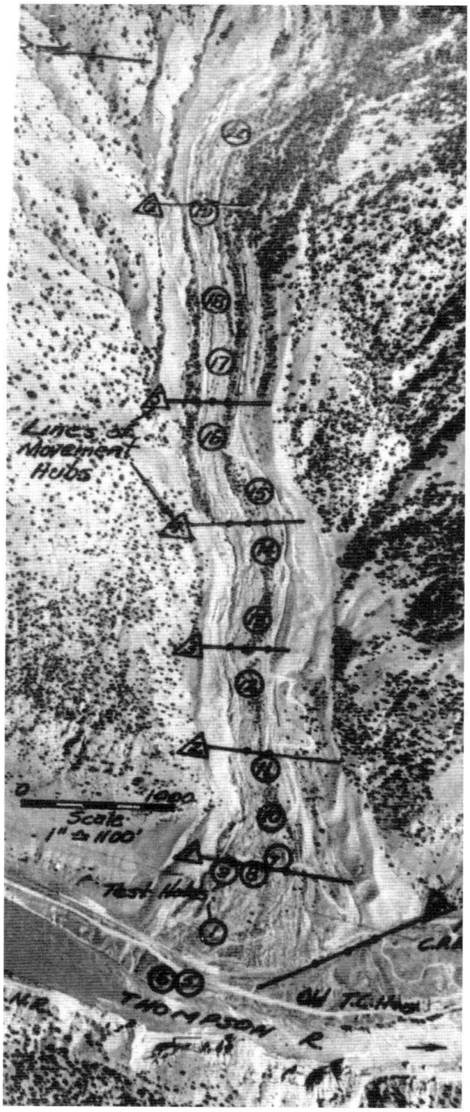

Drynoch Slide near Spences Bridge, B.C. The total movement is 2.2 miles long with annual movement averaging about 3 feet per year. The CP Railway have records of movement dating prior to 1900. An interceptor drainage trench was built across the slide near piezometer 12 and movement line 2.

Drainage trench location near B which extended across and below the failure zone. Perforated 24-inch culverts were placed, and the entire trench was backfilled with sandy gravel. The intercepted water drained northerly to Squianny Creek. The Trans Canada Highway is under construction. The lower slide has been dormant for 40 years since the cross drain was installed.

for the occasion. She looked at me, smiled and said, "Young man, let's enjoy ourselves. Put the top down!" I did. After dinner I drove up Citadel Hill and parked where we could enjoy the view of the city and the harbour. A wonderful lady.

At Nova Scotia Tech, I was the only student under Dr. Geoffrey Meyerhof, an internationally recognized geotechnical professor and researcher. He was always extremely helpful, and to this day I cherish the memories of him, his instruction, and Nova Scotia Tech. He also enjoyed discussing projects that I had worked on in B.C.

To further assist my finances, Dr. Meyerhof arranged for me to instruct in the structural engineering lab. This provided the extra income needed to make up for the higher rent.

After my first year, I returned to Victoria and the Materials Branch for the summer. In my second year, Nova Scotia Tech was short of courses I wanted, so I took two self-directed programs, one in air-photo interpretation and one in Pleistocene (post-glacial) geology. They have both been very useful in my engineering. I believe these courses should be part of every Canadian geotechnical engineer's training.

During the second semester of year two, I was advised that Reg Thurber had resigned to start a consulting practice. I phoned Evan Jones, the deputy minister, and asked if I could be considered for the materials engineer's position. He said they would not fill the position until I returned and was interviewed. That provided me with the incentive to finish my program sooner. Being the only one in the class made this possible. I completed my thesis and exams in mid-March.

On my return to Victoria, I was interviewed and appointed Senior Materials Engineer — four and a half years after joining the Branch. This was April 1958, and, for me, *life-break number three.*

In 1959 I was advised that the proposed Trans Canada Highway location was to cross an active slide area, called Drynoch Slide, near Spences Bridge, B.C. The Materials Branch was requested to evaluate the location and recommend how to deal with the problem.

Background information that was obtained included aerial photographs, topographic maps, geologic maps and climatological data. CP Rail had had problems with their track at this location since before the turn of the century. I spoke to Ken Truman, their regional engineer, who advised that the track moved about 3 feet each year, with the greatest movement in the spring after snowmelt. They had to periodically realign the tracks, and occasionally fix broken rail.

I arranged to travel to the site and inspect the slide area. Upon doing so, I was impressed by the size of the slide, which looked a lot like glacial movement. The upper area extended a full 2.2 miles from the highway. It ranged from about 150 to 400 feet wide, and was confined between two side ridges the full length of the movement.

The slide area was flown over, and up-to-date stereographic air photos taken for three-dimensional review. Topographic maps were developed from the photos to a scale of 1 inch to 100 feet, with a 5-foot contour interval. The slide boundaries were readily identified on the air photographs.

Following the site inspection I drafted a memo to Fred Brown, Chief Engineer, describing the options available. They included:

1. Construct two bridges across the Thompson River to bypass the slide area,

2. Construct a 1,200-foot centre span suspension bridge to cross over the slide,
3. Tunnel under the slide, or
4. Stabilize the slide and construct across it.

Because of the high cost of the first three options, it was recommended that a program be developed to stabilize the slide. This would require an extensive geotechnical investigation. Approval was requested for special funds, and Brown approved $100,000.

Horizontal drains being installed in the upper slope of the trench being cut across Drynoch Slide to stabilize that slope during excavation and prior to backfilling with free draining granular backfill. Note the seepage from the slope.

Ten boreholes were drilled through the slide into bedrock on the centreline of the slide, spaced about 500 feet apart, commencing just above the railway. The soils were sampled every 5 feet and tested in the laboratory. An open pipe, called a standpipe piezometer, with slots, was installed in each hole into bedrock. The depth to the water table was measured with an electric down-the-hole probe, and the depth where the pipe was bent or broken by slide movement was measured by lowering a 2-foot-long piece of reinforcing steel. This rod wouldn't go past a break or moderate bend. Later more sophisticated testing instruments were developed for this purpose.

In a borehole about 2,000 feet above the river, a sample of wood, perfectly intact, was encountered at 55 feet. It was carbon dated and determined to be 6,000± years old. This was interpreted to approximate the original date of the slide.

At the same time, permanent survey stations were located on stable ground on each side of the slide at the "sections" (survey lines measuring the topography) of the boreholes. From these stations, the movement of the top of each borehole could be measured. Readings were taken monthly. The depth to the water table was also measured.

The rate of movement was calculated each month and plotted against the depth to the water table. The movement was greatest when the water table was highest — during the springmelt period. Raising the water table reduced the shear strength of the soil.

The graph indicated that if the water table was lowered a further 6 feet in the fall dry season, the movement should cease. If it could be lowered a further 15 feet, the safety factor (ratio of resisting forces to driving forces) could be increased to be adequate for long-term stability.

The drilling and surveys indicated that the slide was narrowest about 1,500 feet above the proposed highway location, and about 55 feet deep at this location. A design to obtain groundwater lowering there was developed and presented to the Chief Engineer.

The proposal to stabilize the slide involved excavating a deep trench across the slide from the narrowest point to 5 feet below the base of the slide, placing two perforated 24-inch-diameter asphalt-coated metal culverts in the bottom of the trench, and then backfilling the excavation with clean sandy gravel. The pervious trench would intercept all the water seeping down through the slide, and lower the water table. The intercepted water would then flow down a deep creek channel to the north toward the Thompson River.

The contract to dig the trench, place the culvert and backfill the trench was approved and called for the late fall when the water table was lowest. Special horizontal drains were installed to lower the water table and maintain the trench slope stability on the upstream side during excavation. The trench was excavated in two lifts and the culverts placed. The entire trench was backfilled with pervious sandy gravel. The contract was completed in two months.

Movements between the trench and highway stopped before the trench was backfilled, and the entire slide zone between the drainage trench and the river has remained stable since 1959. Lowering of the water table was successful, and achieved at a reasonable cost. The new Trans Canada Highway was constructed across the stabilized slide at a cost of less than 10 percent of the other alternatives. The lesson from this successful project is that, provided the basic fundamentals of landslide engineering are understood, very large scale landslides can be stabilized.

In late 1958 the Department of Highways had decided to construct the Trans Canada Highway into Vancouver from the east as a freeway. Engineers from the suburban areas of Mallairdville and Burnaby advised the Department that they would have to select a route through areas that were unsuitable for commercial or residential development. The areas they indicated as appropriate were swampy, with considerable depth of muskeg or peat. This material is organic and highly compressible — unfavourable as foundations for a major highway.

Fred Brown approached me and asked, "Chuck, will you perform a preliminary site investigation and determine potential options? Try and have information for us in two months."

I answered, "I'll get a program started immediately."

I then arranged for Dick Readshaw, my Assistant, to lay out a drilling and testing program to provide the geotechnical information required.

The results were challenging. Over a length of about 4 miles, weak compressible muskeg existed to depths of 35 feet, with soft clay underneath that.

The typical procedure in the past had been to excavate all the muskeg and replace it with granular soils. The excavation was done by backhoe, draglines or displacement by blasting. Some of the proposed route was adjacent to the Canadian Pacific Railway. Excavation of the deep peat would result in collapse of the railway grade into the excavation.

I was familiar with the concept of "preloading" weak silts and clays to increase their strength and bearing capacity, and to reduce long-term settlement. It occurred to me that preloading may be applicable to muskeg.

Arrangements were made to drill several boreholes to obtain samples for large-scale consolidation tests in the laboratory. The initial results were encouraging. The muskeg consolidated rapidly into a moderately hard mass about

Test section on the Burnaby freeway portion of the Trans Canada Highway to preload up to 35 feet of organic muskeg (peat). Up to 9 feet of hogsfuel or coarse sawdust was placed as a weightless spacer under incremental thickness of granular fill and surcharge.

two-thirds its original volume. The concept of preloading was discussed with the Department's Chief Engineer, and approval to construct a full-scale field test was granted.

An initial test section was constructed on Lulu Island, where the peat was 8 to 10 feet deep. The test plot was 100 feet square. Eight feet of sand was placed over the peat in two lifts. Instrumentation comprising piezometers, settlement gauges and lateral movement gauges were installed. (Piezometers are used to determine water pressure and depth to the water table.) After two months, the settlement totalled about 2.5 feet, and had almost stopped. When 3 feet of sand were removed, a slight

Preloaded peat area with final grade construction underway, Burnaby. This highway has given 40 years of excellent service. Preloading peat is now used internationally.

rebound occurred.

Consolidation tests were then performed on samples in the Burnaby area, where the peat was 25 to 35 feet deep. Up to 9 feet of settlement was predicted from the laboratory tests. Since the natural water table was near the surface, the fill would settle and create a canal.

After much thought, the idea of "a weightless spacer" occurred to me. The Vancouver area at that time had many sawmills with sawdust and coarser hogsfuel (wood chips and bark) as a waste product. Provided the sawdust would remain under water it would last for decades. I thought that sawdust might provide the spacer. In addition to being weightless under water (actually it provides some buoyancy), it provided good trafficability for loaded construction vehicles.

Three test sections were constructed, two in Burnaby and one in Mallairdville near the railway. The test fills comprised a thickness of sawdust equal to the estimated settlement that would occur under the fill loading — 8 feet of sand.

Initially all scrub trees were cut and left on the muskeg surface. They would be below the water table so would not deteriorate. This was followed by installation of settlement gauges, piezometers and lateral movement gauges. The thickness of the sawdust spacer varied from 4 to 9 feet, depending on the depth of muskeg. This layer was then covered by 8 feet of sand placed in two, 4-foot lifts. The second lift was placed after the settlement caused by the first lift had decreased substantially. The muskeg gained strength after placement of the first lift.

The behaviour of the test sections was exciting. Settlement was rapid and practically ceased after four months. At that time 3 feet of sand was removed, which brought the surface to approximate final grade.

The test sections proved that constructing about 4 miles of freeway would be feasible, surprisingly economical, and much faster than total excavation. Fred Brown approved the procedure.

The National Research Council in Canada became aware of this research and invited me to prepare a paper for their annual Muskeg Research Council Conference in Ottawa. The attendance at the conference was international, and the test program received wide recognition and subsequent use. The next year the Highway Research Board in Washington, D.C., requested a presentation. This was given by Norman Lea, who had supervised the field construction, and myself.

In the course of the B.C. government's decision to construct a highway from the northern Transprovincial route near Hazelton, north to the Alaska Highway, it became apparent that a link to Stewart, B.C., on tidewater was essential. This would involve an approximate 40-mile highway from Meziadan Junction, west through Bear Pass.

About a mile west of Bear Pass, a glacier from a side valley to the south filled the main valley through which the new highway would be constructed. In order to bypass the glacier, the route at this point was designed to run along the north valley wall, about 400 feet above the valley bottom. The depth of the glacier in the valley was about 200 feet.

During the first year of construction in 1959, the highway was built in two sections, one east out of Stewart and the other west out of Meziadan. The next spring a serious, short-event flash flood occurred in Bear Creek. It washed out several miles of the newly constructed highway east of Stewart.

Ted Webster, Director of Construction, contacted me and asked, "Would you investigate the problem?" At issue was whether the replacement construction should be relocated to higher ground, which would be much more expensive, and take longer. I agreed to investigate the flood.

The area was first inspected from a helicopter at high, intermediate and low elevation. This was followed by a ground inspection. It was apparent that the glacier that crossed the valley had essentially created an ice dam about 200 feet high. The snowfall the past winter had been well above normal, and spring developed early, with weather much warmer than usual. Above average runoff occurred and a lake had developed upstream of the glacier. However, at the time of the inspection, the lake had completely drained. Nevertheless, the lake elevation and shoreline that existed only 48 hours previously were readily observable. The maximum lake elevation had been about 15 feet below the top of the ice; that is the lake had not overflowed the glacier.

Instead Bear Creek was flowing under the upstream edge of the glacier and exiting from under the downstream edge. The creek channel had been scoured about 15 feet below the base of the glacier. Large boulders were strewn along the channel downstream of the glacier, indicating very high volume and velocity had occurred there very recently.

Realizing that ice is lighter than water, the thought occurred to me that when the elevation of the lake reached about 90 percent of the thickness of the glacier, the glacier would float. As it lifted, a major flow developed below the glacier under an initial pressure head of about 180 feet of water. The velocity would have been very high initially, scouring a channel in the soil under the ice. As the water level lowered, the glacier settled back to its original elevation. However, the scoured channel

Bear Glacier near Stewart, B.C., dammed Bear Valley. A lake developed upstream and lifted the glacier. The lake drained suddenly and washed out several miles of new highway downstream. The trench was developed across the top of the glacier to lower future lake levels. A jökulhlaups-type failure.

remained and the entire lake drained. Gradually the ice deformed to fill most of the channel, but the creek flow continued under the glacier until the next winter.

I realized that if a channel were cut across the glacier about 30 feet deep, any future high-lake level would flow through the channel and not develop enough pressure head to lift the glacier.

Upon my return to Victoria, I looked for references on glaciers that created lakes, and discovered references to "jökulhlaups," a term that referred to precisely the type of occurrence I was postulating. I hadn't discovered something new, but I had recognized the mechanism of the failure. There are several other areas along the Coast Mountains where the phenomenon occurs annually. The best example is the

Taku Glacier east of Juneau.

I reported my findings and conclusions to Webster, recommending that a channel be constructed across the glacier that summer. Webster agreed, set up a special budget and asked me to arrange for the work to be done. I contacted the regional construction engineer in Prince George, Mike Nesbitt, to meet me at Stewart and assist in arranging the program.

Neither of us had previous glacial experience. However, we jointly agreed to try excavating the trench with a D-8 Caterpillar equipped with dual rippers. Excavation proceeded rapidly to a depth of about 10 feet through the weaker upper ice.

Drilling attempted to blast the hard glacial ice to develop the trench. The procedure was a failure. Instead a D-8 Caterpillar with hobnails welded on the tracks and a single ripper were used to develop the trench.

Gradually however, the ice became dense enough that the Caterpillar tracks began slipping on the ice. The ice was now blue in colour, and upon close inspection live ice worms were observed. They were about the size of typical earthworms, but lighter in colour. My family still doesn't believe this!

With the Cat now unable to gain any further traction, the trench wasn't deep enough. We discussed the problem and decided to try drilling holes for explosives, similar to blasting rock.

A blast drill was brought to the site. The drill steel came in 8-foot lengths. The first 8-foot hole drilled rapidly. The drilling was stopped and another length of steel added. We tried to start the drill. It wouldn't move! The drill bit had frozen solid in the hole! We tried several other holes, and the same thing happened. Back to the drawing board.

We put our heads together again. At this time the contractor's foreman arrived on the site to retrieve the Caterpillar, so his voice was added to the discussion. He suggested, "I think the pressure on the tracks isn't high enough. Why don't we weld a two-inch by one-inch piece of steel on each track panel — like hobnails on boots. This will greatly increase the contact pressure and increase traction." Nesbitt suggested replacing the dual ripper with a single ripper to further increase pressure. We agreed that was a good idea. It took a day to weld the "hobnails" on and install the single ripper.

The following morning the excavation was restarted. The redesign worked like a charm. The Cat ripped and pushed the ice from the centre of the glacier both upstream and downstream over the flanks of the glacier. The total length of the trench was about 300 feet. The bottom was excavated on a 1 to 2 percent grade downstream, and about 20 feet wide.

To finish off the trench, several inches of dark sand were spread over the bottom. During the summer, the sun heated the sand sufficiently to lower the base about another 10 feet by melting, which was very cost effective. The design worked and no further flash floods occurred. Over the next four years, the glacier melted sufficiently that a much lower highway grade was feasible. Today, Bear Glacier has receded completely out of the valley and about 1,000 feet beyond.

As the Trans Canada Highway was being developed in the narrow Fraser Canyon, south of Boston Bar, the route selection faced difficult problems. The Canadian Pacific Railway ran on the west side of the river and the Canadian National Railway on the east side. The original highway crossed the river on an old, narrow, low-level suspension bridge. A new crossing was to be located about a mile downstream, crossing the river and both railways.

A.B. Sanderson Consultants of Victoria, B.C., were selected to design the bridge. They were also responsible for arranging the foundations investigation. Their drilling contractor recovered 5- to 8-foot lengths of sound granite rock core at about 80-feet depth, and the drilling consultants interpreted this to be bedrock. Based on this, a rigid steel arch structure with about an 800-foot centre-span was designed, a type of structure which cannot withstand much differential lateral settlement.

The contract was called and awarded. The contractor excavated both foundation sites to the depth of expected bedrock. Instead of hitting bedrock, however, nests of large boulders were encountered at both arch foundation locations. The contractor phoned Sanderson with the unfavourable news. Sanderson contacted the Chief Engineer, Mr. Brown. He phoned me, explained the situation and asked, "Chuck, will you go up to the Alexander Bridge site and evaluate the foundation conditions? Bedrock has not been encountered where it was supposed to be."

I agreed, "I'll go up first thing tomorrow morning." I caught the 7 A.M. ferry from Nanaimo to Horseshoe Bay (near Vancouver), drove about 150 miles and arrived at

West abutment of the Alexandra Bridge under construction over the Fraser River and Canadian Pacific and Canadian National Railways. The arch foundations on sandy gravel were preloaded by applying the total bridge load with tensioned anchor cables to bedrock and reducing the tension gradually as the bridge was constructed.

the site by 1:30 after stopping for lunch.

I asked the resident Bridge Engineer to show me the foundations report, which included the drill logs. The report had indeed concluded that bedrock had been encountered. I went to look at the west-arch footing excavation. Boulders ranging from 6 inches to about 8 feet were exposed in the base, with no sign of bedrock. I inspected the east-arch footing excavation and found the same conditions. I realized that further drilling was required as soon as possible. The construction was delayed until the issue was resolved.

Edco Construction of Vancouver was constructing the new highway near the west end of the bridge. I made contact with the superintendent and arranged to rent an air-track percussion drill and crew from him to drill a hole at the west-arch footing site, and at the thrust angle of the arch. The driller commenced drilling, and I timed the rate of penetration per foot. The rate varied, but penetration continued without much difficulty to about 60 feet. At that depth the drill rate slowed significantly. Drilling continued slowly for another 15 feet. I suspected the drill had finally hit bedrock. The same condition was encountered at about 70 feet at the east-arch footing.

The first 60 to 70 feet below the boulder bed was silty, sandy gravel, old riverbed materials that could settle several inches under the heavy bridge loading or a severe earthquake. It would likely not settle uniformly across the footing. The rigid arch-bridge design became suspect.

On my way back from the site I concluded that the bridge could be built, as designed, if the granular foundation soils could be consolidated under a load greater than the ultimate bridge would provide. The question? How to consolidate the soils? I thought two options were available:

a) Construct the arch foundations and load each with 1,600 tons of steel, half the weight of the bridge. The steel could be railway rails rented from the adjacent railways. The procedure would be to place the steel on the abutments with more rails on the mountain side of the abutments to allow for the nonvertical thrust angle and measure settlement. When settlement stopped, which in the granular soil would be rapid, in the order of weeks, the steel could be removed and the bridge constructed.

b) Drill 6-inch-diameter holes down to and into the bedrock and set a steel casing. Lower high-tensile steel cables into the pipe, 20 feet into bedrock. Fill the lower 20 feet with concrete to anchor the cable into the bedrock. Develop sufficient of these anchors and tension them to continuously apply 1,600 tons of load. Monitor the settlement until it ceased. Then unload the cables (as the bridge was constructed), concrete the anchor holes, and construct the bridge. The soil profile and rock depth would have to be confirmed with standard rotary drilling and sampling.

I discussed these options with Brown and Sanderson. It was a new concept to them and I could see they were uncomfortable with it. So I said, "These are unusual proposals, but I believe they'll work. I suggest Dr. Karl Terzaghi at Harvard University be retained to review the project and my proposals." Terzaghi at that time was the leading geotechnical engineer in the world.

Brown said, "Good idea," and gave instructions to his secretary to place the call.

In a few minutes Terzaghi was on the phone. Brown explained the project and asked him if he would review it and recommend how to proceed. He said, "I'd like to come. I enjoy working in British Columbia, but I've just returned from Brazil and am very tired and have an assignment to finish here. I would suggest you contact Hugh Golder in Toronto. He's just establishing a consulting practice, and has extensive international foundations experience." He gave Brown the phone number.

Brown wasn't able to contact Golder until the next day, but when he did, Golder agreed to leave the following morning.

On his arrival in Victoria, I reviewed the site conditions with Golder, as well as my thoughts on the options. From a political standpoint the bridge location could not be changed. Golder said, "I must see the site." We agreed this was essential and proposed to travel to the site the next day.

Golder was a tall, slim and typically proper Englishman. He had been senior geotechnical engineer for a major civil engineering contractor in England for many years, but had left them to spend time at Harvard University with Terzaghi. Following some teaching, he wanted to return to the practice of engineering. He was friendly, had a sense of humour, appreciated proper English, didn't tolerate fools, and liked red wine and Dewars scotch and soda — without ice. He said, "Why dilute a good drink!"

Golder, Sanderson and I drove to the site and spent about four hours inspecting the excavations and local geology. On the drive up, Golder had brought us up to speed on his experience, and on the way back, we discussed the options for construction. Golder concurred that preloading the arch foundations would be a reasonable solution. He preferred loading with the tensioned anchors into rock because the densification was in the line of the arch thrust, and subsequently wrote

Major slide in weak lacustrine clay which liquefied due to excess fill loading during highway construction. Lakelse Lake area near Terrace, B.C. No geotechnical investigation had been performed. The clay was displaced and a new fill constructed.

a report with this recommendation. Sanderson was still somewhat uneasy with the atypical approach. He would have preferred the footings on bedrock. He agreed to the preloading with the proviso that the bridge design be changed to a three-hinged arch, a type of structure that could tolerate more differential settlement.

I got along well with Golder. Over the next three years I arranged for him to review two other major projects, both landslides. One was a large slide that developed in the Kicking Horse Canyon, where a major highway fill had impacted the CP Rail line. A second project occurred near Terrace, B.C., where the highway construction induced two major soil slides into Lakelse Lake. Golder worked with me on the redesign at both locations. Little did I realize that in a few years our relationship would be of an entirely different order.

Shortly after Alaska had been made a state, the U.S. cabinet arranged for a study to be done on linking the panhandle cities, Wrangell, Petersburg, Ketchikan, Juneau and Skagway, to the north-south Stewart-Cassiar Highway in British Columbia. Brown & Root, consulting engineers from Houston, Texas, were commissioned to perform the study. They sent an experienced engineer, James Burch, to meet with B.C. Highways officials and inspect the potential routes. Realizing that such highway connections would be of great benefit to B.C., Tom Miard, deputy minister of highways, met with Burch.

To facilitate the study, Miard arranged for four B.C. Highways Department engineers, along with Burch, to inspect potential connector routes from the air. They were Mike Nesbit, Regional Construction Engineer; Art Beaumont and Al Tranfield, Location Engineers; and myself as Geotechnical Engineer.

We all flew to Watson Lake at about Mile 600 on the Alaska Highway on Canadian Airlines. We stayed overnight at the Watson Lake Hotel, at the time reputed to be the largest hotel in the world constructed of logs. The lounge was unique. It had a party table, which sat about twelve in a circle.

After dinner we assembled around the lounge table. There were four others there. Before long we began to tell jokes, in order around the table. After about two hours most of us had run out of stories, but not Burch and Fred Baker, an Alaskan from Fairbanks. When Burch began to struggle to remember more, the Texan showed his annoyance by commenting that Baker's jokes weren't as worthy as his own. Baker smiled, got up, walked around the table to Burch, put one hand on his shoulder and said in his quiet Alaskan way, "James, if you're not careful we'll split Alaska in two. That will make Texas the third-largest state in the Union." Everyone had a good laugh. Burch hesitated, then joined in the laughter, stood up and shook Baker's hand. Peace was restored.

From Watson Lake we had chartered Doug Chappel, pilot of Trans Provincial Air, to fly us around the north in a six-passenger Beaver float plane. We took off from Watson Lake and headed south. Burch wanted to see the standard of highway being built south of Cassiar, a road that skirted Dease Lake for nearly 20 miles. We landed on the lake with the completed highway close by, and Chappel nosed the plane up onto the beach. We all got out and stepped along the pontoon to the shore. Chappel tied up the plane and we walked up to the highway. During its early years, the road was intended to be a gravel-surfaced highway, designed for access to natural resources, particularly mining and forestry. In the long term it would be paved.

We walked along the road for about a mile. Burch measured the road and shoulder width, and also the ditch width and depth. He stopped at several fills and culverts and took notes and numerous photographs. He had previously been given a set of plans and sections.

On returning to the plane we enjoyed sandwiches, fresh fruit and coffee. We took off again, heading south to gain an aerial view of the potential location for a crossing of the major Stikine River. But the sky had become cloudy and some low-lying fog began to develop. Chappel said, "The weather is closing in. I'm going to set down and wait for it to open up more." He landed on Mortua Lake, not far from the Stikine. We had planned to fly over several river tributaries toward Alaska that day. It appeared that we wouldn't make it.

We stayed about 500 feet out from the shore for about an hour. The wind would blow the plane toward shore and Chappel would start the engine and taxi back out. After an hour he said, "I've got a collapsible fishing rod on board. Would anyone like to fish?"

Burch responded, with his Texas drawl, "I haven't fished since I was six years old. I'd like to give it a try." Chappel pulled out a sectional rod, put it together, attached a spinning reel and tied a black fly to the line. We had to show Burch how to spin cast. It was slightly awkward, because he had to cast with one arm while holding onto a wing strut with the other, so he wouldn't lose his balance as he stood on the pontoon. The plane was bobbing quite a bit in the wind and waves.

His first two casts were quite a mess. The line became tangled and it took us some time to reorganize it properly. The third cast was quite a good one. Bam! A trout swiped at the lure, coming halfway out of the water. In his excitement, Burch let go of the strut, falling part way into the water. We caught him after one leg had submerged. However, he held onto the rod. We coached him on how to play the fish and gradually bring it toward the plane. Chappel reached down, slipped his fingers through the gills, removed the hook and hit the fish on the head with a heavy screwdriver. This was long before the days of catch and release. Two hours later, Burch had brought in four beautiful trout, ranging from about two to four pounds. He was ecstatic.

Chappel found a burlap bag to put the fish in. Then he noted, "The fog is lifting. I'd like to go up and take a look. I want to get through to Atlin by dinnertime, if possible."

Burch requested, "I have to get a picture of the fish. Otherwise they won't believe me back home."

Chappel agreed, "Okay. Get in. I'll taxi in to shore and you can get some pictures."

On shore, Chappel cut a willow limb about three-quarters of an inch thick and strung the fish through the gills. Burch wanted the plane in the background. A great photo was taken with Chappel and Burch each holding an end of the willow branch. At that moment I think Burch had totally forgotten about his engineering assignment. While on shore, Chappel also cleaned the fish, washed them and returned them to the bag.

Shortly after we took off for Atlin. The weather was still cloudy and windy. The plane bounced a lot, and everyone tightened their seatbelts a little. We flew through several passes, manœuvring around big fluffy clouds and mountain peaks. Near six o'clock we reached Atlin Lake and headed toward the town.

As we came in for a landing I noticed a Royal Canadian Mounted Policeman on the dock, and suddenly realized Burch did not have a Canadian fishing licence. The fine could be as much as $500. I tapped him on the shoulder and told him not to say anything about the fish.

Fat chance! Burch was so excited that when we stepped onto the dock he proudly announced to the officer, "You should see the fish I caught! One of the great thrills of my life!" The Mountie smiled but kept quiet. But that wasn't enough! It's also illegal in Canada for fishermen to catch freshwater trout and have them cooked in a restaurant. Burch next said, "I've got to have fish for dinner tonight. Would the restaurant do this for us?"

Flying in northern British Columbia to locate potential highway connector routes for Wrangell, Ketchikan, Petersburg, Juneau and Skagway to the north-south Stewart-Cassiar Highway in British Columbia. The study was commissioned by President Eisenhower.

Chappel had filed a flight plan for Atlin, and the Mountie knew him well. He also knew that Chappel often flew on government business with visitors. To this day I'm certain that the Mountie suspected Burch didn't have a licence and didn't know the regulations. But this was the North. He grinned and said, "Go up to the restaurant and ask Kim the Chinese cook to prepare the fish. And make sure you give him a big enough *tip* for all the meals!" He almost yelled the word tip. The advice was obvious. We all smiled.

The dinner was great. Burch gave the cook a U.S. one hundred dollar bill, and everyone was happy. After dinner we explained to Burch that he had bent Canadian sport fishing laws beyond what should have been the limit. I'm sure it made a good story for him to tell back in Texas. But I bet the fish were bigger than two to four pounds.

The next day we spent about six hours in the plane, flying over numerous potential highway routes across the Alaska panhandle. We would circle, make notes on bridge sites, culvert locations, and decide which side of the valley the road should be on. Unless the north or drier side was extremely rugged and rocky, it is preferred, particularly in high-precipitation areas. This was not far enough north to be in permafrost.

Late that afternoon we were flying south over the Iskut River. B.C. Highways had a location crew camp at Bob Quinn Lake that Chappel wanted to reach, where we could refuel and stay overnight. He had to fly over a high hill to the east to get there, but as we approached the lake we saw that the cloud had dropped to where the top of the hill was hidden. Chappel began to circle and gain elevation until he was just into the cloud. He then circled back down, slowly losing altitude until we could again see the ground and river below. He next put on a pair of yellowish sunglasses. He repeated the manœuvre twice more, then said sharply, "The lake is open."

He looked at his watch and climbed easterly up through the clouds. We flew for about four minutes in dense clouds. Then he descended rapidly. Suddenly we were out of the clouds and over a small lake. Within a few minutes we had landed.

Afterwards, I asked him what the manuœvring was all about. He said he didn't know if Bob Quinn Lake was clear, so he circled up and down to stir the clouds and try and get a glimpse of the lake. His tactics worked, and once he saw it was clear he timed his flight and elevation to get over the mountain and through the clouds. The sunglasses were special ones that allowed clearer vision in cloudy conditions. Nice flying!

The next morning dawned without a cloud in the sky. The plan was to fly out along the Iskut River to look for a route from near Bob Quinn to Juneau. We took off and were flying about 2,000 feet above the river when, without any notice whatever, the Beaver flipped upside down! Fortunately we all had our seatbelts on tightly.

Almost as suddenly as we had flipped, Chappel righted the plane with a sharp bank and descent. He yelled, "Is everyone okay?" Several articles that had been on the floor had fallen to the upside down ceiling and back again. Everyone was okay. It happened so quickly there wasn't time to be afraid. He explained later, "We hit a wind shear, which occasionally happens in clear air, but that was the worst I've ever encountered." Chappel headed back to Bob Quinn, landed, and did a thorough check of all struts, flaps and control wires. Everything seemed to be all right.

For two more days we flew over potential routes, taking notes and photographs and exchanging thoughts and recommendations. Both days were uneventful. About noon the last day we flew toward Terrace, the end of our aerial inspection.

The weather was clear, with only some forest fire haze. Lakelse Lake, home base for Chappel, was perfectly calm, hardly a ripple on the water. Chappel circled. I was sitting behind him, and he turned to me and said, "Chuck, will you find the food bag under your seat and take out a can of food, any can. Slide your window open and throw the can hard toward the lake when I tell you to."

I found the can and opened the window. Chappel dipped the plane and yelled, "Throw!" I did; he continued on for about ten seconds and then made a steep 180-degree turn and dropped down for a landing. I noticed the water had rippled from the splash of the can in the water.

On landing Chappel explained, "The lake was so smooth I couldn't judge my distance above the water, so I needed you to create a ripple for me. Many pilots have misjudged their elevation under such conditions and crashed."

It was an interesting trip. Good jokes, good fishing, great company, airborne excitement, a new procedure to select highways, and top-notch flying lessons. I'll bet it's a trip James Burch will never forget. I won't either.

In early 1960, oil exploration was developing in the isolated northern corner of British Columbia east of Fort Nelson. The area was uninhabited; a landscape of tundra with extensive muskeg deposits overlying minor-clay, glacial till and bedrock. Surface water was extensive and drainage poor.

Access for heavy drilling vehicles was impossible during summer months due to extremely poor vehicle trafficability, which, if allowed, would cause indescribable environmental damage to the sensitive tundra. As a result, oil exploration had been confined to the winter months when the ground, lakes and swamps were frozen.

In order to bring equipment to the exploration area, it had to cross the Sikanni Chief River, which did not freeze over adequately until about mid-December. This was about six weeks later than vehicles could travel on the frozen muskeg. The industry therefore made a major submission to the B.C. government to construct a bridge, and the government agreed.

Jack Alton, the Highway Department's Bridge Engineer asked me if I would go to Fort Nelson, scout the Sikanni Chief River, locate a bridge site and arrange for a foundation investigation. This was in early August. Arrangements were made to obtain air photos, topographic maps, geologic maps, river flow data and climate summaries.

I flew to Fort Nelson on a Monday. I had decided the best way to inspect the river would be to hire a local resident with a riverboat. I contacted the RCMP and asked them if they knew of anyone locally who could provide a riverboat and act as guide.

Constable Jim Weir said, "I know just the man. Frenchie Thibault has been trapping, prospecting, hunting and logging in this area for years and has a reputation for being able to do almost anything. He lives west of town with an Indian lady in a log cabin. I'll draw you a map with directions."

I drove out and encountered a well-used log cabin with several adjacent lean-to structures. Outside were several boats and a half-ton 4x4 pickup. A dog sled, snowshoes and steel traps hung on an exterior wall. Frenchie was home. I introduced myself and described my project. He was very interested, so I showed him the maps and air photos. The latter intrigued him, so much so that I promised to send him a set when I returned. I needed my set for the field inspection. He agreed to provide a long boat, 25-horsepower motor, gas, his own food, a rifle for our safety against bears, an axe, machete, a compass and long tape. The long boat was required so we would not get caught in whirlpools in the river.

I guessed that we would require at least two extended days. We agreed on $200 per day in cash. I didn't have that kind of money with me so I had to meet with the government agent in Fort Nelson to arrange for payment. The next morning Frenchie met me with his boat and equipment at the floatplane dock on the nearby Muskwa River, into which the Sikanni Chief flowed.

We spent two days along the river, travelling about 8 to 10 miles in each direction. We looked for indications of high water marks along the shoreline. This included sand bars, debris deposits along the banks, skinned bark on trees and particularly debris caught up in the branches of trees. The height of high water could vary at different places because of the variable river cross section, water velocity associated with rapids, river curvature and particularly ice jams occurring during spring break-up.

We initially located three potential bridge sites. The last half of the second day we revisited each. I took numerous ground photos and we ran approximate survey sections. Ultimately I chose the site that was nearest Fort Nelson and slightly narrower than the two other sites. Frenchie agreed. I located the three sites on the air photos.

Frenchie's help on the job was invaluable. He prepared a fire to make coffee for each lunch, and provided a solution to what was our biggest problem — the mosquitoes. He smeared on a waxy substance to keep them away. It smelled rotten. He wouldn't tell me what it was, saying instead, "If I told you, you wouldn't use it." I used it the second day; the mosquitoes were that bad.

In late afternoon of the first day we rounded a bend in the river and encountered six wolves on the bank. Eating a small deer, they looked at us, but paid little attention. Frenchie assured me, "They won't bother us. They're well fed."

Late the next day when we had finished, I stopped with Frenchie at the government agent's office to get his money. I gave him an extra $50 and thanked him.

I stayed overnight at Fort Nelson, planning to catch the daily plane, a DC-3, the next morning at 10:30. Prior to leaving, I dropped in at the RCMP office to thank them for putting me onto Frenchie. Weir said, "There are many stories about his accomplishments. One of the best occurred last spring. Pacific Petroleum had a drill rig and service vehicles east of the river in muskeg terrain. They were nearing their drill target depth when an early spring break-up began. They continued to drill. When they had completed the hole they tried to move the drill rig and vehicles out. They got everything stuck, potentially to spend the summer there."

The drilling manager of Pacific Petroleum advertised over the radio for someone to remove the equipment. Frenchie answered the invitation, but his $7,000 bid was not the lowest. One week later the low bidder returned and advised that he couldn't get the equipment out. The manager went back to Frenchie and said, "Frenchie, I accept your offer of $7,000 to get our drill equipment to Fort Nelson." Frenchie replied, "My cost is now $10,000. It will be more difficult now." The manager objected but Frenchie held firm. Pacific Petroleum had no choice but to accept.

One week later, Frenchie arrived back in Fort Nelson with the drill rig and vehicles. He parked them in front of the Pacific Petroleum office, went in and asked for payment. The manager said, "Come with me to the bank." Frenchie's payment was arranged. On the walk back to the office, the manager said, "Frenchie, well done! How did you do it?" Frenchie smiled slyly and answered in his French-Canadian accent. "I no tell you. Next year, maybe you get stuck again!" Smart fellow.

Later in 1960, the Highways Department Bridge Engineer asked Materials Engineering to perform a geotechnical foundation investigation for a new bridge across the Thompson River about 1 mile south of Spences Bridge. We received a plan of the

The Thompson River Bridge near Spences Bridge under construction. A foundation investigation was performed for each pier.

proposed bridge. There were to be six separate spans crossing the river at a 30-degree angle. The normal practice was to drill one borehole at the centre of each abutment and pier.

Ray Bordeleau, our most experienced driller, was assigned the work. A barge with a drill mounted on it was secured for the river holes. The foundation conditions were typical of a deep river-cut valley. From 20 to 40 feet of sandy gravel with boulders were encountered, and below this a thin boulder layer over very dense glacial till. Two deep holes were drilled to bedrock, which was encountered at between 80 and 95 feet. During each spring freshet, the flow in the Thompson River increased substantially, eroding much of the loose sandy gravel. As the flow reduced during the summer, the sandy gravel was slowly replaced. This layer is described as the active scour layer.

Foundations for the bridge had to be founded into the dense glacial till for long-term stability. Steel H-piles were recommended and selected. Special points welded to the pile tips were developed so the piles would penetrate the till. To resist the lateral pressure of the water and ice against the piers, the four downstream piles at each pier were battered downstream at an angle of 1 horizontal to 6 vertical (1H:6V).

During construction, I inspected the pile driving for the central river pier. The piles under the downstream half of the pier were driven first, from the centre of the pier downstream. About half the upstream piles had been driven when I arrived, and I watched as the next was driven. It had soon penetrated about 35 feet of sandy gravel, each blow of the hammer driving the pile about 3 inches. At 35 feet the driving became very difficult, obviously in the glacial till. The pile was driven about four feet into the till to "practical refusal."

The next pile behaved differently. It was still going at 50 feet. A 30-foot length was welded on to the top. It drove another 10 feet before refusal. We assumed it had hit a scour hole. The next pile also drove 50 feet, and so a 30-foot section was again welded on. This drove the full depth. Another 30 feet was added. This also drove that added depth. The total length was now 110 feet, and the driving resistance remained reasonably similar. We didn't believe the pile had broken, which could occur at a weld. After more discussion, we decided to add another length. Driving recommenced.

Suddenly, the site engineer yelled, "Look upstream; there's our pile!" It was slowly coming up above the surface of the water! The pile had done a 180-degree bend! It had apparently hit the edge of a deep scour hole, bent and followed the scour hole along the surface of the dense till or bedrock and continued up until it broke the water surface.

Rather than trying to pull the bent pile out, it was decided to have a diver cut it off at the existing riverbed. The rest of the upstream piles were driven next. They hit the till or bedrock at 70 feet. The intermediate piles had sharper and longer drive points welded on, and these were successfully driven to solid foundation. During the next freshet, when the fast water had re-scoured the sandy gravel, several hundred tons of hard large riprap were dumped into the scour hole around the piles to provide long-term stability.

Based on that experience, for future major bridge foundation investigations, wherever scour holes could develop, I instructed that boreholes be drilled at each end of the piers.

In mid-1963 an opportunity came out of the blue. I was in Terrace, B.C., working on a landslide problem. I was finishing up the fieldwork and called in at the District Highways office to update the District Engineer on the stabilization. There was a phone message there for me to call Hugh Golder at the Vancouver Hotel.

I placed the call. Golder answered and asked when I would be back in Vancouver. I replied, "I have a plane at five this afternoon and will arrive in Vancouver at seven P.M."

"Good," he said, "I'll meet your plane and we can have dinner together." I thought it was simply a courtesy client contact.

He met me and we drove to Hy's Steak House for dinner. Halfway through dinner he indicated, "Golder Associates has developed into a moderate-sized consulting group in Ontario. We've just been retained as geotechnical consultants to assist CBA Engineering on the design of the High Arrow Dam on the Columbia River. The dam is part of the Columbia River Treaty Agreement to regulate the Columbia River and increase electrical generation at Grand Coulee Dam in Washington.

"This job will open the door for more work in B.C. As a result, we'd like to open a consulting office in Vancouver. After discussing this with my young partners, it was decided we'd approach you and offer you a partnership in Golder Associates to open that office for us."

I had to take a deep breath. This was totally unexpected.

He continued, "We know you better than anyone else in the west. I know you and have worked with you. Larry Soderman and John Seychuck, two of my partners, went to university with you in Winnipeg and they're in favour. We propose a salary of $12,000 per year [I was earning $9,600 at Highways]. We'll provide you with a free automobile and operating costs, and pay your moving expenses to Vancouver. Provided the first year is successful, we'll give you a five percent interest in Golder Associates. If this is agreeable to you, Larry Soderman will come out to decide on other details such as benefits with you. We'd like you to start about October 1. Any questions?"

I answered, "This has come as a major surprise." I wanted to say yes immediately, but something inside said hold on a bit. I said, "I'm not familiar with your consulting procedures and reporting. Could I spend a few days in Toronto to meet the others and gain some familiarity with your programs?"

He answered, "That's a good idea. Victor Milligan acts as Chief Engineer. You should spend time with him." I had other questions regarding staff and procedures, and finally asked, "How would you propose to finance the office?"

Golder answered, "I'll discuss that with my partners and let you know."

After ordering dessert and coffee, Golder asked, "What do you think; are you interested?"

I said, "Yes I am, very much."

We were quiet for a bit. I was thinking. The salary was good, a car and expenses, a future partnership, I knew Golder, Soderman and Seychuk.

I looked at Golder and said, "Dr. Golder [I didn't have the nerve to call him Hugh yet], I accept your offer. Would you put it in writing for me when you return to Toronto?"

"Certainly," he answered.

Life-break number four.

I had originally planned to catch the 9 P.M. ferry to Nanaimo and drive home to Victoria that evening, but I missed the ferry. Part of me was glad to stay over in Vancouver; it gave me more time to think. I didn't know how Pat would take the news. We had moved into a new architect-designed house less than a year before. Life was going well; we had many good friends in Victoria. It would mean a whole new start.

After I told her, for half a day she was very upset. Then suddenly she said, "If that's the way it is to be, let's go to Vancouver and enjoy it."

We had enjoyed Victoria and its slower lifestyle, but in hindsight Vancouver was a good move for us. It opened up our opportunities and interests. In Victoria we used to say, "When the dogs chase the cats, they both walk." Now we had a chance to run.

3

Introduction to Consulting

Introduction to Consulting (1963-1970)

H.Q. Golder, internationally recognized geotechnical engineer, established a consulting company in Toronto. Several years later he offered the author a partnership to establish an office in Vancouver. The early history of this office provides an excellent lesson in the successful development and expansion of an engineering consulting company.

Early major geotechnical projects which are described include the foundation investigation for the nearly instant Simon Fraser University in Vancouver, and Canada's first use of roller-compacted concrete to stabilize the Nakusp waterfront. The author participated in a think-tank study to develop a method which would allow Humble Oil to drill for oil offshore in the Arctic Ocean. The program involved dredging sand to form islands which then froze to become drill platforms sturdy enough to resist the considerable forces of wind-blown ice.

The importance of rock mechanics and geotechnical engineering in mining was recognized in 1966 following a rock slope stability study for Kennecott Copper at its open pit molybdenum mine in northern B.C. This led to an investigation and stabilization by the author of two major slides at Anaconda's 1,200-foot-deep Anamax Copper Mine south of Tucson. A dewatering program using a drainage tunnel was developed. A noteworthy project which included special blast design near the final wall at Gortdrum Mine in Ireland is described.

The first International Conference on Stability in Open Pit Mining was organized by the author in Vancouver in 1970. This was the first of ten Mine Geotechnical conferences. The author was involved in editing all proceedings.

Chapter 3
Introduction to Consulting

I provided the Department of Highways with one month's notice of my leaving, even though I was only required to give two weeks. I felt it desirable to leave on as good terms as possible.

Three weeks before leaving Highways I spent two weekend days in Toronto going over business supply needs, deciding on a letterhead, and reviewing consulting reports. I was advised that $25,000 would be deposited to a corporate bank account in Vancouver. The goal was for me to use that to establish sufficient business that Golder Associates would not need to supply more before I developed income which exceeded expenses. As it turned out, I did not require additional financial assistance. I was very proud of that.

Larry Soderman, a partner in Toronto, came out during my second last week with Highways. We had lengthy discussions during the evenings about potential clients. Since I had extensive experience in materials testing, I proposed that we offer that service as well as geotechnical engineering. He agreed. We also discussed business operations and business development. Larry was very helpful.

On my last weekend with Highways, Victor Milligan, Senior Operating Principal partner, came out from Toronto and we looked for potential office space. In adding materials testing to our services, we would need laboratory facilities, and given that, we agreed that a finished basement or space on the first floor was preferable. It was also agreed that the location should be within several miles of downtown Vancouver. After looking at three locations, we inspected a basement office that had been used by a small printing operation. Within fifteen minutes we agreed that it would be suitable. There was room for a laboratory, front office and my office. Extra space was available behind, on which we took an option. The location was on West Broadway, which was convenient to downtown Vancouver.

I hired a secretary, an attractive middle-aged widow who was a good typist and could take shorthand. A telephone arrived early the first business morning in October 1963. I had arranged for good second-hand furniture and a new typewriter.

Hugh Golder had rented a furnished apartment that he would use when he came to Vancouver. I bunked there during the week and commuted to Victoria on weekends. This lasted for several months until we sold our home in Victoria. In Vancouver, we decided to rent for one year before buying a house. That gave us a chance to decide where we wanted to live in the Vancouver area. The assumption was made that the office would be successful.

After 10 A.M. on the first day, I sat at my desk waiting for the phone to ring. By noon I realized no one in Vancouver knew about the new business. I awoke to reality. That afternoon I made up a list of potential clients and drafted an announcement notice. The list included structural and civil engineers, architects, city engineers, BC Hydro, CP Rail and CN Rail, forestry companies and construction contractors.

While with Highways, I had come to know many structural engineers who would use geotechnical engineers to perform foundation engineering studies. I called a number of them and tried to make appointments to meet them at their offices. Only about one out of three agreed to see me. It didn't take long for me to realize they were busy and weren't too happy to waste time on what actually was a salesman's call — a consulting engineering salesman. I decided to change tactics. I began to invite them to lunch. They had to eat! This was accepted by all I invited. I did the same with architects. It was a better atmosphere than the office, and turned out to be a successful approach.

In the first month I landed three small jobs. They were foundation studies for two houses and a small apartment building. The next month was the same. That was followed by a study for a small bridge in North Vancouver. In the fourth month I invited architect Geoff Massey to lunch, and he accepted. Prior to the invitation, I had explained who I was. He was related to Vincent Massey, the former Governor General of Canada, and a partner of Arthur Erickson, the internationally known Vancouver architect.

On my arrival at his office, I could hear a loud argument transpiring behind closed doors. After a few minutes, a gentleman walked out, very upset. I knew him — he was a local geotechnical consultant — and we nodded. Massey then came out to the foyer, saw me and invited me into his office. We shook hands.

He said, "Have a seat." We chatted for about fifteen minutes, then he said, "I've just fired the geotechnical engineer on the proposed new Simon Fraser University. If you can give me a reasonable proposal by tomorrow noon, you've got the job."

The premier of the province, W. A. C. Bennett, had decided a new university was required in the Vancouver area. The top of Burnaby Mountain was selected as the site. A statement of interest was called for from internationally respected architects. From that interest, six architectural firms were short-listed to prepare detailed submissions, which included the development of models. Because of the high cost involved, all were to be paid a set fee for that preparation. It had been decided that several buildings would initially be developed, including the Administration, Science, Library, and Theatre buildings, as well as a sports complex.

Gordon Shrum, a well-known retired judge and past president of MacMillan Bloedel, the forestry giant, was selected by Bennett to coordinate the entire project. The timing was unreal. Bennett wanted the university ready to open in just eighteen months!

Erickson Massey were selected as the overall university architects, and they also designed the sports complex. Other Vancouver architects were selected for the remaining buildings. Massey showed me the overall site model, and it was impressive. A structural engineer was selected to design a water tower. All these structures, including their access roads, required geotechnical studies.

I prepared a proposal and hand delivered it the next morning. Massey reviewed it immediately and said, "Chuck, it looks good. When can you start?"

I answered, "As soon as we receive a set of site plans and the names of the other architects."

He said, "I can get all that for you now." Within an hour I was back at the office. I was so excited I phoned Golder to tell him the good news, and he congratulated me.

Simon Fraser University campus at the top of Burnaby Mountain overlooking Vancouver. The foundation investigation was my first major consulting project.

I spent the afternoon planning the program, then spent one hour on the site. An old access road to the top of the mountain existed, and general clearing of the site had been completed. A site architect showed me where the stakes were for the various buildings.

From observations of the vegetation, topography, and a map of the surficial geology of Vancouver, I concluded that bedrock on the site — sandstone and shales — would be shallow. Hence, I decided that a test-pit program should be phase one. I rented a track-mounted backhoe that could dig to a depth of 12 feet.

It is standard practice for a technician or junior engineer to supervise the digging of all field test pits, describing all the soils or rock encountered at various depths, and taking photographs and samples for testing. For the project, I would need a field technician, as well as a laboratory technician.

I advertised locally for the field technician on the basis that he would be hired only for the length of the project. I interviewed and hired a retired civil engineer who wanted to remain active. Mike Kendall, the head soils technician with the Materials Branch, had told me just before I left Victoria, "If you ever need a senior soils technician, please let me know. I'd like to work for you." I called Mike and offered him a job. He accepted immediately.

The project went very well except for two problems. While test pitting for the sports complex, a fault zone was encountered about 10 to 12 feet wide, running north–south under the proposed swimming pool area. Two angled holes were drilled 50 feet from the surface exposure to intercept the fault at depth. The fault was nearly vertical. I advised the architects of this and recommended the building be moved 50 feet west. They were adamant it not be moved. They said it would compromise "the feeling" of the project. I'm not sure what that meant, but my recourse was to recommend that the floor and beams that were to straddle the fault be strengthened significantly. Fortunately, to date there has been no serious earthquake at the site.

The second problem involved the library. Since it was designed with two levels of basement, a test-hole program was performed for the structure. This resulted in about 20,000 cubic yards of excavation being required. The average geotechnical profile consisted of the following:

0 – 5 feet — loose to medium-dense sandy gravel

5 – 11 feet — weakly cemented gravelly sandstone

11 – 18 feet — soft to medium-hard sandstone

18 – 25 feet — medium-hard sandstone with hard sandstone and shale lenses

I concluded that all this material could be excavated with standard excavation and ripping equipment, and that no blasting was required. I therefore recommended to the library architect that the excavation be bid as one material. By our providing the detailed drill logs, the contractor could estimate the variable diggability and bid one average unit rate. I was not asked to review the design drawings and specifications. In fact, I did not see them until after the contract was awarded and the contractor had nearly completed the excavation. After removing the upper 5 feet on average, the contractor had then drilled and blasted all the remaining material. Upon becoming aware of this, I asked to see the specifications.

The specifications defined the material as follows:

1. Overburden — can be dug with a hand shovel
2. Sandstone (soft to medium-hard) requires a small backhoe to excavate
3. Sandstone (medium-hard) requires a large backhoe to excavate
4. Sandstone (hard) requires drilling and blasting to excavate

The architect estimated 5,000 cubic yards of each material to provide a basis for bid comparison. The contractor bid low on the first three items, and high — $20 per cubic yard — for the hard sandstone. This was an unbalanced bid, normally unacceptable practice. However, a clause to deal with this had not been included in the contract documents.

After excavating the overburden, the contractor had in fact drilled and blasted the entire remainder of the excavation, claiming it was too hard to excavate with a backhoe. This meant 15,000 cubic yards excavated at $20 per yard, or $300,000. Dr. Shrum reviewed all cost items over $5,000. He noted this cost and questioned the architect. I was asked to attend the meeting and brought my relevant files, including my report, which recommended a unit bid for all excavated material.

The architect said the contractor had told him it would take twice as long to excavate the material as compared to blasting. Shrum didn't buy this. I didn't either. Shrum gave the architect a severe tongue lashing for not bringing this to his attention. I subsequently have learned that many contractors bid low in order to land contracts, and then claim extras for as many items as they can make an extra payment argument for. To counteract this, the designer must give very careful thought to contract requirements and wording. At the same time, he should develop specifications that are fair and practical. It is unfair to throw all the risk on the contractor with what I call "weasel clauses."

During my early business development efforts, I received many comments to the effect that Golder Associates was an Ontario (foreign) company, and that the potential client preferred to work with a local firm. I always emphasized that I was a local British Columbian, but this didn't seem to matter. I discussed this with Golder, Milligan, and Soderman. They recognized the issue. On a later trip to Vancouver by Larry Soderman, we again discussed this situation.

At lunch Larry suggested, "What do you think the reaction would be if the name were to be changed to Golder Brawner Associates?" I was surprised. My first reaction was that I didn't think Dr. Golder would approve. Larry said, "Leave that to me!"

Soderman was an excellent engineer who had another great asset. He was a people person, an outstanding communicator and an excellent teacher. Unfortunately, he had a heart problem, an abnormally rapid heartbeat, and he died from two heart attacks in his early forties. A great loss.

The Board of Directors accepted the name change for the Vancouver office. Personal

Roller-compacted soil-cement slope stabilization at the Nakusp, B.C., waterfront to control slope stability and erosion. The Arrow Lakes were raised about 90 feet by construction of the Keenleyside Dam near Castlegar, B.C., the first such slope in Canada (1968).

announcements were sent out, as well as a paid announcement in the local newspapers. There was an immediate change of attitude. Almost overnight Golder Brawner became accepted as a local company.

In the fall of 1965, I received a call from Jack Milligan of BC Hydro. A storage dam was being designed on the Columbia River near Castlegar. The dam would raise the high water level on the Arrow Lakes by about 90 feet. This would result in major flooding along much of the shoreline, extending over 200 miles. Farmland, beaches, houses, highways, parks and several small communities would be below the high water level.

The town of Nakusp, population about 2,000, was located on a terrace above the lake. The rising water level would impinge on the terrace slope along about 3,000 feet of shoreline. This terrace faced south, overlooking about 6 miles of Arrow Lake. Winds from the south could create waves 6 to 8 feet high. The terrace contained a major sandy gravel deposit, and had a slope about 60 feet high, with an average slope face angle of 35 degrees. The high waves would cause major scour of the slope, and over many years would eventually undermine numerous houses and the historic Leland Hotel. The slope required protection from this potential erosion.

Milligan asked me if we would develop a design to protect the slope with rip rap — coarse blasted rock — in order to control the potential scour. He asked for a specific proposal outlining our approach, staff available, schedule of fees, and a potential time scale. The contract was to include a geotechnical investigation, location of a rip rap quarry, field surveys, design of scour control protection, and supervision of construction.

The workload in Vancouver had increased substantially and so I had hired Robert "Bob" Wilson and Glen Gilchrist as new senior engineers. Bob was designated project engineer. One aspect bothered me. Placing rip rap would severely restrict any use of the slope. It would be dangerous to try to access the beach across it, and I also didn't feel it would be attractive.

I contacted the Waterways Experimental Station in Vicksburg, Mississippi, the U.S. Corps of Engineers, Tennessee Valley Hydro Authority, and the Bonneville Power Authority in Washington, enquiring about slope protection methods. Rip rap, I was told, was the common procedure. However, one reference was made to soil-cement slope protection at small dams in Arkansas and Oklahoma. I contacted the Portland Cement Association for information.

In the central area of the United States, quality rock for rip rap is not readily available. On the upstream slope of several water storage reservoirs, cement was mixed with the soil, placed in layers and compacted to a high density. The soil cement was spread with a paving machine in 6- to 8-inch horizontal layers and compacted with rubber-tired rollers. Each lift was stepped back to develop an average 2.5 horizontal to 1 vertical slope. The final slope had a stepped appearance. The result was an erosion resistant layer equivalent to "medium-strength sandstone."

I proposed this slope protection alternative to Hydro. The initial reaction was negative. I felt sufficiently strong about the more favourable appearance, the ready slope access, and the potential winter use after moderate snowfall, that I tried a little professional psychology — I offered to pay for two Hydro engineers to inspect the dams in the U.S. Several days after this offer, Milligan phoned and agreed to making the inspection but said, "Hydro is not that poor. We'll pay our own way." The inspection trip was arranged.

The trip to inspect the dams and upstream soil-cement facing was uneventful. At the sites, the stepped appearance was quite uniform and attractive. Milligan and Graham Morgan of BC Hydro were impressed. One condition, however, was unexpected — about every 20 feet a crack had formed running down the slope. It was concluded that the cracks were due to the soil cement shrinking. Actually this would be beneficial, as it would allow drainage through the facing. Today this facing is called roller-compacted concrete.

Back in Vancouver, Hydro decided to call a public meeting to present the two erosion control options to the residents of Nakusp. The meeting was held in the local school auditorium. Milligan introduced the subject and emphasized the purpose of the meeting — to obtain input from the residents as to which option they preferred. I was then introduced and described each option. I had arranged for a series of colour slides to show both the rip rap facing and the soil-cement slope. The people asked a number of questions and then voted. The soil-cement design was preferred by most. Hydro announced they would follow the residents' wishes.

The stabilized slope about 35 years after construction. The alternative protection of rip rap was considered to be unacceptable. The area is now a park with easy access to the lakeshore.

The project was designed and plans and specifications were completed by late summer. A cost estimate was also arrived at. The work was reviewed by Hydro and approval was obtained to proceed. We advised them that soil cement should not be placed in the winter, because the soil cement could freeze and be damaged before it had adequately cured. The tenders were not called until late April. Six contractors bid, and the lowest bid was 6 percent over our estimate. Allowing for 6 percent annual inflation, our estimate, which was prepared the previous fall, was on the money.

An earthworks contractor by the name of McKnight was awarded the contract. The work went extremely well. Initially the terrace was cleared of vegetation and sloped uniformly, top to bottom, at 2.5H:1V. A layer of select sandy gravel, 5 feet thick, was placed on the slope to allow the slope to drain when the reservoir was drawn down. Drain holes were also constructed through the soil cement to further facilitate back drainage. To prevent under-scour, a rip rap toe berm was placed at the transition of the soil-cement toe and the beach. The rip rap voids were filled with sandy gravel. Four sets of concrete steps down the slope were constructed, spaced at uniform distances along the slope. Following completion, the town of Nakusp developed an attractive parkway along the top of the slope, which included lawn, flowers, trees and benches.

I inspected the waterfront in July 2001, about thirty years after construction. The slope was in excellent condition. To my knowledge, this was the first use of roller-compacted soil-cement slope stabilization in Canada.

Business development at Golder Brawner was an ongoing priority. Following my success with muskeg preconsolidation for Highways, and my writing several papers on the projects, I realized that writing technical papers that emphasized practical experience provided excellent personal and corporate exposure. I decided to write two to three articles for publication in trade magazines every year. These publications were always looking for practical engineering articles. They normally do not pay for the article, but I learned they would provide 200 reprint copies free of charge. When I received the reprints, I would send them to a select list of clients or potential clients with my card. This was very effective in opening doors to the services Golder Brawner could provide. I speak to many engineers and new consultants who wish to enhance their professional exposure and their careers, and I always recommend this practice to them all. Reference to such publications enhances one's résumé.

In my second year at Golder Brawner, it was decided to hold a cocktail party before Christmas at our newly constructed home in West Vancouver. This was a new endeavour, and I was unfamiliar with the protocol for such an event. I hired Mrs. Vera Vaux, local specialist in organizing social parties. She recommended specially designed invitations, serviettes, matches and decorations inside and out. She also recommended a specific theme for the party — "travel" — and so our invitations looked like airline tickets. She further suggested we hire Bill Conrad as lead bartender. He had the reputation of remembering everyone's name, as well as the drink they had ordered years before. He subsequently proved that this reputation was deserved.

Invitations were sent to sixty clients, potential clients and their wives. Thirty-five couples attended. I believe it was a success. Over ten new clients were secured as a result of the party.

This became an annual event for ten years. At that time the business had grown to a point that I discontinued holding the parties. They had been fun and the senior staff, as well as Pat and I, made many new friends. I nevertheless continued to invite potential consultant clients to lunch, and to meet with government and corporate officials at their offices.

In 1965 I made an appointment to meet Ken Truman, Regional Engineer, and Jim Cherrington, Assistant Regional Engineer for Canadian Pacific Railway (CPR) at their offices. During my time with Highways, I had become aware that the railway had to deal with periodic landslides in a transportation corridor they shared with B.C. Highways through the Fraser, Thompson, and Kicking Horse canyons, as well as Rogers Pass through the Selkirk Mountains. Since I had extensive landslide experience while with Highways, I met with them to outline that experience.

During our meeting, Truman commented that Dr. Golder had acted as a specialist witness for the Canadian Wheat Board against CPR on two accidents where debris flows had washed out small bridges. In each case, a loaded train carrying wheat was derailed and many carloads of wheat were lost. Hence the Wheat Board sued the Railway. In both instances, Golder gave evidence that there were indications on aerial photographs that such events could occur. He performed field inspections of the upper gullies and located the source of the flows. He stated that electric warning fences placed across the gullies would have warned of the washouts in advance. (The flow debris would break the wires and trigger a warning in a nearby CP Rail office.) In both cases, the court held CPR unobservant and awarded damages to the Wheat Board.

Truman challenged, "Golder cost us a lot of money. Why should I hire you?" Without really thinking I countered, "Mr. Truman, if you can't beat them, join them!" What nerve!

In hindsight, he would have been justified in telling me to leave. There was quiet for a few seconds then Cherrington chuckled, "So you think you might save us money?"

I answered, "Mr. Cherrington, we would sure try. We have an excellent team, and experience. I'd like a chance to prove it to you." I left on what I felt were reasonable terms, but wasn't sure if they considered I had been too outspoken. I haven't been that aggressive since.

Three weeks later Cherrington phoned. "Chuck," he said, "we have a project for you. We're going to build a new rail-line to a coal project that Cominco is building in the Fording River valley in southeast B.C. We'd like you to perform a preliminary geotechnical survey to assist us in selecting the best location. There are some difficult areas — deep soil cuts, some very steep rocky locations and a number of swamps. Would you come to our offices so we can decide on details, timing and costs?"

What an opportunity. We arranged to meet that next day.

At that meeting, I was provided with topographic maps of the entire valley, and overlapping stereo photographs of the potential route. In addition to the difficult areas Cherrington had mentioned, I noted several possible snow avalanche and debris flow areas, and a number of old landslides. The east side of the valley was along the west side of the Rocky Mountains, which reached an elevation of nearly 10,000 feet.

The new railway would be about 32 miles long. Many fills, culverts and several bridges would be required. The geotechnical investigation was performed over one and a half years. CP Rail established a location and survey crew camp about midway along the route. It involved several camp trailers and a dining trailer.

On one visit to the site, Cherrington, his field supervisor Milt Page and I worked through the lunch hour and arrived at the dining room unexpectedly, after all the others had eaten. There were adequate leftovers, however, so we gulped these down and then left to go back to the field. After inspecting several high rock areas and selecting a preliminary centreline, we returned for dinner, this time with the survey crew. As we were finishing, the bull-cook (camp chief) brought a record book and asked us to sign for the dinner. I had assumed a contractor was providing the camp. Above our names and at the end of the lunch signature list he had written, "Three big cheese — no pay." He wasn't about to let us get away with a free lunch.

Jim Hawthorne of Humble Oil in Houston, Texas, called in 1966 to advise that his company was interested in pursuing a drilling program for oil in the Arctic Ocean north of Aklavik on the MacKenzie River Delta. They realized that standard drill rigs which rest on the ocean floor would not be feasible, given the wind-driven motion of the arctic ice. The forces generated by the moving ice are so large that drill rigs simply cannot withstand them.

Hawthorne indicated that they were looking for a Canadian consulting geotechnical engineering firm who could perform a think-tank type of study to determine possible exploration options available to them. I said that the first concerns we would have would be the depth of water and the seabed soil conditions. He asked if Golder Brawner had the capability to perform such an investigation. I answered, "Yes, we do. Would it be worthwhile for us to come to Houston to meet with you to learn more about your program, your timetable, particularly the details of your drill rigs, and to describe the services we provide?"

He answered, "That's a very good idea; when could you come?"

I replied, "I'd want to have Dr. Hugh Golder, our senior principal come with me. I'll contact him and phone you back. And by the way, there will be no charge for our visit."

I managed to contact Golder, and he agreed to go to Houston the next week, preferably Wednesday. I phoned Hawthorne and he confirmed the date.

In the meantime, I contacted the Canadian Hydrographic Survey, who said they would forward their depth survey maps of that area. They advised me that the water generally varied from only 40 to 120 feet deep for over 50 miles off the coastline. They also advised that the sea bottom there was primarily sand. According to past seismic surveys, the thickness of sand was variable.

Golder came out to Vancouver and we flew together to Houston. It was July. What a shock. The temperature was about 110°F and the humidity high. Fortunately, the autos and offices were air-conditioned.

Hawthorne took us into a conference room. He explained that their subsidiary, Imperial Oil, had performed offshore geophysics programs, and believed there to be a reasonable chance of oil. He showed us photos, and provided details of the drill rigs they were using in the Caribbean Sea.

During our discussions, we broached the idea of dredging sand from the sea bottom and creating islands to drill from. He said, "That's an interesting idea. How would we determine if it could be done?"

Golder answered, "A geotechnical soil drilling program could be performed from the ice to sample the seabed materials — determine the type of soil, gradation, density, thickness, *et cetera*." Golder described a dredging project he had supervised in the Caribbean. Further details were discussed until late afternoon.

Hawthorne then invited us to dinner. Over the meal Golder and I described some of the other projects the company had been involved with. When we were finishing, Hawthorne asked us to provide a general outline of the program we would propose, and the people who would be involved. We said we would detail a phase one program, with cost, that would determine the site conditions before any further investigation should start. On unusual, difficult or large projects, a two-stage investigation has much merit.

The next morning we flew back to Vancouver. Golder and I agreed that Dave Campbell, an Associate partner in Vancouver, would be the ideal person to lead the study. We spoke to Dave and he agreed.

The program involved boreholes drilled through the ice on a grid within an area 10 to 40 miles offshore. A drilling technician, Wilf Rosenberg, was sent up with the drill and drill crews to oversee the work. We contacted the RCMP in Inuvik, the largest local town, to let them know of the project, and to determine if they could provide any advice on logistics.

The RCMP were very specific. They said, "Unless you have experience on the offshore ice, arrange for assistance from the Eskimos. They know the ice and weather behaviour, particularly the wind and whiteout conditions. They can also provide transportation to and from the drill and set up wind protection." Campbell and Rosenberg arranged all the site requirements with the Eskimos.

The project went reasonably smoothly. The seabed proved to be primarily sand, which could be dredged to create large islands from which the oil drills could then operate. The surface of the islands would freeze naturally to provide good support, and thermo piles, which use cold internal fluid convection, would freeze the island at depth.

In this area, the ice can be about 6 to 8 feet thick, and during arctic storms the entire sheet will move several feet per day. The island had to withstand this massive thrust. Northern experience indicated that if the beach angle was about 7 degrees, the ice would then ride up on the beach, break, and move around the island. This is indeed what occurred, and the procedure was subsequently used for the first several years of offshore oil exploration.

In 1966, I received a call from Carl Broadbent, Mining Engineer with Kennecott Copper Ltd., Salt Lake City, Utah, saying that they were developing a molybdenum open pit mine near Alice Arm, B.C., north of Prince Rupert. "A serious rockfall has occurred from a rock slope above the construction site, directly adjacent to proposed crushers and a major mill building," he added. "Is there any chance you could go up to the site with me early next week?"

I thought quickly, concerned that I did not yet consider myself a rock mechanics engineer. I had gained some rock experience while with Highways, but I was still hesitant to accept a consulting assignment in that particular field. I said, "There's an engineering geologist in Vancouver who has more experience than I have. I suggest you contact him. His name is Dr. Victor Dolmage." I gave him the phone number.

About twenty minutes later Broadbent called me back. He said, "We didn't communicate well together. I'd appreciate it if you could come up with me."

I agreed, "Okay, I'll go with you on the understanding that if I recognize the problem, I'll develop recommendations. If I don't understand it, I won't charge you for my time or travel costs."

"That's more than fair," he replied, and the trip was set up.

We flew to Prince Rupert, then caught an amphibious airplane to the site. We disembarked at a dock that had been constructed on a creek delta where a townsite was to be constructed. It was a several mile drive up a creek valley to the construction site near the top of a mountain.

The mine was to be an open pit. To reduce the hauling costs, the crushers, concentrators, warehouse, and mine office were to be constructed near the pit on a level area excavated in the mountainside, which was all rock. The excavated slope was steep — about 75 degrees — and up to 220 feet high.

The rock was a bedded limestone. The bedding dipped out of the slope at 40 to 48 degrees. About 1,800 cubic yards of rock had slid on one of the dipping beds, creating what is called a planar failure. The rockfall had destroyed an 80-foot length of footings and foundations that had only been poured several hours earlier. It was fortunate no one had been killed.

The rock face contained adverse bedding — shaley partings which felt slippery to the touch — over about 40 percent of the slope. Hence, the friction angle along the bedding was much lower than the dip angle. As a result, the safety factor (forces resisting failure divided by forces causing failure) was less than 1.0, the condition required for failure.

I said, "Karl, wherever the bedding dips out of the slope, the rock above will have to be excavated below the lowest bed which daylights the slope, or the rock will have to be held in place by tensioned grouted rock bolts." He understood what I said. I continued, "If the site surveyors can survey cross sections every twenty-five feet, I can map the areas that require removal or bolting, and develop the design for the entire slope." He agreed.

The survey and design took two days. Broadbent was pleased with the recommendations. Golder Brawner supplied an engineer to supervise the work.

Several months later Broadbent phoned and said, "Chuck, we appreciated your help at B.C. Moly. We have rock problems at Bingham Canyon near Salt Lake City, at Ely, Nevada, and on a project in Chile. How soon will you be able to help us?"

Bingham Canyon is the largest and deepest open pit mine in the world — 2,300 feet deep! I didn't feel ready for this, but I really wanted the work. I thought quickly and said, "Karl, I'm very heavily committed. Is there any possibility you can wait two months?"

"Chuck, I really don't want to wait that long. How about one month?"

"I'll call you back tomorrow," I answered. "I'll look at my commitments and see what I can do."

He said, "Okay, but the sooner the better."

I stayed up half the night thinking. The fact was that I could go much sooner than two months, but I was still wary of my lack of rock experience. I decided I would stall as long as I could. In the meantime I would go to the University of British Columbia library and find whatever literature I could on rock mechanics.

The next day Broadbent agreed to a six-week delay. After spending eight hours in the library, all I could find were eleven relevant articles on surface rock stability. Remember, this was 1966. There were several books and many articles on underground rock mechanics and tunnels, but these were of no use to me.

I reviewed all eleven articles, and noticed that several names came up more frequently: Dr. Don Coates in Ottawa, Dr. Evert Hoek in England, Dr. Leopold Mueller in Germany, and Dr. Manuel Rocha in Portugal. The U.S. Bureau of Mines also had several references. I decided that if I could spend some time with any of them, I would arrange a rush trip. I phoned each man. Coates, Hoek and Rocha agreed. Mueller wasn't fluent in English, but suggested Dr. Klaus John, an Associate, who agreed. The U.S. Bureau of Mines also agreed. I left the next week.

I was away two weeks. Not only did I spend time personally with each man, but several of them supplied me with notes and a number of research reports. On my return, I wrote an article entitled, "Rock Slope Stability in Open Pit Mines." It was published in the *Western Miner Magazine* in October 1966.

In combining my own experience with my UBC review, and what I had learned from each man during my trip, I concluded that mining engineers and geologists were making several major errors.

1. Water is not a lubricant in rock slopes, which many claimed.
2. Final wall slopes are damaged by excess blasting forces, which was not recognized.
3. Drainage of rock slopes increases stability. Many engineers did not recognize its importance.
4. Stable slope angles in rock will not have to be flatter than the angle of repose of broken rock (37°). The author of at least one article did not recognize the low strength of clayey rocks.

Commencing in six weeks time, I successfully assisted Kennecott on improving stability at the projects Broadbent had mentioned. Word of my success reached the Anaconda Mining Company, who also had headquarters in Salt Lake City. They contacted us regarding a project in Arizona, which is described later. In addition, Cleveland Cliffs Mining Company and Hannah Mining Company, who have open pit iron mines in Michigan, made contact, looking for assistance. Note that most of our early work in rock mechanics was in mining in the United States, not in Canada. I even obtained a work permit for the U.S. during that time.

I couldn't handle all this rock mechanics work alone. While in London at Imperial College with Evert Hoek, he had introduced me to four graduate students who were studying for their doctorate degrees. I phoned him and asked about them, "When will they graduate and do they have jobs?"

Hoek replied, "They'll finish in four to six months and no, they haven't accepted jobs yet."

I offered all of them jobs working with us in rock mechanics out of Vancouver. David Pentz, John Sharp and Brian Stimpson accepted and joined our staff that year.

Within two years we had one of the leading surface rock mechanics teams in the world. Projects developed both in open pit mining stability and transportation rock stability. While working on open pit stability, we also became aware of the need for geotechnical engineering for tailings dams and waste piles. We quickly added these areas to our consulting services.

In addition, Pentz got us involved in computers. At that time computers were very crude, requiring a temperature and humidity controlled room and special punch cards. The cost to set up a system was about $1 million, a sum well beyond our means. Then IBM came out with the 360 model, and the University of British Columbia installed one. Pentz suggested we should try to rent time. The university was approached and said something could be worked out.

Pentz soon said, "I can't do all the computer work. We need to hire someone to look after the computer programs." I asked him if he could try to locate someone. Three weeks later he said, "Chuck, I've got him. He's coming in this afternoon. But I want to warn you; he's different! Please don't act surprised."

Pentz had spent several sessions with the candidate and was convinced he could get us into the computer age. On a July afternoon, Pentz brought his recruit to our offices. His name was Walter Rae. He wore a wool toque over neck-length hair, had a moustache, wore a logger's shirt with the top three buttons open, a pair of old blue jeans (they were clean), and a pair of high-top boots. Normally this is not the appearance one looks for in an engineering applicant.

I remembered that Pentz had said don't be surprised. I smiled, put out my hand and said, "Glad to meet you. You come highly recommended by David. Have a chair."

While his appearance was unusual, his handshake was firm. That was a good sign. I began, "David has convinced me that we need to get into the computer world. We've developed a great group of young engineers here, and to meet today's and future challenges we have to maintain a leadership status in our consulting practice. David has described to you the type of work we do. I'd like you to tell us what we have to do to take advantage of computer technology, and how you can help us do that."

Rae spoke for over an hour, describing the type of problems that could be dealt with, the major benefit of reviewing more options in less time, and he gave us a number of case examples. He was a good communicator. Rae convinced us he would be a valuable addition to our group. I asked him what salary he would be looking for, and he cited a figure that was acceptable. I outlined our benefit package and hours of work. He said, "I have a different work schedule than most. When I am turned on by a project I sometimes work for twenty-four hours at a stretch. However, you'll get at least a hundred and fifty hours of work every month."

I realized that he would have little if any contact with clients, so his appearance or work schedule could be accepted. I stood up, "Jim, if you're agreeable, the job is yours." I put out my hand.

He was surprised and stood up slowly, saying, "Great, I can start whenever you say."

"How about Monday?" I asked.

He agreed and we shook hands.

Pentz and I needed a few days to prepare the staff. I had learned that some staff do have prejudices, and that you could no longer surprise employees or put round people in square boxes. The workplace was changing.

The transition to computers was successful and this new capability soon led to us obtaining a number of major international projects.

Paul Hodges, former Manager of the major Palabora open pit mine in South Africa, had become Mine Manager at the Anamax open pit mine south of Tucson, Arizona, in 1970. He arrived to a situation where very large rockslides had developed in the south and north zones of the pit. Each slide, involving millions of cubic yards of rock, was moving at a rate of about ½ inch per day. The slopes were about 1,200 feet high. Several hundred employees were working in the pit and would be in danger if either slide took off — which was a genuine possibility. A major slide involving an 800-foot highwall at Chuquicamata, Chile, had failed dramatically only a few years previously. Fortunately, monitoring had offered sufficient warning that no lives were lost.

David Pentz had visited Hodges at Palabora and discussed rock slope stability with him. Hodges phoned him again at Golder Brawner and asked him about performing an inspection of the pit slopes. Pentz spoke to me after the call, and we agreed that we should go together. He called Hodges back and made arrangements for our visit.

The most important factor influencing rock slope stability is the structural geology of the rock, that is the orientation of the discontinuities in the rock — faults, shears, joints or bedding — in relation to the orientation of the rock face. If a through-going discontinuity or discontinuities dip out of the slope face at an angle steeper than the angle of friction along the discontinuity, failure must be expected along that zone of weakness. The structural geology is most important because we cannot change it; we must determine its existence, shear strength and orientation. We must live with the geologic model that nature has provided, and account for its influence in the design.

The two other very important factors are groundwater pressures in the slope (these may vary), and the impact of seismic forces due to heavy blasting near the final wall. Surprisingly, intact rock strength is usually not critical for stability in most surface mining projects.

Upon arrival at the site, and after introductions with Hodges and mining and geological staff, we asked to see geological maps (plans and cross sections, perpendicular to the pit slopes). The most notable conditions observed were the presence of near vertical faults, usually at right angles to each other. Displacement along the faults had resulted in rock being ground down to clay sizes with low permeability, which greatly reduces water flow through the rock. Water pressure was compartmentalized within fault-bound blocks.

Practically all rock mechanics studies which had been performed by Anaconda, the owner, dealt with the strength of intact rock. One smaller rockslide area had been stabilized using high-strength anchor cables which were tensioned and then grouted. The grout was to hold the tension in the cables, to anchor the end of the cables below the failure zone, and to reduce corrosion of the cables. This stabilization was not carried down the slope as the pit was deepened. At greater depth, the high rock stresses extended deeper, and the slide depth ultimately extended beyond the original grouted anchors. A larger potential slide volume developed.

Wireline monitoring of pit slope movement at the Anaconda Twin Buttes copper mine, Arizona. A limit switch is tied into the wire system which triggers the flashing light and a siren if the movement exceeds one-half inch during an eight-hour shift. The pit was 1,200 feet deep.

The area south of Tucson is considered a desert, with only 6 to 10 inches of precipitation per year. As a result, the conclusion is often reached that open pit mines there will not encounter water at depth, but this is a fallacy. Even in the Atacama Desert in Chile, which sees only ½ inch of rain each year, the water table is encountered at a depth of 400 to 600 feet. Mining extends much below this depth.

After lunch, we drove down into the pit. At the bottom I observed standing puddles on the mine floor. In the area where blast holes had been drilled, I dropped pebbles down several holes, and heard a splash, indicating they contained water. Some holes were even flowing water (artesian) near the toe of the slope. It was concluded that the mine was deeper than the water table, and that water was seeping through the rock toward and into the pit. Hence, water pressure and seepage forces could be the primary cause of the two slides.

Back in the office I said to Hodges, "Water pressures could be the cause of the movements. If so, drainage of the slopes using horizontal drains or drainage tunnels would likely stabilize them."

Hodges appeared startled and reflected, "Anaconda have been studying the rock conditions and stability here for many years. I find it hard to believe that they didn't consider the impact of groundwater."

I said, "Paul, we need to drill some holes and install piezometers to determine the water table near the pit walls." Pentz agreed. A program was developed and agreed to by all.

A week later three holes had been drilled and it was determined that, within 100 feet of the slope, the water table was about 250 feet above the base of the pit. There was no doubt that water pressure in the slopes was largely responsible for the slides. It was concluded that effective drainage of the lower slopes would stabilize the movements.

Hodges was initially very surprised, but then very complimentary. His reaction was not unreasonable. Most mine managers in the 1960s had never studied soil or rock mechanics at university. It was a new specialty for many mining engineers.

It was now a question of how to lower the water table in the lower slopes. In order to improve slope stability as the pit was being developed incrementally another 100 feet deeper, the water pressure should be lowered at least 300 feet back from the pit face. One option was the installation of horizontal drain holes from the toe of the slope. The holes would be drilled using 4-inch drill bits on the end of drill casing. Drill holes, 350 feet long at a 2- to 3-degree upward slope were proposed. The bits would wear out before each hole was finished, so the casing had to be pulled and a new bit installed periodically. The final bit was left in the hole by reversing the casing direction a small amount. Special 2.5-inch-diameter steel or plastic pipe with drain holes in it would be installed full length inside the casing. Then the casing would be pulled and the perforated drainpipe left in the hole. The drainpipe was

Drainage tunnel (a) extends 300 feet into the pit wall and circles the pit. The drainage of the slopes halted two very large pit wall slides. Note the water in the base of the pit.

required since the drill hole has a tendency to cave whenever heavily fractured rock or fault gouge is encountered. The drains would be required about every 50 feet around the pit.

The second option was to install a 300- to 350-foot drainage tunnel perpendicular into the slope at the toe, and then develop a tunnel in each direction all around the pit. At selected intervals along the tunnel, percussion drain holes could be drilled vertically and horizontally around the pit to increase the effectiveness of the drainage tunnel.

Anamax selected the second option. The program was successful in stopping both slides. In fact, ongoing dewatering with depth allowed steeper slopes to be used as the pit deepened.

Another benefit of dewatering is to drain the rock so blast holes are dry. In dry holes, inexpensive ammonium nitrate (ANFO) can be used, whereas in wet blast holes expensive slurries or emulsions are required. The difference in cost of the two types of explosives can be several million dollars per year for a large mine. A further problem is that blasting energy impacts saturated slopes more adversely than unsaturated rock; sometimes called hydrodynamic shock. Blasting shock can be doubled below the water table compared to above it.

Our success at Anamax enhanced the reputation of Golder Brawner. At mining conferences, company staff inevitably tell others of their successes. This led to our involvement in an increasing number of national and international projects.

On many projects that I have been involved with, serious instability had already developed. My experience has been that many consultants would outline a geotechnical investigation which would include drill holes, core recovery and logging, rock testing, piezometer installation and monitoring of movement. This would be followed by stability analyses and the preparation of a report. This process often meant that three to four months passed, while the slide continued to move. With greater movement, the risk of exceeding the peak shear strength greatly increases. If this occurs, there is an increased danger of a complete failure. If failure doesn't occur, greater cost will be required to restore that lost shear strength.

In the majority of open pit mine failures, water pressures contribute to the movement. The most effective procedure to restore stability rapidly is to install horizontal drains. The program can commence almost immediately. In many instances, large slides have stopped moving with only 5 to 6 drains installed. To enhance the stabilization, a vacuum can be applied to the drains. At the same time as lowering the water table, I will recommend a typical investigation. This will determine how much stabilization is required. Clients look for rapid response, and that is where experience is so important.

Upon returning from one of my business trips, Bob Wilson, a principal partner at Golder Brawner, advised me that a representative of the Teamster's Union, Ken Steele, had been in the office to advise Golder Brawner that his union was going to attempt to sign our technical materials testing staff to a union contract. Bob expressed his concern to me, saying that the union representative had been very aggressive. To add to his aggressiveness, Steele was a big man, which in itself can be very intimidating.

We had developed a reasonable capability in materials testing, and had laboratory and field technicians, as well as secretarial staff who could be involved. The Teamsters had already organized most other materials testing firms in Vancouver, but I wanted no part of becoming unionized. It would destroy the flexibility we, as well as our staff, then enjoyed. I advised Bob I would not allow unionization to occur and that I would look after it.

After thinking about the issue overnight, I decided to discontinue materials testing, except for geotechnical testing. I contacted Steele, said I wished to meet with him, and asked him to come to our offices. I wanted the meeting on our turf. In the meantime, I contacted Terra Engineering Services and asked if they would be prepared to take over our materials testing contracts. They agreed. I then called our existing clients, advised them of the union pressure and asked their approval to the change in provision of testing. I had decided I was going to deal with Steele decisively, leaving no doubt that we did not intend to unionize.

My secretary advised me that Steele had arrived. I asked her to please bring him in. He entered the room, I introduced myself but did not shake hands. I said, "Mr. Steele, please have a chair." He sat down. I stood directly in front of him and said forcibly, "Mr. Steele, Golder Brawner does not intend to be unionized. I have arranged for all our materials testing contracts to be taken over by Terra. All work that we will do in the future will be professional geotechnical engineering, which comes under the purview of the Association of Professional Engineers of B.C., and outside your area of unionization. I've found your aggressive approach to us to be very repulsive. I do not expect to see or hear from you or the union again. Mr. Steele, this meeting is concluded! Good day." I waved him to the door. He was so taken by surprise that he said nothing, rose from the chair and left. We never heard from him again.

In the early fall of 1968, I was contacted by Reg Gagnon, Chief Engineer of the Quebec Cartier Iron Mine in Gagnon, Quebec, near the Labrador border. He indicated that they were having problems maintaining their haul road into the pit due to movement of the pit wall. He asked if I could come to their mine and perform an investigation to determine the cause of the movement, then develop a plan to stabilize the slope and haul road.

I flew to Montreal, changed planes to Quebec City and on to Gagnon. Reg met the plane and took me to the local hotel where we had dinner together. He was French-Canadian, bred, born and educated in Quebec, a very friendly and dedicated engineer. He reviewed the history of the mine for me. They were mining 60,000 tons of iron ore and about 120,000 tons of waste rock per day. Iron ore is extremely heavy, and the loaded trucks travelled slowly up the 10-degree ramp. He told me that they were considering using electric assist to increase the speed of the vehicles. (The trucks used diesel engines to generate electricity for electric motors on the wheels.) The concept was to construct an electric line over the haul road that an electric bar on the top of the truck could make contact with — similar to a trolley bus. Obviously they couldn't consider this unless the haul road was stable. He added that they were reducing production until the problem was resolved, and the implication was clear — an early solution was imperative.

The next morning Reg took me out to the mine. The open pit was about 1 mile long, 2,000 feet wide and 600 feet deep — a very big hole! The access ramp was on the east wall of the pit, and was wide enough for two mine trucks to pass, each hauling 150 tons. One-third of the way down the ramp a wide crack had been filled with a grader. Near the bottom of the ramp the road had heaved several feet. We drove out to the centre of the pit and looked back at the wall. I could trace the continuance of the upper crack, as well as the heave, along the toe area. An area of the wall over 800 feet longitudinally and 400 feet in height was involved.

I said to Reg, "It's necessary to determine the structural geology in the slope, the zone of failure, and the water table within the slide to perform a stability analysis. This is required to assess the potential benefit of various stabilization alternatives. In addition, a plan and cross sections of the pit in that area are required. Finally, hubs should be placed on either side of the upper crack to determine the rate of movement of the slide. If the movement is accelerating, the stability is reducing and the road may have to be closed."

"We can go back to the office and arrange to get that information for you," he replied.

Back at the mine office, I was introduced to the mine geologist, who brought me up to date on the geology of the mine area and the east pit wall. The upper crack appeared to coincide with a fault zone. The location at depth was uncertain. No information on the water table in the slope was available, except where seepage exited from the slope.

I advised Reg, "We need a cross section down the centre of the slide area showing the geology, rock types and boundaries and the fault zone, the upper crack and the zone of heave, and the phreatic water surface. This will require two drill holes on the section to obtain rock core and samples of the fault material. Also, we need to install standpipe piezometers to determine the water table, and locate the failure zone as identified by a bend or breakage of the standpipe. With this information we can perform a back analysis assuming a safety factor of 1.0, and calculate the average friction value along the failure surface. We can then determine a change in safety factor for any change in the water table or in the surface geometry."

Reg asked, "Can you arrange the drill holes quickly?"

I answered, "Yes. I'll make some phone calls if you can let me use a phone." He led me to an adjacent vacant office and showed me how to make a credit card phone call. I called the Golder office in Toronto and spoke to John Davis, who agreed to arrange for the drilling and a site inspector.

The drill arrived on site three days later. It took six days to complete the holes and place the standpipe piezometers. I arranged to return to the site when the second hole was being completed.

The water table was sufficiently high that I concluded slope drainage with horizontal drains would halt the slide. To gain additional stability for the longer term, it appeared reasonable to reduce the weight of the upper slide area by lowering the elevation of the haul road.

Upon my return to Vancouver, I performed stability analyses and prepared a report with specific drainage and slope unloading dimensions. The construction of another access haul road was strongly recommended. Quebec Cartier completed the work as designed, and the movement stopped. Gagnon was pleased with the final result and recommendations.

Gortdrum Mine, Ireland. This is one of the earliest mines where controlled blasting was used to improve slope stability and safety (1969).

In May of 1969, I received a phone call from the General Manager of Gortdrum Mines in Ireland inquiring if Golder Brawner could visit in the near future to provide a review of their pit slope stability and design. I had never been to Ireland and indicated that I could be available in about a month.

"Can you mail me information on your mining program, site conditions including geology and hydrology, and instructions to get to your mine?" I asked.

He said, "Yes, we'll do that tomorrow. Please give me your address."

I gave it to him and asked, "What's a nearby town where you would recommend I stay?" He suggested Tipperary.

That evening I mentioned the project to my wife, and asked her if she would come with me. We could spend several days in Ireland, also in London, and then fly home from there. It was a chance to have a short holiday together.

She had to check on live-in babysitting first. In the meantime I queried my travel agent about flight routes. He recommended Vancouver to Montreal to Shannon, Ireland.

On that trip I made a travel mistake I now try not to duplicate. Pat and I arrived Monday morning in Shannon, after an overnight flight, rented a car and drove to Tipperary. I had agreed to meet Stewart, the General Manager, after lunch, spend the afternoon with the mine staff, and then have dinner with Stewart. Halfway through dinner, I found myself having great difficulty staying awake! I went to the washroom, washed my face in cold water and ran on the spot for about two minutes. That helped to wake me up, such that I could again communicate sensibly.

Gortdrum had unusual geology. Minerals mined there included copper, lead, zinc, silver and mercury. Daily tonnage was reasonably low— about 6,000 tons of ore per day. The impressive feature was the excellent blasting program they had developed. Construction of the final wall used a procedure called controlled blasting (closely spaced holes with light explosive charges). It was recognized that production blasting (heavier explosive charges) near the final wall would damage that slope and result in extensive ravelling rock. This would be dangerous to all operations in that area and would require flatter slopes.

Their production blasts utilized 6-inch holes spaced 12 feet apart in a square pattern and drilled 32 feet deep. They were loaded with about 500 pounds of explosive per hole and generally detonated ten holes at a time (per delay). At the final wall, they drilled the 6-inch holes just 5 feet apart with 25 pounds of explosive spread down the hole. This line was drilled at 75 degrees below the horizontal. Only ten holes were detonated at one time to create a pre-shear crack along the final wall. Two further rows (buffer rows) 7 feet and 14 feet out from the final wall were drilled and also loaded lighter than normal. A typical final wall blast involved detonating the pre-shear line first, followed by the production holes, outer buffer line, and finally the buffer line nearest the pre-shear line. An excellent, stable final wall was achieved. Most of the half rounds of the line drill holes were observed in the final wall after excavation. This is a measure of good final wall blasting, with only minor damage to the wall rock.

The structural geology was generally random jointing, with only two unfavourably oriented faults that dipped into the pit. At these locations the pit slope was set back behind the faults. At other locations the overall pit slope angle was 45 degrees, with catch benches every 30 feet.

There had been no pit slope stability problems to date. As a result, the manager asked me, "Can we steepen our slopes? This could save us several millions of tons of waste removal."

I had stated previously, at an AIME meeting in Minneapolis, that if there had been no pit slope instability at a mine, the slope angles are excessively conservative. I described this to him and added, "I would prefer you to keep your final bench face slope design. However, the slopes can be steepened by double benching." That is, delete every other bench. This will allow you to steepen your slopes to about fifty-two degrees."

It should be noted that benches to catch ravelling rock have become standard practice in open pit mining. They are generally 20 to 25 feet wide, depending on the amount of ravel that is expected. Where ravel is minimal, the bench width can be reduced and double benches can be designed, as in this case.

He was very pleased and said, "I'd like you to describe this to my engineering staff."

At the time of my visit the pit was about 300 feet deep. Ultimate depth was to be about 600 feet. No significant seepage or wet blast holes had as yet developed. I advised, "If you encounter considerable water as you deepen the pit, you will likely have to develop a dewatering program to minimize the cost of explosives and maximize slope stability. Water reduces stability by reducing shear strength, creating seepage forces, filling tension cracks and increasing seismic forces (water hammer effect) below the water table. At that time I can help you develop the drainage design."

Many mining engineers still treat blasting as a "bombing" operation — production, production, production! They don't appreciate that controlling blasting with less seismic energy near the final wall can often allow steepening of the final slopes by 5 to 7 degrees. That represents savings of many millions of dollars, and improves stability. I still have difficulty getting this message across.

I drafted recommendations on Tuesday afternoon and finished about four o'clock. I then gave a verbal summary of my site inspection, conclusions and specific recommendations. I particularly complimented them on their blasting program. Even by today's standards, it is one of the best programs I have encountered anywhere in the world.

I drove back to the hotel and tidied up for dinner, then Pat and I went to a local restaurant. As it was getting quite dusky I had my headlights on dim. In the centre of town I was stopped by a policeman, who said, "You'll have to turn your lights off and use only your parking lights!"

I replied, "Why? It's getting dark and it's dangerous to drive without lights."

Belligerently he retorted, "We have a county law that requires the use of only parking lights in the town at night. Either turn your lights off or park your car!"

I nodded and said, "I'll turn them off." Later I wondered what he would do if he caught me speeding. I didn't intend to find out.

The next day we went sightseeing and stopped in Galway to visit a rebuilt portion of the town that was over 500 years old. An attractive young Irish lass showed us around for about an hour and gave us a great lesson on that time in history. She seemed embarrassed when I gave her a tip at the completion of the tour. However, she had earned it.

We drove for several hours passing through small villages, a coastal town, farmland and peat farms. The peat is cut by hand into blocks, laid in the sun to dry and later burned for fuel and cooking. It is a major industry in Western Ireland.

At one small village we stopped for gas. A young girl came out and I said to her, "Would you fill it up please?"

She stared at me briefly, then turned around and ran into the gas station. Moments later another girl came out. I asked her the same thing and then queried, "Did I say something wrong? Did I frighten her?"

"Oh no; she can't speak English! She only speaks Gaelic. She was embarrassed." This was a great surprise to me. I couldn't help thinking later that she had a narrow life in front of her.

Late that afternoon we turned inland and began to look for a hotel for the night. We noted a sign on the side of the road advertising accommodation at a renovated castle five miles ahead. Minutes later we turned onto the access road to the castle. It was well maintained with beautiful green manicured landscaping. We parked and went in to enquire about a room. The check-in counter was small, set in the centre of a display of old British furniture.

The clerk advised us he only had a room with twin beds left. The price was modest, about $20 Canadian. I signed in, after which he gave me a pamphlet that described the castle's history. It had been constructed by an Irish lord in the thirteenth century and continuously occupied since that time, except for several decades in the early 1900s. An American entrepreneur had bought it ten years earlier and turned it into a hotel with fifteen rooms, a dining room, and an Irish pub. It was very charming. A bellboy took our bags upstairs to a medium-sized room with a very high ceiling and two windows up high. He informed us that the bathroom was down the hall and showed us a washbasin and water jug on a side table, adding, "I'll bring up some water shortly." He gave us information about the dining room and pub and then asked, "Would you like a wake-up call in the morning?"

We said, "Yes, seven-thirty please." He nodded, smiled and left.

Dinner that evening was pleasant and we crawled into bed early. As I slid under the covers my foot encountered an object — a hot water bottle! It had been placed there during dinner. We still joke about the "warm" Irish hospitality. The next morning our wake-up call was a knock on the door and a waiter bringing us tea, a nice custom in the rural parts of the United Kingdom.

The following day we drove to Dublin Airport to return our car and catch a flight to London. Along the way, we were amused by some of the Irish road signs. The one we enjoyed most was "Lay by," which indicated a parking area along the roadside. We joked, "It isn't very private."

This was the only time I had flown on a British Airways Comet. This ill-fated early jet was taken out of service after two in-flight explosions, which resulted in many lost lives.

In about 1972, White Pass and Yukon Railroad had signed a contract to haul lead-zinc concentrate from the new Cyprus Anvil mine near Faro, Yukon, to Whitehorse by trucks. From there the concentrate was shipped on the railroad to a new storage facility and deep-sea dock in Skagway, Alaska, for shipment overseas. White Pass had a fleet of special trucks designed and built to carry 50 tons of concentrate. The distance was about 200 miles. After only six months the trucks were being shaken apart by the severe washboard that had developed on the gravel road. The fenders were tearing from the frame, the radiator bolts were shearing, and the doors and locks were loosening. The trucks had to be provided with special air cushion seats for drivers to survive the six-hour drive. At each terminus the trucks had to be washed to remove a substantial layer of dust.

A. Freisen, President of White Pass, contacted me and asked me to meet with him to discuss the problem. At his office he showed me photos of the metal failures. It was obvious that the cause was metal fatigue due to severe vibration. Based on my experience with highway grade design and stability, Golder Brawner was retained to evaluate the problem.

It was indicated to Freisen that the problem was likely related to the heavy truckloads overstressing both the road foundation and the road surface layer. Based on a major Canadian highway grade strength research program in the late 1950s, which included British Columbia involvement, it was determined that road surface deflection and rebound under legal axle loads was related to strength and life expectancy. An instrument called a Benkelman beam was placed between the dual back tires with the truck stopped. A dial gauge read rebound to the nearest thousandth of an inch when the truck was driven away. For asphaltic pavement, if the rebound is less than 0.05 inch on removal of the truck, the asphalt pavement would withstand about one million truck passes with 18,000 pounds single-axle dual wheel loading, or 32,000 pounds dual-axle dual wheel loading, without failure.

During inspection of the highway surface condition, it was noted that numerous sections of the gravel surface were in good condition, while others were severely rutted or wash-boarded. Benkelman beam rebound tests were performed at many good and bad sections. Where rebound values exceeded 0.06 inch under the loaded concentrate vehicles, the road surface was deteriorating. Where the rebound exceeded about 0.08 inch, the damage was severe. At each test location a borehole was drilled through the road-surfacing layer, the grade and into the sub-grade. Evaluation of this data indicated 24 inches of clean granular sub-base and crushed base gravel was required over a strong sub-grade — i.e., compacted till. And 36 inches of clean granular sub-base and crushed base gravel was required over moderately weak sub-grade. This study developed the foundation base criteria.

To provide resistance to washboard, which results from shearing and shoving under heavy wheel loads, the upper 9 inches was selected as ¾-inch maximum-sized crushed gravel with between 10 to 12 percent passing the No. 200 sieve. Four percent calcium chloride was added by weight to water. Water was added so the moisture content would be within 2 percent of modified proctor optimum moisture. This mixture was spread in a 12-inch loose lift and compacted with eight passes of a 50-ton rubber-tired roller.

The clay and calcium chloride assisted in greatly increasing cohesion and dust resistance. The final grade developed with this procedure typically would last about three months. As rutting or washboard developed, a heavy-duty grader equipped with surface ripper as well as a blade ripped and regraded the surface. A water truck applied a calcium chloride mixture and the new surface was recompacted. The calcium chloride causes corrosion of metal, so White Pass built a truck-washing facility in Whitehorse to wash the vehicles after every trip.

The mine was of sufficient economic benefit to the Yukon that a program was commenced in the second year to surface the grade with asphalt. After the new highway was constructed from Whitehorse to Skagway, truck haul replaced the train haul.

By the beginning of 1970, it had become apparent that major opportunities for consulting geotechnical engineering existed in the mining industry internationally. Golder Brawner had developed a world-class group of geotechnical engineers with skills in rock slope stability, tailings dam technology, and waste pile stability. It had also become apparent that it was easier to perform one large consulting project than ten small projects. Contracts with mining companies were usually large and had the potential for ongoing involvement. Project overhead per man-month was generally less. It was decided to embark upon a major marketing program.

This included development of a seminar on mining geotechnology to be presented to mining companies. This usually took one-half or one full day. The factors that influenced stability were described, investigations to evaluate stability were outlined, and analysis of stability was reviewed. Ongoing evaluation of stability, which emphasized the observational approach, was described, and methods to develop or maintain stability were presented. This was followed by presentation of case examples of failures, and the stabilization program that was then developed. Coloured slides were used throughout the seminars. I would include the offer to present such a seminar in most proposals submitted — at no cost. I attribute our early high success rate in obtaining project contracts to the success of the seminars.

In early 1970, I decided to organize an international conference on "Stability in Open Pit Mining," to be held in Vancouver. By this time I had become aware of the major players in the industry — mining companies, consultants, university professors and researchers. An editorial committee and sponsors were established. Chapter topics were developed and potential authors were considered. The intent was to invite specialists to prepare a specific book chapter and present it at the conference. All specialists that were invited accepted the invitation. The topics and authors selected for each chapter were as follows:

1. *Introduction* – C. O. Brawner
2. *The Role of Slope Stability in the Economics, Design and Operation of Open Pit Mines* – Richard M. Stewart and Bruce A. Kennedy
3. *Geologic Factors Controlling Slope Stability in Open Pit Mines* – F. D. Patton and D.U. Deere
4. *Influence of Rock Structure on the Stability of Rock Slopes* – Evert Hoek
5 *The Influence of Groundwater on Stability* – Norbert R. Morgenstern

6. *The Influence and Evaluation of Blasting on Stability – Alan Bauer and Peter N. Calder*
7. *Influence of Earthquakes on Stability* – Robert V. Whitman
8. *Methods of Analysis of Stability of Rock Slopes* – David L. Pentz
9. *Field Instrumentation for Rock Slopes* – K. Barron, D. G. F. Hedley, and D. F. Coates
10. *The Stabilization of Slopes in Open-Pit Mining* – H. Q. Golder
11. *Design and Construction of Tailings Dams – Leo Casagrande and B. N. McIver*
12. *Case Studies of Stability on Mining Projects* – C. O. Brawner
13. *Panel Discussion* – K. Barron, C. O. Brawner, L. Casagrande, H. Q. Golder, B. A. Kennedy, and D. L. Pentz
14. *Conference Summary* – Evert Hoek

A registration fee was established to cover all costs. The Canadian Institution of Mining Engineers was approached to publish the proceedings. They rejected the idea! Naturally I was quite upset at this. I had felt it would have been a great project to develop exposure for the Institute. However, I was also a member of the American Society of Mining Engineers (SME), and they had published a number of mining books. The head of the Publication Department was Miss Marianne Snedeker. I phoned her, and she was most interested. Based on her interest, I decided to act while the "iron was hot." I arranged to meet with her the following week in Denver.

Miss Snedeker, an extremely professional woman, was very helpful. She described the publication process and provided me with a set of author instructions. Victor Milligan, Senior Operating Principal of Golder Associates in Toronto, was co-editor with me.

The conference was a great success. Over 250 mining and geotechnical engineers from all over the world attended.

Based on this success, I was involved in nine further mine geotechnical conferences in Vancouver up until 1988. The proceedings of all the conferences were published, most by the American Society of Mining Engineers, with two by Miller Freeman Publishing. Following the 1970 conference, an arrangement was made with SME for further conferences whereby they would provide a full page of advertising for four consecutive monthly issues of *Mining Engineering*. Announcement of the conference and call for papers, followed by the details of the program and registration details, were included. In return for free advertising, SME could sell the proceedings and keep all profit. The financial return for SME was good.

Organizing the conferences involved a lot of work. In addition to the SME advertising, the program was sent to every mining manager in the free world. Address labels for the mailing were purchased from the *Engineering and Mining Journal Mine Directory*. Golder Brawner secretarial staff, and later Mrs. Penny Manfredi, provided me with tremendous assistance. I am a strong advocate of the philosophy that the efficiency of a manager is directly related to the efficiency of his secretary or administrative assistant. Four of the proceedings received publication awards from SME.

The overall results were beneficial for all. The texts form a basic library of mining geotechnology, and the exposure for Golder Brawner was excellent.

University engineering programs generally contain no instruction in business-related subjects. Initially, this hampered me. I was fortunate to receive an unsolicited brochure from the American Management Association in New York that described numerous specialty business conferences. They lasted five days and were usually held at a tourist resort. The program involved morning presentations by top business leaders, an afternoon of relaxation or play, a dinner together and round-table discussion with the instructor in the evening. Wives were invited.

The first conference I elected to attend was "To be President of a Small Business" at Jasper Lodge, Alberta. It proved to be very beneficial. I attended two other such seminars — "Financial Management" and "Personnel Management" — which were also worthwhile.

To anyone who aspires to become management or to develop a consulting company, I recommend either a graduate degree in Business or a series of specialty management courses early in your career.

Dr. Hugh Golder was from England. He held the strong conviction that if you provided top-quality engineering, the firm would be successful, technically and financially, and would prosper without significant business promotion. Consequently, Golder had little involvement in business development. He acted primarily as a senior engineering review consultant. Many of the older Canadian engineering firms were well established. Golder Associates and Golder Brawner Associates were a new group trying to become established. A major problem was that most of the larger projects that were being considered required competitive proposals. Usually, as many as six consulting groups would be requested to develop submissions. On large projects these proposals could cost tens of thousands of dollars. With a statistical chance of success of about one in six, considerable financial risk was added to the operating cost and the success of an engineering firm.

I didn't like the odds. As a result, I attempted to pursue projects on a sole source basis. In addition to the procedures described earlier — lunch invitations, Christmas parties, specialty in-house seminars, mailing of technical papers, organization of international seminars and participation in geotechnical conferences — I became heavily involved in the development of specialty corporate brochures. One brochure, specifically on mining geotechnology, was developed in the Vancouver office for corporate use.

As a result of this approach, more than half of the geotechnical engineering projects in mining that Golder Brawner obtained were negotiated without competing proposals. This resulted in reduced lead-time, savings to clients, and improved corporate profits.

It took many years for Dr. Golder to accept the benefit of this proactive business development approach. Some senior staff with British or Australian backgrounds never adjusted to business development or marketing while I was with Golder

Brawner. Occasionally this led to internal friction. Today the competitive nature of the engineering industry has become standard to such an extent that engineering fees are lower than they should be for the quality of work or ingenuity that is generally provided.

Our consulting business developed rapidly as the mining geotechnical work exploded internationally. Senior staff were added including: Peter Stacey from the Iron Ore Company; Norman McCammon and Richard Butler from Toronto; Bruce Kennedy and Richard Talbot from Anaconda; Ken Rippere from Kennecott; Kevin Rosengren from Mount Isa Mine, Australia; as well as, Dr. Milton Hsu, Bill Hodge, Russ Hammett, Herbert Hawson, Bjarn Carlsen and Norm Norrish. Offices were opened up in Calgary, Kelowna, Kamloops and Edmonton. A U.S. office, initially supervised from Vancouver, was opened in Seattle.

Space quickly became a premium. Mortgage rates were low. Around 1970, it was decided, with approval from Toronto, to buy property and build an office building with downstairs laboratory facilities. The new building, situated at 780 West Broadway in Vancouver, was designed and constructed by Beedie Construction, a contractor who specialized in small offices and warehouses. The building had been added to three times by 1978.

4

Canada

Canada
(1963-2003)

The author was involved in numerous interesting surface mine stability projects in Canada, including an investigation into the failure of a high waste pile at the Kaiser Resources coal mine in B.C. which killed two people. A recommendation to the B.C. Ministry of Mines resulted in new guidelines for waste piles in that province.

A 33-million-cubic-yard open pit slope failure occurred at the Canadian Johns Manville mine and damaged buildings in Asbestos, Quebec. The author established a stabilization program involving the unloading of the slope, the installation of horizontal drains and deep drainage from underground exploration drifts. The use of the observational approach, recommended by Dr. Ralph Peck, was successful at the Steep Rock Mine at Atikokan, Ontario. This allowed mining of iron ore to continue below an 800-foot-high open pit slope where pit wall movement averaged ¼ inch per day.

The most challenging mining project in the author's experience was at Syncrude Canada where he joined the Geotechnical Review Board. He has been on the Board for thirty years. Challenges have included the development of numerous dragline options, highwall slope monitoring, and geologic evaluation using a borehole dipmeter. Further challenges were the development of 300-foot-high tailings dams on glacially shoved Clearwater clay shales, and a 250-foot-high divider dyke in the pit at temperatures down to –40°C. A provincial highway was developed on top of the dyke.

A rockfall derailed a freight train near Spences Bridge, B.C., which killed the two operators. The investigation led to the first rock slope stability classification program in North America and an ongoing rock slope stabilization program for the Canadian Pacific Railway.

Chapter 4
Canada

Jim Morris, Manager of Kaiser Resources' coal mine in the Crowsnest Pass area of southern B.C., called me at home on a Sunday, about noon in November of 1968, and said, "Mr. Brawner, we've had a very serious spoil pile failure. It has closed Highway 3 (the southern trans-provincial highway) and killed two people who were caught by the slide in their car on the highway. We require your help immediately. Can you come right away?"

In view of the emergency, I considered going to be essential. I answered, "I'll catch the first plane that I can this afternoon or evening. I'll be there tomorrow morning at eight." I phoned my secretary at her home to cancel all my early week appointments, and then called Pacific Western Airlines for a late afternoon reservation.

That evening I arrived at Cranbrook by 7:30, rented a U-drive, drove to Fernie and by 9:00 P.M. had checked into a motel. I arrived at the mine office at 8:00 the next morning. I had to pass through the slide area, which had by then been cleared enough to allow one-way emergency traffic.

When I saw him, Morris said, "Thanks for coming so soon. Henry Kaiser is here, as well as the mines inspector. They'd like to meet with you." He took me down the hall to a conference room where plans had been spread over a table. Morris made the introductions. Kaiser queried me on my background and then said, "The sooner you see the slide the better. They're still clearing the highway and only allowing emergency traffic through under escort. They've recovered the car and two bodies. Highways have decided to construct a Bailey bridge over the river to detour traffic until the area is stabilized. We should probably go to the top first and look down. It'll be safer." I asked Morris if he had a pair of binoculars. He did.

The mine had a four-wheel-drive vehicle, which we took up the switchback access road. Fortunately, there was no snow on the ground. Crows' Nest Resources was originally an underground coal mine which had operated since near the turn of the century. By the 1960s, it was struggling to stay open. At the time, the Kaiser group from San Francisco was looking for mining projects, and had explored the potential of developing the property as an open pit operation. Their evaluation was optimistic, and so they purchased the property. Initially a small pit near the top of the mountain was developed on property west of Highway 3. Kaiser had contracted out the mining, but performed the mine design with its own staff.

The coal seam being mined dipped into the mountain. The open pit wall above the seam (called the "highwall") contained shales, siltstone and limestone, and was up to about 150 feet high. Waste rock was blasted, loaded and dumped down over the mountain slope. The overall pile was about 800 feet high at its natural angle of repose. A near-vertical slide scarp, about 120 feet high, had developed where the slide has slipped away from the rest of the waste pile. Below that, the slide surface appeared to follow the original slope configuration. However, the surficial soils had been scoured off. In the lower centre of the slide, a moderate seepage flow had developed.

The waste material had developed into a slide flowing down the lower slope and across the highway, stopping at the edge of the river. It had also severed the

Waste pile failure which crossed the southern provincial highway near Sparwood, B.C. The toe of the pile encroached on weak glacial lake silts and caused the foundation to liquefy. Two occupants of an automobile were killed.

trans-provincial power line. It was the north half of the spoil (or waste rock) pile which had failed, but a major concern was that the south side might also go — with little warning.

I asked Morris, "I note the water flow; was there a small creek and gully there originally?"

He confirmed, "Yes, there was. We assumed the dumping would develop a zone of coarser rock at the toe of the pile as it got higher and larger, and that would act as an underdrain." He added, "The dump was deepest there. The south half of the spoil pile is not as deep."

There were no tension cracks behind the scarp, nor at the top of the south half of the pile. I said to Morris, "I'd like to speak to the dump operators as soon as possible. I believe there had to have been some warning of this failure."

He answered, "Okay, I'll arrange that. You might also be interested to know that yesterday morning, when the slide occurred, one of our miners who lives across the river from the slide was dressing and looking out his window. He watched the slide as it developed and came across the highway."

"Yes, I'd like to interview him," I answered.

We drove to the bottom and the south side of the slide. Two highway officials were there reviewing further action. We were introduced, and they indicated to

Morris that they expected him to make the site safer as soon as possible. Morris said, "We've hired Golder Brawner to advise us on that. As soon as we get Mr. Brawner's recommendations, we will advise you."

I indicated to Morris that I wanted to climb up the south edge of the failure to observe what I could and to look at the exposed section of the central scarp from that vantage. Morris said, "I'll go with you." Kaiser declined.

The lower half of the slide was quite wet and slippery, which made climbing difficult. We stopped at several locations and used the binoculars. About midway up the slide, climbing became easier. I could see no seepage issuing from the central scarp of the slide, so I decided to venture across its bottom to obtain a better look. Morris came part way, acting as a spotter for me, looking for signs of instability as I continued.

As I approached the scarp, I could see that about 5 to 10 feet of the natural ground had been scoured away. Within the intact dump, a layer of coarser rock was visible on top of the original topsoil, and the lower 2 to 3 feet of this layer was blended with rock fines, ranging from a fine silt to coarse sand. This zone was apparently blast and crushed rock fines that had been washed down through the coarse dumped rock by rain and snowmelt. Given its comparative impermeability, this layer would act to reduce seepage from the mountain and build up water pressures behind. About 15 feet above the coarse basal layer, I also observed a 1- to 2-foot zone of very compact snow, indicating that dumping had taken place after a major snowfall — a poor practice. I noticed no seepage from the scarp face. I took photographs, but it was a cloudy day, and the contrast of materials was poor.

We returned to the office to meet Kaiser again. Morris had arranged lunch to be brought in. Kaiser asked, "What do you think?"

I explained, "The zone that has failed appears to be stable, including the toe area. The upper slide has scoured out all the weak material. The spoil on the road is up to about twenty-five feet thick. To ensure that no further slumps will occur, the new back slope being excavated should not be steeper than 3 horizontal to 1 vertical. This material is loose and wet. The major concern is the stability of the south half that is still there. I recommend that several horizontal drains be installed directly under the lower third of the south portion. To ensure safety of the drill crew, the drains should be installed from the side of the spoil pile. Highways officials should complete the Bailey bridge and detour all except construction traffic until further notice. BC Hydro should also detour their power line until the area is stabilized."

Kaiser stated, "Your recommendations sound reasonable. When should we speak to Highways and Hydro?"

"I'd first like to interview the dump and Cat operators who worked on the top of the dump last week, and the worker who observed the failure."

Morris said, "They'll be here shortly."

I interviewed four Cat operators. Three had observed nothing unusual. The fourth one said, "I came on to the early morning shift. There had been no dumping for several days, but I was told dumping would start that morning. I walked my Cat to the centre dump area and there was a crack about one foot wide, and the crest had dropped a little."

I asked, "What did you do?"

He replied, "I filled it in!"

"Who did you tell?" I queried.

"I didn't tell anyone," he responded.

Today he would have had explicit instructions to immediately call a supervisor. Prior to spoil pile failures, there is almost always a history of cracking and movement.

Shortly later I interviewed the employee who had watched the slide develop. He recounted, "I was looking out my bedroom window while I was buttoning my shirt, looking directly at the mountain face. Suddenly I saw the lower area start to move, and then a wide crack across the slope opened up about a hundred feet above the bottom. This was followed by a second crack above, with movement below that crack. This cracking and movement continued working its way up the waste pile. About that time I saw the slide come across the highway and catch the car. A few seconds later, the upper part of the waste pile started to move downhill. It was all over in about one minute. I ran to the phone and called the mine telephone operator."

His description indicated the failure had initiated in the toe area and retrogressed up the slope as each new failure removed support for material above. I asked Jim Morris, "Do you have any overlapping air photos of that area prior to the mining?"

He said, "Yes, I'll get them." The use of stereo air photos is an extremely useful tool in geotechnical engineering. Stereo photos for three-dimensional viewing can also be obtained on the ground by taking two photos of the same area 10 to 12 feet apart.

Morris returned shortly with the photos. I had a pocket stereoscope with me, and so I laid out the photos and selected a stereo pair of the waste pile area. In the upper area, the vegetation was primarily fir trees. In the lower area there was a marked change to a preponderance of deciduous trees, what appeared to be poplar, birch and willow. The line of change was continuous and appeared to follow a topographic contour. This change in vegetation meant a change in subsurface moisture content, or soil type, or both.

I was generally familiar with the glacial history of that area. A glacier had dammed the Elk River Valley at Elko, west of Fernie, and a glacial lake developed — Lake Fernie. The lake filled until the elevation reached that of Crowsnest Pass and water then flowed eastward into Alberta. The elevation of the vegetation change coincided with that of the Pass. Below that elevation, weak lacustrine soils were deposited in a lake environment. It was obvious the waste pile toe moved downhill as dumping continued, until the toe encroached enough on the weak lacustrine soils to overload them and induce failure at the toe.

No geotechnical investigation had been performed for the spoil pile. A review of the air photos and several test pits would have been sufficient for a qualified geotechnical engineer to identify the potential problem. Unfortunately, this had not been done.

Before leaving the mine, I recommended that about 12 waste pile mounds, 15 to

20 feet high, be constructed above the highway and in the lower third of the failure, the mounds to be staggered across the slope. Any further spoil movement would be deflected several times and likely stop. (This technique is also used to stop snow avalanches.) Following that, some 100 feet of the top of the south pile could be pushed by Cat into the scoured out failure zone to reduce the load in the south area.

No further problems developed. Kaiser chose to close that small pit operation. Within the year, the B.C. Ministry of Mines developed preliminary guidelines for waste dump design, construction and monitoring in British Columbia.

As I was preparing to leave, Kaiser asked how I was going to get to Vancouver. I said, "Drive the rental car to Cranbrook and catch Pacific Western Airlines."

He suggested, "How about coming with me on our corporate jet to Vancouver?" I accepted immediately.

The plane was a Gulfstream jet with two pilots. The cabin had folding tables and very comfortable seats. The captain came back to the cabin and asked Kaiser, "Cranbrook airport is uncontrolled air space, and I've never had the opportunity to do a full power climb. We're to get up to 39,000 feet. I'd like to really test this baby. Do you mind?"

Kaiser said, "Go for it! Is that okay with you, Chuck?"

I nodded, but with a lot less enthusiasm. The captain warmed the jet up on the apron for at least five minutes before taxiing to the end of the paved runway. "Fasten seat belts for a thrill ride," he said over the intercom. The engines roared, and I was forced back in my seat. Within what appeared to be only seconds, we were airborne and rapidly picking up speed. Suddenly we were in the steepest climb I have ever been in; it felt like at least 45 degrees. The noise was deafening. Then, just as suddenly, it was quieter; the plane levelled off. About five minutes later, the captain came back, a large grin on his face, "Wow, that was great, just under two minutes. She behaved beautifully."

With the Crow's Nest failure, the Aberfan spoil pile failure in Wales that killed many school children, and the earthquake-induced El Cobre tailings dam failures in Chile in the same year, geotechnology suddenly became an essential part of mine waste facility design internationally. Most government regulatory agencies responded quickly with new guidelines or regulations that were more rigid, and required more thorough geotechnical site investigations.

In late 1968, Sam Smyth, Exploration Manager of Kennecott Exploration in Vancouver, phoned to say, "Mr. Brawner, we have a copper prospect in west-central B.C. that we're exploring on Galore Creek, north of Hazelton. We've completed an adit [an underground mining tunnel] about 1,200 feet long into the ore body to obtain large-scale samples for metallurgical testing. All the drill core is available at the site. Our Salt Lake City office has recommended you to assist us in evaluating potential pit slope angles required for our feasibility studies. Would you be available in the next few weeks to go with me to the site to inspect rock conditions in the adit and to map the rock core?"

I answered, "Mr. Smyth, I would be pleased to go with you. Can you give me more details?"

He said, "I'd like to go up the second week in October. The plan would be to fly to Terrace on Sunday, overnight there and charter to the site Monday morning. I'll also bring a geologist with me. We have a camp there with a caretaker-cook. I would plan to return to Terrace Wednesday afternoon. I don't want to leave it any longer, as there is danger of being caught by a heavy snow. I'll arrange for your tickets."

I then requested, "Can you bring site maps, geological reports, air photos and drill hole logs with you? I'd like to study them on the way up."

He replied, "Good idea. Bring warm clothes with you. They've had freezing weather there already."

The trip to Terrace was uneventful. Monday morning we left Terrace in a single-engine de Haviland Otter aircraft. A supply of food, gasoline, oil and drill supplies had been loaded earlier. (The Otter has replaced the Beaver, the workhorse of northern bush flying, to carry heavier loads. It could carry up to about twelve passengers and baggage, or about 1.5 tons of freight.) It was about a two-hour flight to a dirt landing strip on the bank of the Scud River, near the mouth of Galore Creek. The Scud River is a tributary of the major Stikine River.

The weather was sunny and the flight was smooth. We crossed several rivers, mountains and the Nass River the first hour. The scenery then changed to an area of inactive volcanoes and volcanic flows. I had never been over this area before, and I kept my nose to the window for over thirty minutes. We then entered the Stikine Valley and followed it to the Scud River, where the pilot turned the plane and dropped to a lower elevation. Within minutes we landed on the narrow, bumpy airstrip.

Smyth had contacted the camp by radiotelephone before leaving, and so the caretaker-cook met us with a four-wheel-drive power wagon. It was several miles to the camp, up an access road that had a number of very steep sections, and traversed several narrow ledges above Galore Creek. The camp was a combined cookhouse, dining room, bunkhouse, storage building and separate core storage area.

On the way up I learned that the deposit contained at least 120 million tons of copper ore grading over 1.2 percent copper. However, a portion of the ore body was covered by the lower reaches of an active glacier. This condition indicated early development would have to be an open pit, changing to an underground operation under the glacier. There was evidence that the glacier was retreating about several hundred feet per year.

The first afternoon I spent four hours with the geologist in the adit, mapping structural geologic features — joint patterns, shears and the occasional fault zone. Samples of fault gouge were taken and placed in small sample jars for later testing. After dinner that evening I scanned through drill hole logs.

Mapping the adit was completed by noon the next day. After lunch I went to the rock core storage area, which was partially under cover. I inspected and photographed rock core and compared the drill logs to my own mapping.

By midafternoon dark clouds had moved in and it started to snow. By dinnertime six inches had fallen. Smyth tried to contact the aircraft charter company to check

Snowbound for five days at the Kennecott Galore Creek copper prospect in northern British Columbia. Chuck feeding and playing with a black bear cub who had lost her mother.

on our pickup the next day, but couldn't get through. There was only static. The next morning he tried again, and again had no luck. The low cloud continued and it snowed periodically throughout the day. We were stranded! The bad weather continued for the next three days, which were spent in the core shed and the adit, or with Smyth in the bunkhouse discussing rock stability, pit slope design, and weather.

One amusement existed. A black bear cub, born the winter before, stayed around the camp. The mother had broken into the camp earlier and had to be destroyed by a conservation officer. I guessed the cub weighed about forty pounds. He was provided with a sparse diet of lettuce, vegetables and the occasional small fish, and had become quite playful and tame. The caretaker advised us to make sure we wore leather gloves when we played with him. His teeth were quite sharp. He had been taught to "come," and to stand on his hind legs.

Finally, on Saturday the weather cleared, and Smyth reached the charter company with the camp radio. They would send the plane in that afternoon, provided the weather stayed clear. A tentative arrival time of 3 P.M. was agreed upon. He was to come in on skis.

By then 3 feet of snow had fallen. The caretaker cleaned the snow off a D-6 Caterpillar, fuelled it and made two passes to open the road to the airstrip. We left the camp at 2:30 by truck. The steepest parts of the road were extremely slippery, and so I got out of the cab and climbed into the back of the truck. I wanted to be able to jump off if the truck started to slide near the cliff edge. Nearing the airstrip, the caretaker appeared to lose control of the truck. I jumped into the inner snowbank, sinking to my hips. The truck came to a stop, almost sideways, only a foot from the crest. Slowly the driver applied forward power and straightened the truck out, continuing to the bottom of the hill. I followed on foot.

At about 3:15 the Otter came into sight and landed. So much snow swirled around the plane you could hardly see it. The pilot turned the plane around and stopped but kept the motor running. He said he wanted to taxi back and forth several times to pack the snow for takeoff. We were a light load, so takeoff was short and uneventful. The plane landed at the Terrace airport on its wheels — small hydraulic cylinders raise the skis to where they are above the base of the wheels.

The next summer we flew into the site again to inspect the area for a mill site, tailings dam and small hydro development. The feasibility study indicated that the most desirable route to access the property would be via a 10-mile tunnel to the Stikine River, and then a barge to Wrangell, Alaska. The high cost of this route has shelved the project to date, but it will be developed some day. High-grade copper deposits are rare in Canada.

In 1970 Golder Brawner became involved in one of the world's truly challenging geotechnical engineering projects — the Syncrude tar sands development. Atlantic Richfield (ARCO) of the U.S. was one of the early major partners. Their lead engineer on the project was Peter Zambus, a mining engineer by training. Zambus retained Golder Brawner to specifically evaluate pit slope stability and design.

Early stability studies by others were based on the tar sand behaving like a uniform material, in which case failure of excavated slopes would be expected to be curvilinear. Following our inspection of selected core materials, a shallow test pit, riverbank outcrops and the already-operating Suncor project, Dave Pentz and I concluded that the most likely failure mode would be along clayey beds which dipped out of the highwall. As movement developed, the shear strength would reduce and planar failures could occur quite rapidly. Zambus advised Syncrude of this potentially serious failure mode. Shortly after that, ARCO withdrew from the Syncrude consortium to pursue its major Prudhoe Bay oil discovery in northern Alaska. Nevertheless, Ernie Pitchall, Manager of Engineering at Syncrude, asked me to be part of an advisory team.

Syncrude and each of its owners already had other advisory consulting engineers. The advisory team's first major decision was to determine whether bucketwheel excavators should be used, or whether to pursue the use of large draglines. The bucketwheels excavated benches about 50 feet high, whereas the draglines would mine the entire thickness from a top bench. Suncor, several miles away, was already in production using bucketwheel excavators to extract the tar sand, but, on that project, a number of problems had developed. These included the inability to reject waste or low-grade material, excessive wear on the bucketwheel teeth, requiring frequent replacement, and poor equipment trafficability on the tar sand due to compression under heavy wheel loads.

All the advisory geotechnical consultants were assembled in November 1972 to assist Syncrude in this major review and decision. The advisory group became known as the Geotechnical Review Board (GRB). The draglines had potential benefits in that they could mine the entire depth (150± feet) of tar sand, and could separate waste zones within the sands, but big questions remained as to the stability of the tar sand slopes on which the dragline would be operating, and the ability of the draglines to chop-cut the overburden.

The original Review Board was comprised of Arthur and Leo Casagrande, U.S. consultants; Elmer Brooker and Dave Devenny of EBA Consultants; Gerry Kendall of Imperial Oil; Herman Knight of Peabody Coal; M. Brashears and Russ Slayback of Leggett, Brashears, Groundwater Consultants; Frank Patton, Geologist; Karl Taylor, retired from Bechtel; Robert Hardy of R.M. Hardy Ltd.; Fred Matich of Geocon; Norbert Morgenstern of the University of Alberta; and myself of Golder Brawner Associates. This Board provided advice to Syncrude for four years, until 1975. An early major recommendation made by the Board, accepted by Syncrude, was to use a dragline to excavate a large test pit in order to evaluate diggability and slope stability. This recommendation resulted in Syncrude purchasing a 17-cubic-yard dragline to dig the test pit, a machine that came to be fondly known as "Little Beaver." This trial proved to be fortuitous. The test pit design, to be full depth, called for 70-degree slopes. The test pit walls in the tar sands began to slab and sluff almost immediately due to excessive steepness and methane gas exclusion. And as the pit neared its base, the floor began to heave due to excess pore water pressures in barren water sand below the tar sand. Both conditions required a major change in design. Pit slope angles were changed to 50 to 55 degrees, and major basal dewatering was required in advance of mining.

The GRB also grappled with the question of highwall failure with a dragline on the crest. (Risk analysis techniques were not yet available.) Herman Knight, dragline specialist with Peabody Coal, quickly settled the issue. He said in his Texas drawl, "Assume that you'll lose one dragline every ten years and build that into your economics. If you can't afford that, you shouldn't go with draglines."

After twenty-five years now, and some 50 block slides in the highwall, no dragline has been involved in a failure. Over 6,000 miles of highwall have been mined safely in that time, and over 1.2 billion barrels of oil have been produced. The GRB takes personal pleasure in that success and major financial saving.

As soon as the major dragline decision was made, Pitchell decided to reduce the GRB to five members, each with a special expertise: Karl Taylor with large-scale industrial geotechnical experience; Robert Hardy, tailings dam and tar sand engineering properties specialization; Frank Patton, geotechnical instrumentation and clay shale geology; Russ Slayback, groundwater hydrology and dewatering; and myself for open pit mining and dragline experience. (By this time I had extensive experience in dragline coal mining in Canada, Australia and the United States.) From that point on, the GRB has met twice per year for a week at a time. At the meetings, Syncrude staff provide a general update of the project, followed by specific updates on many geotechnical problem areas. They make detailed presentations and solicit discussion with the GRB, usually presenting questions on each major issue for the Board to address. The GRB then completes its review report before it leaves the project. This is a procedure recommended to all review boards. The mining concerns reviewed include highwall monitoring and movement criteria, tar sand behaviour,

A 90-cubic-yard dragline digging tar sand to a depth of 155 feet. Stability of the slope below the dragline was essential. Major instrumentation of the slope and visual monitoring was performed hourly. Syncrude Canada mine near Fort McMurray, Alberta.

geological structural predictions prior to mining, basal dewatering, blasting of hard siltstone layers, tar sand windrow stability, highwall slope angle design, haul road design, *et cetera*.

Tailings disposal has also become a major stability issue. Two major tailings storage dams were required at Syncrude, for disposal of over 300,000 tons of tailings per day. Both are founded on glacially sheared Clearwater clays, and both required major stability studies. Side slopes as flat as 8H:1V to 20H:1V have been required at some locations.

Monitoring movement and water pressures, along with the observational approach, have become a way of engineering life at Syncrude. Allowable rates of movement, in combination with standard stability analysis and back analysis of all failures, were used extensively to develop empirical stability evaluation techniques. The GRB provided many benefits; these included:

1. Worldwide experience
2. Unbiased recommendations
3. Latest technology transfer
4. Engineering rapport with Syncrude staff
5. Focus and recommendations on major issues
6. Engineering continuity

Thin clay layers within the tar sand dipping out of the slope were the main stability concern. Visual monitoring observed the development of cracks and alerted the operator. The dragline immediately began to move from the danger area. The clay bed dipped at about 17 degrees. A resistivity dipmeter was developed to locate such beds before mining.

7. Confidence for government regulators
8. Emphasis on risk management (after about 1990)
9. Increased confidence (for Syncrude) in engineering design
10. Excellent training for Syncrude staff

Typical of the major contributions the GRB has made to Syncrude are the following:

1. Initial endorsement of draglines for mining
2. Recommendation for a mining scale test pit
3. Recognition that the primary highwall failure mechanism would be structurally controlled block failure on clay layers dipping out of the highwall
4. Development of facies modelling of the tar sand influenced by mode and sequence of deposition — facies were related to engineering properties
5. Development of deep basal dewatering of barren water sands below the tar sand to maintain highwall stability
6. Recognition of failure by plastic bulging in high-grade tar sand — in excess of 12 to 13 percent bitumen

7. Development of a slim line resistivity dip meter tool, in conjunction with Schlumberger, to measure dip and dip direction of clay layers whose orientation could lead to block failures
8. Development of monitoring procedures and limiting movement criteria to walk draglines away
9. Development of directional mining procedures to mine through areas where block slide potential was high
10. Development of blasting techniques to break up hard siltstone layers, which interfered with dragline mining
11. Ongoing evaluation of the behaviour of the tailings dams, and development of the observational criteria to control the rate of dam construction
12. Modifications to the design of drop structures in a major drainage interception ditch
13. Development and test review of composite tailings methodology
14. Development and review of winter construction of a 180-foot highwall berm as a licensed dam to separate in-pit tailings storage (to temperatures of –40°C)
15. Development of criteria to safely develop major waste dump construction

Over the years, several board members have resigned or passed away, and new members have been added. The present members (2003) are C. Brawner, Chairman; Dr. "Nordie" Morgenstern, Fred Matich, John Sobkowicz and Ed McRoberts.

The names of Karl Terzaghi, Arthur Casagrande and Ralph Peck stand out as world leaders in the development and application of geotechnical engineering. Nordie Morgenstern, in my opinion, has now joined that elite group. He is a brilliant engineer, grasps the issues faster than any engineer I know, and has extensive international experience. He has presented many keynote lectures and has been awarded many prestigious honours. He also has been president of the International Society of Soil Mechanics and Foundation Engineering. Fred Matich is one of Canada's most experienced geotechnical engineers. He is quiet and reserved, but extremely thorough in all investigations and evaluations in which he is involved. He has maintained detailed records on every one of his past assignments, as well as a photographic record of most projects. Ed McRoberts is a senior consultant with AMEC in Edmonton, Alberta, and John Sobkowicz is a senior consultant with Thurber Associates in Calgary. They are recent replacement additions to the Review Board who have extensive tar sands experience. For further information on the GRB and its effectiveness, the reader is referred to an excellent article by Gord McKenna, Environmental Engineer at Syncrude, entitled, "Celebrating 25 Years — Syncrude's Geotechnical Review Board," published in the September 1998 issue of *Geotechnical News*.

In September 1974, an empty unit coal train was travelling eastbound near Spences Bridge, enroute to the southeast British Columbia coalfields. The train rounded a curve

Rockfall on the Canadian Pacific Railway (CPR) caused an empty coal train to derail near Spences Bridge, B.C., in 1974. This accident led to a Canadian Transport Board hearing and subsequently to development of a priority stabilization program on the CPR all across Canada by the author.

at 35 miles per hour and struck a rockfall on the track. The two lead locomotives and several coal cars were derailed and plunged over a 100-foot-high, near-vertical rock face down onto the Trans Canada Highway. The two locomotive operators were killed.

The accident blocked the main CP Rail line, where daily traffic amounted to about thirty large-unit trains hauling commodities such as coal, wheat, potash, sulphur and petroleum products. Closure of the track gradually backed trains up all the way to Vancouver and Winnipeg. Early opening of the track was imperative.

Ken Truman, Regional Engineer of the Canadian Pacific Railway, phoned me and said, "Chuck, we've had a terrible accident near Spences Bridge due to a serious rockfall. Two men have been killed. I'm leaving immediately by car. Is there any chance you can come with me?"

I thought quickly then answered, "Ken, I'm not dressed for fieldwork but I have boots in my car. I can be ready in thirty minutes at my office. Can you pick me up on your way to the freeway?"

He replied, "Great, I'll see you in half an hour."

I called my secretary into my office, told her about the problem and my need to rush. I asked her to cancel one appointment I had for the afternoon, advising her, "I probably won't be back until late tomorrow. Would you run over to the drugstore and buy me two rolls of 35mm colour film?"

I didn't have field clothes, but there was a Coast Mountain Sports store three blocks away. I hurried over there and purchased a light jacket, outdoor shirt, hiking pants and a baseball hat. Spences Bridge could still be very hot that time of year. I returned to the office and changed. I put my camera and a field notebook into my briefcase. I was ready when Truman arrived.

As we drove Truman said, "The priorities are to determine the cause of the rockfall, evaluate the existing stability of the rock slope, and determine whatever measures are required to stabilize the slope to allow early reopening of the main line track." On the trip to the site, which took about three and a half hours, we also discussed implementation of a stability inspection and evaluation of all other rock slopes on the CP Rail system, including those in central and eastern Canada.

The site inspection indicated the rockfall involved about 20 cubic yards of rock. The failure was bounded by two near vertical joints with an intersection angle of 120 degrees, and an irregular sand-filled basal joint which dipped about 40 degrees out of the slope. Railway maintenance staff advised there had been very heavy rainfall earlier that day. They also advised that a loaded coal train had passed the site about twenty minutes before the accident. That sparked my interest.

I had become aware that the number of rockfalls had increased following the implementation of single commodity trains. Several CP Rail roadmasters had con-

The second locomotive destroyed the lead locomotive and killed the two operators.

firmed that when I questioned them. Also, on one occasion during a stability inspection, I had become caught in a rock tunnel when a coal train passed through the tunnel. I pressed myself tightly against the rock to be as far away from the train as possible, and, as the train passed, I observed that vibration in the tunnel wall increased steadily until the caboose finally passed. I experienced this again on several later occasions, bringing me to the inevitable conclusion that a person's ass is very sensitive to vibration.

An inspection of the Spences Bridge rock face also revealed parallel joints behind the failure zone. This rock could also fall. An immediate decision was made to install tensioned grouted bolts, followed by the drilling of drain holes into the rock face to relieve any water pressure. Truman arranged for Emil Construction Company to immediately mobilize a stabilization crew to install the bolts and drains. This was completed the next morning, and train traffic recommenced by noon.

Truman asked, "What do you believe caused the rockfall?"

I answered, "I believe that several factors contributed to the failure. The adverse structural geology allowed the failure to occur — the failure was along the discontinuities in the rock, not through the intact rock. I also believe the heavy rainfall that occurred before the accident caused the joints to become filled with water, creating water pressure in them, and that seepage flowed outward through the sand-filled lower joint. I suspect that the vibration created by passage of the earlier loaded coal train may actually have been the triggering condition, with the rock falling just after the train passed."

To develop a better understanding of the potential impact of vibration in rock cuts, a program was developed the following year to measure the magnitude of train vibration. The study indicated that, just as I had observed, vibration created by unit trains continued to increase as the train passed the seismic geophone until the last car went by. This is analogous to infantry marching in step across timber bridges during World War I, and collapsing due to the build-up of vibration. (Troops eventually learned to 'break step' as they crossed bridges.) Over the years, rock has occasionally hit a passing train, and vibration would certainly have contributed to such a rockfall. It was also determined that vibration near rail joints was about three times greater than that between the joints. Originally rails were 39 feet long between joints. This subsequently changed to 78-foot rails, followed by continuous welded rail, and this has greatly reduced vibration.

On the way back to Vancouver Truman said, "I believe this accident is sufficiently serious that the Canadian Transport Board will call for a special hearing. This process usually takes about six months. Over the next week I'd like you to propose a program for CP Rail to institute as soon as practical to evaluate rock stability conditions along our railway. Based on this study, also propose stabilization of rock slopes that you consider to be unstable."

The program proposed was to develop a comparative classification system to identify rock slopes believed to be at greatest risk of rockfall potentially causing train derailment. The classification included details relating to location, site visibility, alignment, climate, train frequency, slope dimensions, structural geology, history of past rockfall, evidence of seepage, workspace available, and the stabilization measures recommended. A simple site rating system was developed, from A — serious potential of rockfall of sufficient size to derail a train in the near future, to E — negligible risk of rockfall even during extreme events such as major precipitation occurrence, spring thaw, freeze-thaw cycles or an earthquake. All rock cuts on the railway in Western Canada were inspected and rated in the fall of 1974, and across

COPY

Figure 18 - DETAILED APPRAISAL STABILITY ASSESSMENT

STABILITY ASSESSMENT

PRIORITY RATING A

Region Pacific Subdivision Cascade Date Apr. 10/75

Mileage 24.9 Train Traffic Very Heavy

Alignment slight curve Sight Visibility 400 ft. West; 800 ft. East

Average Climatic Conditions High winter snowfall, hot dry summers

Past Stability Record Several local blocks

DESCRIPTION OF SITE

Cut [x] Fill [ ] Other [ ]

Height 50 ft. Length

Rock [x] Soil [ ] Other [ ]

Geologic Description Blocky faulted granite - 3 major faults. Random joints.

Evidence of Water Slight seepage

Work Space Available Limited

NOTES

DESCRIPTION OF POTENTIAL INSTABILITY - (TYPE, MAGNITUDE, SERIOUSNESS)

Very large block could fail at east tunnel portal. Also - second block 60 ft. eas of portal.

Tunnel has thin roof - some rockfalls will cause daylight

DETAILS OF RECOMMENDED STABILIZATION

Put in 8 - 1-1/4" dowels at lower edge of rock - shotcrete fault around dowels - 4 - 1-1/4" dowels at second block.
Shotcrete entire tunnel roof and side and tunnel portal

PHOTOGRAPH OF SITE

the rest of Canada in 1975. Initially the rock cuts were rated travelling by track motorcar and walking through the cuts. All rock cuts were photographed. In later years helicopter inspections were added to examine the rock slopes from above. Tension cracks were sometimes observed from the air which were not evident at track level. At some locations, instability was considered to be sufficiently imminent that urgent stabilization was recommended, as soon as practically possible. In two instances the potential of rockfall was considered to be so serious that train traffic was halted until the insecure area was stabilized.

Based on the 1974 inspections, the railway allocated emergency funds to develop a stabilization program for ten rock slopes where the stability rating was considered to be most critical. The program included rock scaling, increasing catch-ditch dimensions, installation of rock bolts, dowels or cable mesh, application of shotcrete, and the construction of buttresses or installation of warning fences.

Approximately six months after the accident near Spences Bridge, the Canadian Transport Board held a hearing into the accident at Vancouver. Their report expressed concern over the mishap, but commended CP Rail on its quick and professional response. The major recommendations of the Board were the requirement to inspect rock cuts annually, and to continue the stabilization program.

For over twenty years I participated in annual rock slope inspections along the CP Rail lines from Vancouver across Canada. This gave me the opportunity to enjoy Canada's beautiful scenery from a great vantage point — a Canadian Pacific Railway motorcar — an outstanding opportunity that few have experienced. This program for inspection and classification of potential rock slope instability was the first of its kind anywhere in the world in the transportation field. Duncan Wyllie, a young Associate partner, and I expanded it further. In the United States, the program was expanded for use in highway engineering, particularly by the state of Oregon under the direction of Larry Pierson, and more so by the state of Washington under the direction of Steve Lowell, as well as by the state of Colorado under Rick Andrews.

It was a Saturday afternoon in February 1968 when I received a phone call from Jim Hutcheson, President of Canadian Johns Manville, Asbestos, Quebec. "Mr. Brawner, he said, "we've had a bad slide in our open pit which threatens our primary crusher building, several streets and many houses in the town. We need your help immediately. We can have our corporate jet in Vancouver early tomorrow morning to pick you up and fly you to Sherbrooke, Quebec, near our mine. This is extremely critical. I hope you can come."

I thought quickly. I had nothing terribly urgent early in the week so I answered, "I'll be ready at the south terminal of the Vancouver airport at the executive aircraft offices at seven tomorrow morning. What is the weather like there?"

Hutcheson answered, "It's quite cold, minus fifteen degrees [Fahrenheit], and we may have snow tonight. Bring warm clothes."

I asked further, "Could you send the following with the pilots: geological reports, climatic data, the latest mine plans and the area of the failure, any geotechnical reports that you may have, as well as details of the movement? I can review them on the flight to Quebec. And can we have a meeting late tomorrow afternoon when I arrive?"

"Yes, that's great," he said. "We'll get the information for you, have the plane there about seven A.M. tomorrow, and be ready for you on your arrival. Thank you very much."

I realized I would need more money than I had on me so I scurried to the bank (no bank machines then). We had a theatre date that evening so I packed before dinner and laid out my travel clothes — 5:30 A.M. would come early.

I arrived at the airport just as the jet was finishing refuelling at the service centre. As the pilots came in to make a phone call I grabbed a cup of coffee. By 7:00 we were all on the plane. I was given a large file of reports, documents and plans, and directed to a seat where a tabletop had been folded down for my use. A waiter from the cafeteria came into the plane with cold lunches, drinks and coffee. Within ten minutes we were taking off, a quick turnaround for the crew. I busied myself going over the reports, summarizing the documents as I read. By the time we were nearing Sherbrooke, after five hours in the air, I had scanned the documents and prepared eight pages of notes. Just prior to landing, I read the notes quickly to refresh myself on the reports' contents.

We arrived at 3 P.M. Quebec time and Jim Hutcheson met me. It was nearly an hour's drive to the mine. En route, Hutcheson said the mine had begun about 1900 as an open pit mine, and the town developed about a mile away. The ore body (serpentine asbestos ore) dipped at about 55 degrees under the townsite. The company constructed and owned most of the stores and houses. When the mine deepened, the west pit crest began to encroach on the town. Management therefore decided to change to an underground mine. Two shafts were constructed and underground workings were developed, but the serpentine ore is very weak and slippery, with low friction, and almost as soon as underground openings were excavated, they commenced to squeeze. Many types of underground supports were tried to hold the tunnels and adits open, with little success. In addition to the squeezing, groundwater seeping into the openings created difficult working conditions. After two years, management gave up on underground mining and reverted back to an open pit program. The underground workings and shafts filled with water.

The surface mining required a crest setback every two years, encroaching on the town. Each setback involved removing houses and stores for one street. At the same time, new buildings were developed on the other side of town. Some people decided to move away from town and build their own homes. By this time the pit was about 1,200 feet deep.

Several days before the slide developed, the main waterline into the town broke and drained a 70-foot-high water tower. This water saturated the sandy overburden and bedrock below. Over a period of one day, the top of the slope, extending back from the crest for over 300 feet, dropped 10 to 12 feet. Many houses and several stores were seriously damaged. Brick facings fell off, doors buckled and windows broke. Fortunately no fires broke out. The upper crack came within 50 feet of the

Major slide involving about 33 million cubic yards of soil and rock at the Canadian Johns Manville asbestos mine in Quebec. The slide was caused by major seepage from a broken water main in the town. Stabilization was developed by dewatering and unloading the slope.

crusher building, and the skipway from the ore dump, midway up the slope, to the crusher building was pulled apart. This halted production until the skipway could be temporarily repaired.

It had snowed Saturday night, covering the bench tops in the open pit and making it more difficult to observe the slide area. We spent about two hours before dark observing the damage in the town, and the cracks and scarps that had developed in the haul roads. Learning about the waterline break, we drove into the pit and looked for evidence of seepage from the slope. We found the upper zone about 400 feet above the pit bottom. Because experience told me that practically all major open pit slope failures are impacted by subsurface water, I asked that the depth of water in the mineshafts be measured. This revealed a higher water table beyond the pit crest, indicating seepage flow toward the open pit.

Groundwater influences stability in several ways:

1. The buoyancy created by water reduces the vertical weight of the soil and rock, reducing the shear strength.
2. Increasing moisture content reduces cohesive strength in clayey or weak rock, such as serpentine.
3. Water pressure in tension cracks increases as the depth squared (D^2).
4. Seepage forces usually act in the direction of failure.

The slide extended into the town of Asbestos and damaged many stores and houses including the Continental drug store.

I concluded that slope dewatering would be beneficial and improve stability.

The company owned the Iroquois Club, the local hotel. I stayed there while at Asbestos. Monday morning was spent with the geological staff reviewing the geology, rock types and conditions and hydrogeology. During the afternoon, meetings were held with planning and mining staff. Monday evening I was invited to have dinner with Joel Edy, the Chief Engineer, and his wife. They had also invited several other couples. French-Canadians are very musical, and after dinner they sang local songs. My high school French wasn't of much help in understanding their singing, but it was enjoyable.

Tuesday morning I spent reviewing the information provided to me. I learned they were bringing in four new 180-ton trucks that they were planning to use for removal of waste rock. I asked if they could use them to unload several million cubic yards off the upper slide area. They agreed that they could. This would be beneficial in reducing driving forces within the slide.

I also learned that the old underground workings were directly below the toe of the slope. If they were still partially open to one of the shafts, it might be possible to lower the water table in the slide by dewatering the shaft. With seepage exiting from the toe of the slope, horizontal drains would be effective in reducing water pressures.

By Wednesday a program to stabilize the slope was developed. The mine assembled senior geological, planning and operations personnel. For about two hours I systematically presented the program I had developed. It comprised three separate procedures:

1. Unload about 1.8 million cubic yards of overburden and rock from the top of the slide — this was estimated to take about two months.
2. Dewater the shaft and the top underground drift directly under the slide. From the drift, drill about 15 percussion drill holes 50 feet long, fanned out upward to increase the effective drainage area.
3. Drill twelve horizontal drain holes and install perforated plastic pipe 300 feet long at a 2 to 3 percent upward gradient, below the toe of the slide.

Within two weeks all proposals were well underway, and surface movement monitoring revealed the rate of movement was slowing. Two months later the slide movement stopped. Evaluation of the geology and movement directions at various locations on the slope indicated the slide had involved about 30 million cubic yards of material.

When Canadian Pacific Railway signed a contract to haul coal from the southeast B.C. coalfields to tidewater near Vancouver, their proposal involved establishment of unit trains dedicated to hauling only coal. The trains were comprised of two lead locomotives and two slave locomotives in the centre of the train controlled from the front end. Generally 110 cars, capable of being dumped while moving, carried the coal. The split locomotives were required for better power, traction and braking. This configuration was capable of hauling a loaded train on grades up to about 1 percent. However, three locations had grades approaching 2 percent. At these locations, four to six extra "pusher" locomotives were required. The locations were from Rogers to the Connaught Tunnel in the Selkirk Mountains, several miles west of Revelstoke, and near Tappen, west of Salmon Arm, B.C.

CP Rail decided to design and construct new alignments with a 1 percent grade at each location, such that pusher locomotives would not be required. The new alignment would be over 300 feet below the elevation of the existing Connaught Tunnel, which is 5.5 miles long. With the lower elevation, a new tunnel about 10 miles long was required. The Special Projects Division of CP Rail developed the initial alignment up the south side of the Beavermouth River Valley, with a tunnel more than a mile south of the Connaught Tunnel, and a second possible alignment up the north side of the Beavermouth Valley.

I was called to the CP Rail offices, presented with a review of the two options, and asked to develop comments comparing the two alignments. They had complete air-photo coverage of both routes. Back at my office, I took about three hours to study the air photographs in three-dimension using a stereoscope. I arrived at two major conclusions:

1. The north side of the Beavermouth Valley was a major preference. My experience is to stay on the "sunny side of the street" wherever possible, since it will usually be drier and more stable, with fewer drainage, culvert and bridge location problems. The south sides of mountain valleys in B.C. are usually wetter and have more

snow, soft ground and landslide potential.

2. The new tunnel should be located under the existing tunnel as much as possible in order to take advantage of the drainage afforded by that 80-year-old tunnel. The new tunnel would cross under Mt. Macdonald, where melting glaciers above created a source of seepage, which could cause serious tunnelling problems. One of the tunnel engineer's greatest worries during construction is to encounter unexpected large water volumes under high pressure. An example was the Trans-Andean Railway tunnel between Argentina and Chile, which had serious construction problems and substantial loss of life due to an unexpected water inflow.

West of Revelstoke, approximately 2 miles of 2 percent grade were modified by crossing the valley three times, lengthening the alignment to 4 miles and reducing the grade to 1 percent. The three crossings required fills 100 to 120 feet high, each with 10-foot-diameter multi-plate culverts to carry creek flood flow. The foundations in the centre of the valley were comprised of compressible saturated silts. Undisturbed soil samples from boreholes were subjected to consolidation tests in order to calculate potential settlement of weak silty sediments under the fills. Settlement of 4 feet was estimated under the centre of the fill, but just 1 foot near the toe. To provide the necessary foundation support, 6 feet of soil was sub-excavated under the culverts and replaced with compacted sandy gravel. The culvert was then constructed with the central portion 4 feet above culvert grade, and the entrance and exit just 1 foot above grade. Differential settlement of 3 feet could overstress some of the connector bolts, so the connector bolts were only snugged initially. After two years of settlement, loose bolts were tightened. By this time the culvert grade was almost uniform.

At the west end of this alignment change, a 25-foot-deep cut was required in loose saturated organic silt. Geotechnical testing indicated that very flat cut side slopes of 8H:1V would be required for stability. At the north side, this would undercut the Trans Canada Highway, and on the south side the valley creek. Several options were considered, including sheetpile walls, tie-back walls, and electro-osmosis. A preliminary design and cost estimate indicated that the latter would be the most economical.

Dr. Leo Casagrande, specialist in electro-osmosis from Boston, and Dick Loughney, specialist in well point dewatering, were consulted. The concept includes driving a row of sacrificial anodes, in this case scrap steel railway rails, and about 15 feet away a parallel row of well points acting as cathodes. Direct current generated by a 2,500 hp diesel electric locomotive stationed at the site provided the power. The electric current created a hydraulic pressure gradient in the water, resulting in subsurface water flowing toward the well points, from which water was extracted. Each side of the cut was treated similarly. The electro-osmosis lowered the water table, increased the effective stress, and caused consolidation and strengthening of the organic silt. After about a three-month period, the organic silt was further strengthened by point-to-point electrochemical hardening. After four months the cut was excavated at a 2H:1V side slope. The silt had become much drier and was very firm. To provide a foundation for the railway ballast and ties, 6 feet of compacted sandy gravel was placed in the bottom of the cut. The side slopes were covered with 3 feet of gravelly sands to control surface erosion.

After thirty years, the cut slope and foundations are behaving satisfactorily. One interesting observation was that, after the power was turned on, earthworms in the area came to the surface and formed concentric circles on the surface around the anode rails. Anyone walking in the area had to wear rubber boots to keep from getting "tingled." This was an interesting project, using an unusual and specialized technique, which was completed successfully.

In September 1972 Bob Stark, Chief Engineer of Steep Rock Mines near Atikokan, Ontario, phoned and said, "We have movement which has developed in our open pit highwall at our iron mine. It started about one month ago. We installed hubs to measure the movement — the upper crack has opened about one foot and is widening about one-quarter inch per day."

Steep Rock open pit iron mine near Atikokan, Ontario. The Algoma Central Railway is in the background. A crack developed at the end of the pit indicating the development of a major slope failure. The open pit was 800 feet deep.

I flew east the following week to Thunder Bay, Ontario, rented a car and drove to Atikokan. On arrival at the mine, Stark showed me air photos, geologic sections and mine plans. Not only was the highwall stability of concern, but also the safety of the Algoma Railway line, which was only about 150 feet from the pit crest.

Removal of waste rock to expose new ore at the mine had gotten behind schedule. The only ore available was directly below the moving highwall. Personnel operating in the mine were aware of the movement and had reported the problem to their union. The local union leader advised Steep Rock, "We want you to call in specialists to review the movement and advise what to do. If you don't, we'll close the mine for safety reasons. We're also notifying the mine inspector in Thunder Bay." Threatened with mine closure, which would affect about 200 staff, Stark called Golder Brawner.

After the initial meeting, we went out to the open pit, which was impressive — about a mile long, 1,200 to 2,000 feet wide, and up to 850 deep. The host rock was very hard granite. The iron ore was a deposit that infilled a bowl in the granite, and the ore-granite contact was sharp. In the area of the movement, the contact dipped at 70 to 75 degrees into the pit. The upper 400 feet of the open pit was steep, parallel to the contact. The slope in the iron ore averaged 45 degrees, and was comprised of a number of benches.

We drove to the pit end where the major crack existed. It extended along most of the end of the pit, about 75 feet beyond the crest. I looked down the crack and could

The crack being monitored twice per day. Average movement of about 0.25 inch per day was occurring. The pattern of the movement indicated a toppling-type failure. Mining continued below for three months.

see it was at least 60 feet deep. I dropped a small rock down and could hear it bouncing beyond the limit of my vision. I concluded the crack was occurring along a continuous shear zone.

Back in the office, I looked at the plot of movement versus time. It was nearly a straight line. Larry Lamb, the mine manager, asked me, "What's your reaction?"

I replied: "The movement is not accelerating, so there is no immediate danger of the slope collapsing. I have no doubt, however, that it will fail. It appears to be a toppling failure. Because of the large volume, I expect that, before total failure occurs, there will be considerably more movement, and a period of many days of accelerating movement. My immediate reaction is that you have three options:

1. Close that portion of the mine and mine waste in other areas to expose new ore. The mill will have to close until new ore is exposed;
2. Load the crack with many tons of explosive and induce failure. After failure, remove the slide debris to expose ore. This will also result in mill closure, but probably for less time than option one; and
3. Mine and monitor the movement. Based on experience elsewhere, failure of the slope will not occur instantaneously. There will be days of accelerating movement before failure. I believe that you can continue to mine below the highwall as long as the slope is monitored continuously, and the rate of movement does not exceed about one to one and a half inches per day. I can recommend an expanded monitoring program to you."

I then described the experience of Anaconda at the massive Chuquicamata Mine in northern Chile in 1968, where monitoring was used to actually predict the date of a major wall failure three weeks in advance of its collapse. Mining continued below the unstable slope until three days before the slide. The movement was first observed six months prior to failure.

Lamb said he would think about the options overnight and we could discuss them in the morning. I was bunked into the guesthouse, which was very comfortable. Bob Stark invited me to dinner that evening.

The next morning I met with senior staff, and a major discussion of the three options occurred. By late morning, the group decided to try and gain approval for the "mine and monitor" approach. First the union had to be onside. Secondly, the regional mines inspector must be advised and his approval obtained.

The union was approached first. The union head listened attentively. He said he would like to approve the procedure, but he wanted the mine crews to be in on the decision. He said, "It's an important decision. It could mean loss of over one hundred jobs for three months." He requested that Stark arrange for a meeting of most of those working in the mine — drillers, blasters, surveyors, loader operators, truck drivers, Cat operators and mechanics — for 4 P.M., and that I describe the proposed plan with background history to those present. He would then call for a vote.

The group assembled in a large meeting room in the Administration Building. I was introduced and my background in mine stability was presented. I described the tension crack, the movement monitoring, the three options, and my proposal to "mine and monitor," along with a summary of previous experience. There were only a few questions from the mine staff.

Lamb said, "We want you to let us know if you're prepared to work in the mine as long as ongoing monitoring is provided. Anyone who does not want to will not be penalized." He then asked the question in an unusual way. "Will those of you who are *not* prepared to work in the mine with ongoing monitoring please raise your hand?"

The employees looked around at one another, some fidgeted, but not a hand was raised.

The union leader stood up and said to Lamb, "You've got the staff approval.

However, I have one proviso — one of our union employees must be on the monitoring team full time." A good recommendation.

The mine manager agreed. He said to me, "Now I'll call the mines inspector to come here early tomorrow."

The mine inspector arrived at about 11 A.M. Prior to his arrival I had developed the monitoring program. The primary system was to install three sets of tripods. For each set, one tripod was established on each side of the crack, each with a pulley hung from the top. A steel fishing line was tied to a pin grouted into the rock on the slide side of the crack, strung through the two pullies, and then tied to a 10-pound weight. This hung from the pulley on the stable side of the crack. As the slide block moved the weight would raise an amount equal to the movement. A limit switch was installed in such a manner that if movement exceeded ½ inch, a siren and flashing strobe light would be activated. The limit switch was reset every 8 hours.

If the rock slope were to fail, it was estimated that the rock would run out a maximum of 800 feet. A series of green markers, called the green line, were placed 1,000 feet from the toe. If the warning siren or light went on, *all* persons at that end of the mine were to walk to beyond the green line and await instructions. They were to leave the equipment.

A second system was planned. A Laser Ranger EDM survey unit was to be set up on a stable location on the pit crest where the entire face that was moving could be seen. Twelve reflecting prisms were installed on the face of the mine slope, four near the crest, four near mid-height, and four near the base of the granite exposure. The unit was to be set up in a protective shack. The Laser EDM had an accuracy of 0.5 inch based on averaging five separate readings, one after the other. A young Golder engineer, David Woods, was to be the monitor. The union also supplied a man.

This program was described to the inspector. He said the decision was so important that he wanted to consult the Chief Inspector from Toronto, whose name was Cruden. He was contacted and agreed to come the next day.

Following his arrival, inspection of the pit and crack, and my description of the proposed program, Cruden said, "I want a second opinion."

"That's reasonable," I replied. "Who do you have in mind? He should be someone familiar with mining and monitoring and rock stability."

He answered, "I'd like to make a phone call before I commit myself."

Stark took him to a vacant office to use the phone. When Cruden returned he said, "I want a geotechnical engineer I know to review this. His name is Morton. He can be here tomorrow." It was not a name that I knew. As it turned out he was a soils engineer. Later that day Cruden also contacted Dr. Peter Calder, a Queen's University mining professor.

The next day Morton arrived first. He spent several hours in the pit. By that time Dr. Calder had arrived, and they spent several hours together. We all met at about one o'clock. Morton wanted about six additional tripod monitors, and Calder wanted to bring in a noise seismometer. When rock fails there is a pattern of cracking noises. The greater the number of noise bursts, the greater the movement. Cruden said if their recommendations were followed, he would agree. Lamb had no recourse but to agree.

I asked Morton, "Why do you want so many monitors? The block is moving as a unit."

"I want to be sure," he answered.

I then mentioned to Calder, "Don't you think the noise in the mine will drown out the geophone pickup capability?"

"I recommend it be tried," he responded.

It was overkill, but it wouldn't make anything worse, just cost more.

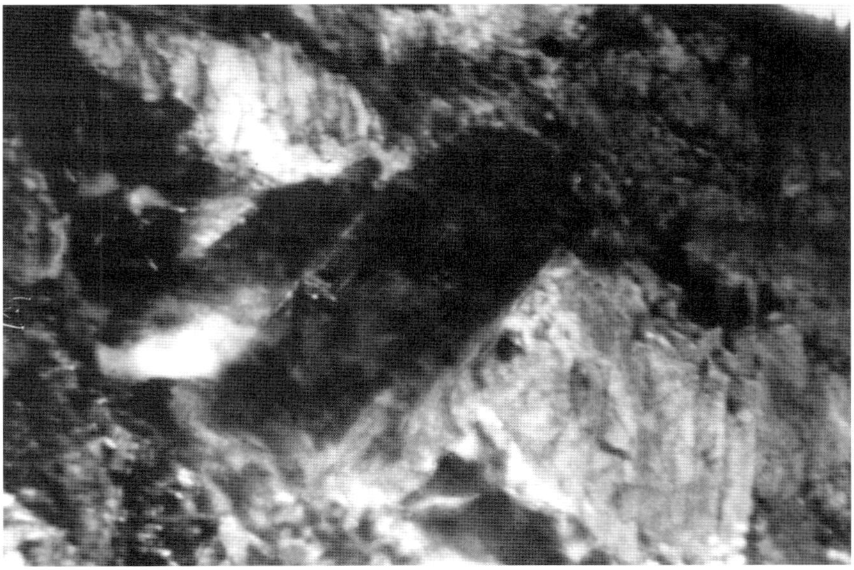

In the spring, precipitation and snowmelt combined to cause failure after mining was completed in that area of the pit. The "mine and monitoring" approach was successful. 400-foot-high toppling block at the Steep Rock Mine.

The mining program proposed to recover all the ore near the unstable slope involved extraction of three additional benches, a depth of 90 feet. This would add additional stress in the slope due to removal of the existing toe. This made the monitoring more important. Prior to leaving I made a further recommendation to Lamb: "Before commencing each new bench, I recommend the line holes all around the pit wall be drilled and blasted first. The movement gauges and laser readings should be taken before, and every five minutes after the blast, to determine whether the blast had destabilized the slope. If movement accelerates after the initial movement, directly due to the blast, and continues to accelerate, cease mining until the movement pattern stabilizes. If it continues to accelerate, close that part of the mine and call me."

The monitoring program continued except for the geophones, which, as I suspected, couldn't recognize rock noise with all the noise in the mine. Mining continued for three months, until all the ore was removed at that end of the pit. In the meantime sufficient waste was excavated elsewhere to expose additional ore.

One amusing incident occurred. A monitoring inspector was resetting the limit switches when he fell over one of the wires, triggering the siren and strobe light. Dave Wood in the Laser EDM shack saw men running in the pit. He said, "I think they all broke the record for the hundred-yard dash." He realized that only one switch had been set off so failure was not actually developing.

Monitoring continued periodically to ensure the adjacent railway would not be in danger. The following May, Bob Stark phoned and said, "Chuck, the slope has moved sixteen feet total, and yesterday moved about six inches. It looks like it will fail tomorrow!"

I said, "Bob, I want photos of the failure. I'll send an engineer down to take care of that. There will be no cost to you." He agreed.

A sequential series of colour slides were obtained of the failure. About 600,000 cubic yards of rock toppled, over in about 20 seconds. The photos are dramatic. The rock failure moved out about 600 feet. The green line was justified, and the "mine and monitor" program was successful! Steep Rock were very appreciative.

Henry Ewanchuk, President of Mascot Gold, called me in 1986 and said, "We're looking at developing an open pit to mine down through the underground workings of the old Mascot Gold mine. If we can mine with 55-degree slopes we'll have a viable mine. Anything flatter is uneconomical. I'd like you to do a study for us to determine the steepest slopes that we could design for. When can you come to our office to discuss such a study? At that time we can review the old mine plans, geology and rock information that we have."

The Mascot mine was near Hedley, B.C., but Ewanchuk's office was in downtown Vancouver. "How about next Monday afternoon?" I queried.

"Great, make it about one thirty," Ewanchuk replied.

At the time I was on a retainer with R.M. Hardy Associates and teaching Geomechanics in the Mining Department at the University of British Columbia. Hardy had a young but experienced rock and stability engineer, Drum Cavers, whom I had worked with on several previous mining projects, and who would be ideal for the project.

The following Monday we met with Ewanchuk and his staff, including Paul Saxton, Vice President. The underground mine had closed in the early 1950s, but most of the workings remained open. Two very large "glory holes" existed underground. They were more than 200 feet long, 100 feet wide and 250 feet high. Many underground adits, inclines and smaller mined-out openings also existed. Mine plans and geologic maps were very comprehensive, drawn on linen with black ink! Geologic mapping of discontinuities was extensive, and some old drill core was still available.

Ewanchuk showed us a preliminary pit plan with 55-degree slopes. They were up to 800 feet high. Such steep, high slopes had not been developed in B.C. before. I indicated to Ewanchuk, "This design wall likely requires a variance from the ministry of mines. It's steeper than regulations now permit."

He wasn't surprised, saying, "If so, we will need your assistance."

A week later we went to the mine. A major exercise was to determine the accuracy of the structural geological mapping. If it was accurate and we could use it for design, there would be considerable savings in time and cost. We inspected about six adits and performed structural mapping.

One of the adits led to one of the glory holes. The only lighting we had was miners' lamps and a powerful flashlight. A narrow ledge circled the glory hole, with no railing to hold onto. It was nerve wracking! We walked slowly, with one hand on the vertical rock wall. Our lighting could not reach the ceiling. At several locations we measured the geologic structure. Partway around the glory hole we came to an old load-out area that led to an incline to the surface. This was where some of the ore had been removed from the mine. We climbed the incline to the daylight.

Back in the office, we compared our mapping with the old maps. Agreement was good. Use of this old data probably saved Mascot Gold about $40,000 in geologic mapping and new drilling to obtain oriented core, which would also require logging.

The structural geology was plotted onto stereonets to develop the geologic model of the rock. That data was assembled into domains of similar geologic structure. Potential failure models were determined, and these were inspected for planar, wedge, toppling or buckling-type failures. No large-scale failures controlled by the structural geology were determined. Hence, the pit wall stability would be controlled by the rock strength, the mining method and blasting procedures, and the maximum slope angle allowed by the ministry of mines.

To evaluate rock strength, samples were obtained from four core holes. Strengths were very high, ranging from 25,000 to 60,000 psi. The general host rock was a skarn. The major mining significance of the hardness would be high-blast drill bit wear and high-crushing costs, increasing overall mining costs. This was emphasized to Ewanchuk.

The ground surface above the ore body sloped, so that the highest wall was along one side. To reduce waste rock removal, we recommended that the haul road be developed on the walls of lower height.

Mining regulations in British Columbia required that benches be incorporated in the wall of a mine to catch rockfall, and that bench height should not exceed the reach of loading equipment plus 5 feet. Under special justifiable conditions a variance could be obtained.

It is noteworthy that benches were not originally designed to catch rockfall. The most efficient method to drill blast holes is to drill vertically. To develop an overall slope, considering stability as well as efficiency, drill holes were offset, necessitating the creation of benches. Recognizing the potential to catch rockfall, regulators stepped in and made benches a design requirement. In recent years, the concept of spacing the benches farther apart vertically became acceptable, provided adequate final wall blasting and scaling of the loose rock were performed.

Ewanchuk was asked, "Would you consider using controlled blasting along the final wall, followed by scaling, particularly at the bench crest? This could allow wall steepening of five to seven degrees. It will also improve mine safety."

He thought for a bit and answered, "Yes, that would be a major benefit."

With those instructions Cavers and I developed pit wall designs. For the highwall we developed a slope angle of 68 degrees! Maximum pit angles in the past had been 50 degrees. For better grade control, Mascot had selected 20 feet as their bench height. Bench height of 30 to 40 feet is more common in open pit mines.

We were informed that they proposed to use 5.5-inch-diameter blast holes. I asked Ewanchuk how deep he could drill a straight hole with that equipment. He said, "I'll have to get back to the drill manufacturer."

Later he advised, "I am told eighty feet in good rock."

We therefore proposed a bench height of 80 feet — quadruple benching! This design was unheard of in Canada previously. We suggested to Ewanchuk, "We should have a special meeting with the geotechnical engineer of the mines ministry to determine if they'll buy into this design." Ewanchuk agreed. The meeting was arranged at his office, because the plans, maps, reports and photographs were bulky. Three ministry officials came.

Ewanchuk described the geology, details of the ore body and ore grades, and the proposed preliminary mine plan. Cavers and I described the rock conditions, structural geology, stability of the rock in the glory holes, where some slopes were vertical for nearly 200 feet, and the use of controlled blasting. When our presentations were complete, the ministry geotechnical engineer said, "That's awfully steep. I'm not in a position to approve that slope design. Both the bench height and the slope angle exceed regulations. If an accident occurred there could be a lawsuit."

It was obviously going to be a hard sell, but I was confident in the design and pressed on. I said, "I'd like to review for you the conditions at the Republic Iron Mine in northern Michigan, where one highwall is being developed at eighty degrees." They were surprised, but listened to my review of the rock and mining procedures there, including controlled blasting.

Still they weren't convinced. They talked amongst themselves for a considerable time. I had saved a final point for last, if needed. I said, "Not only are there millions of dollars at stake, but the potential of a new mine which could employ a hundred to a hundred and fifty personnel for ten to fifteen years as well. That's a lot of tax income. Would you consider granting approval to develop one upper bench face of the proposed design? That way you could observe and monitor the face as it is developed. If that bench is acceptable, then you could approve a second eighty-foot face below the first bench, and so on."

"Give us more details, please," they asked.

"Okay," I replied, "the final wall would be developed using five-and-a-half-inch holes, spaced five feet apart, eighty feet deep with explosives located throughout the holes. These would be detonated, ten holes at a time (per delay) as a pre-shear line full depth. Two buffer lines out seven and fourteen feet would be drilled, and production holes would start out twenty-eight feet from the final face, spaced fourteen feet on the square. Production holes would be blasted first and delayed separately within sixty feet of the final face, followed by the outer buffer line, and then inner buffer line. The buffer holes and three lines of production holes would all be delayed, drilled twenty-two feet deep, allowing for a sub-drill of two feet. Each twenty-foot bench would be excavated before the next bench was drilled. When the face at the final wall was exposed, this wall would be scaled of all loose rock before drilling the next bench commenced. The ultimate blasting plan would be reviewed and approved by Lou Oriard, specialist blasting consultant from Huntington Beach, California. Oriard is one of the most experienced blasting consultants in the world. Each pre-shear blast would be photographed with a high-speed camera and played back with a slow-motion video to evaluate the blast. Once several benches were exposed in the highwall, the slope would be monitored by visual observation for cracks at and near the crest, and prisms would be installed on selected sections and monitored with an EDM survey unit. Faces below would be developed in the same manner."

It was now time for lunch. They decided to go alone so they could discuss the proposal amongst themselves.

On the return from lunch one of the officials said, "We can't give you approval now, but your proposal for a variance has merit. Please document it in detail and send it to us, and we'll discuss it with senior ministry staff and get back to you."

The proposal was sent the next day. Approval was provided about two weeks later, with the proviso that a report was to be forwarded to the ministry as each high face was developed. The ministry would also inspect the wall periodically.

The program went well for the life of the mine. No serious rockfall incidents occurred. A second mining problem, however, required a practical solution. Mining in the pit typically required large equipment. As the pit bottom lowered, it came closer to the glory holes and one major inclined opening. The initial question was as to how close to the large openings mining could occur without failure of the roof. Drilling obtained undisturbed core for shear strength and modulus testing. Stress models were developed for different crown pillar thicknesses. For the largest glory hole, a roof thickness of 80 feet was determined as necessary to support the trucks, dozers and drill rigs with an acceptable safety factor. As the thickness neared this amount, settlement measurements were instituted, along with visual inspection for cracks.

In order to mine the areas surrounding the large openings, it was concluded that they should be filled prior to nearby mining. Where the walls contained no ore, they would be filled with blasted waste rock. Where the walls contained ore, they would be filled with low-grade ore, which could subsequently be mined economically.

The question was then how to fill them. A drop raise procedure was developed. Four drill holes 8 feet apart on the corners of a square were drilled, followed by three centre "burn holes." Explosives were placed at a depth of every 10 feet in the holes. Between each charge, a delay was placed. Blasting commenced sequentially, from the bottom up, with a delay between each blast. At each depth, the burn hole was detonated slightly before the corner holes. The result was a clean open square hole from the pit bottom to the top of the glory hole! Several drop raises were developed for each major opening. A bumper beam was installed at the top of the raises for the trucks to back up against to safely dump ore or waste rock. All large openings were filled sufficiently for mining to be performed above and beside them. The procedure was successful.

As the mine was nearing completion, Mascot wished to extend the centre of the pit two benches deeper. Steepening the lower haul road significantly and reducing the amount of ore in the trucks accomplished this.

This project illustrates that successful mining can present very unique problems, which require flexibility and ingenuity, in this case not only from the mine staff and consultants, but also from government agencies.

One situation in the United States is disturbing. MSHA — Mine Safety and Health Agency — is responsible for inspecting mine operations periodically. My experience is that their inspectors are often not professionally qualified mining engineers. In one recent situation, MSHA required an unnecessary change in design that increased costs and contributed to mine closure. The inspector had been a schoolteacher! In Canada, mine inspectors are required to be registered professional engineers.

5

Alaska

Alaska
(1966-2001)

Early experience with rock mechanics in mining led to rock mechanics assignments on many transportation projects. Failure of a retaining wall along the White Pass Railway in Alaska, famous for its place in history as part of the Klondike gold rush, caused a train loaded with silver concentrate to cascade into the Skagway River. The author was called in to evaluate and reconstruct this section, as well as to design other rock slope stabilization and snow avalanche avoidance programs.

When rock design recommendations were not followed, a rockfall occurred in Keystone Canyon causing the tragic loss of life. On the new Skagway to Chilkoot Pass Highway, rock slopes were designed with variable rock slope angles. The project won a national award. Major reconstruction from Anchorage along Turnagain Arm involved five separate projects beside the Alaska Railway. Many sliver cuts up to 250 feet high required special pioneer drilling, controlled blasting and rock slope stabilization.

Other interesting projects described include rock slope stabilization to safeguard outdoor fences of the maximum security jail at Ketchikan, protection of salmon smolt from blasting using an air bubbler system at the Snettisham fish hatchery, and the rock mechanics evaluation for a "run of river" power project at Tazimina Falls.

Working in Alaska wouldn't be complete without salmon fishing. A successful trip to Sitka, Alaska is related.

Chapter 5
Alaska

My first involvement in Alaska was with the White Pass and Yukon Railway in 1966. With headquarters in Vancouver, they operated a container and petroleum service using the double-hulled ship *Frank H. Brown*. I understand it was the first ship of its type in the world. The ship travelled from Vancouver to Skagway, where extensive dock and petroleum storage facilities were available. Petroleum products were pumped some 100 miles to Whitehorse, Yukon, by a high-pressure, small-diameter pipeline. Containers, general freight, ore concentrate from the north, and passengers were transferred by railway between Skagway and Whitehorse. A trucking line also operated from Whitehorse serving northern B.C. and Yukon.

Mr. A. Friesen was president; a congenial, efficient and no-nonsense administrator. In October of 1966 he telephoned me to say, "Mr. Brawner, we've had a failure of a retaining wall that has closed our railway service to Whitehorse. I'd like you to meet with me to discuss your services in determining how to solve the problem and evaluate the stability of twelve other comparable walls, some of which are much higher. Our requirements are urgent. When are you available?"

I have learned that rapid response to a request for meetings or services is essential to the development of a consulting engineering company. Prompt, reliable, professional service must be a top priority.

I answered, "I can be available at your office today within half an hour, if you're free."

He said, "Great, I will see you here."

After introductions, I provided him with a background of our experience, personnel, availability and fee structure. I indicated I could leave for Alaska late that afternoon and be available in Skagway the next morning to inspect the site. By late afternoon I could contact him with my initial interpretation of the problem, potential alternatives and timing to get the rail line back in service.

He was pleased with the rapid and positive response and said, "Based on your comments and verbal proposal, I hereby authorize you to proceed. We have a geologist on staff, Mr. Charlie Brown, who knows the railway, the operations and our personnel. I'll arrange for him to travel with you or meet you in Skagway tomorrow. I would appreciate it if you would prepare a one-page summary of your proposal. Can you telex that to me before you leave?"

I agreed, "Yes, I will do that. To assist me I would appreciate any railway plans, retaining-wall plans and designs, site photographs, aerial photographs and geology reports that you can make available to me. If they could be ready by midafternoon I can pick them up on the way to the plane." I would review these on the airplane and get a head start on the problem.

He replied, "I'll put together what we have readily available."

As I drove back to the office I reflected on the meeting. I concluded that my immediate, positive response to the problem had contributed to his speedy authorization. I gave major credit to my past membership in the Victoria, B.C. Beaver Toastmasters public speaking club, again concluding that the ability to communicate is essential to developing and providing satisfactory consulting services. Every consultant should make this a priority and take a public speaking program early in his career. You have not fine-tuned your skills until you can present a lecture, seminar or technical paper *without* notes, relying only on slides, overheads or electronic projection as a prompter.

I picked up a package of information at the White Pass offices, caught a flight to Seattle at 5 P.M., and connected to Juneau on Alaska Airlines two hours later. Upon arrival I took a taxi to the historic Baronoff Hotel. The next morning I had an early breakfast, caught a 7:30 flight on a small six-seat Cessna and arrived at Skagway about forty-five minutes later, after a very rough 90-mile flight. Skagway is always windy. Charlie Brown met me, and we drove over to the White Pass Railway station. There he brought me up to date on the railway's history and the wall failure.

During the original construction of the railway at the time of the Yukon Klondike gold rush of 1898–1900, timber trestles were used extensively to cross creeks, deep gullies or steep-sided hill slopes. The trestles were founded upon untreated timber piles, five across down to the rock surface or concrete footings below, spaced about every 20 feet along the track. A heavy timber beam (sill) was placed across the top of the piles to support the ties and rails.

The timber piles were not treated to reduce decay, as they would be today. Within about ten years they had begun to rot. White Pass therefore decided to construct concrete walls up to 30 feet high outside thirteen of these trestles. The walls were very thin, only about 3 feet wide at the base, 1 foot wide at the top, and they were not tied to the bedrock behind or strengthened with reinforcing steel. White Pass backfilled behind the walls and around the trestle posts with free-draining sandy gravel. The walls had weep holes so water pressure would not build up behind them.

The retaining wall that failed was on the outside of a curve, where the train created an outward pressure on the wall due to the centrifugal force. The wall failed during passage of a train loaded with silver concentrate from the United Keno Hill Mine, Yukon. The train cascaded about 400 feet down to the canyon bottom, resulting in serious injury to the four crewmembers. All the silver concentrate was dumped over the slope below.

The failure occurred near Mile 9 outside Skagway. Brown suggested we inspect the track up to Mile 22, near the Canadian border, to familiarize myself with the rail line, and then return to Mile 9. The inspection was made using a motorized track car up to Mile 9. There we walked around the failure and caught a second track car that had been kept in White Pass, near Mile 22.

This was my first trip on this narrow-gauge railway (only 39 inches between rails), a trip I will never forget. It was stunning, with the rapidly flowing Skagway River below, cliffs and canyons along the track, and snow-capped mountains above.

Commencing at Mile 7 in the main valley, the track clung to very steep rock cliffs for over 2 miles. In this 2-mile stretch there were three deep gullies, with a timber trestle crossing each one. The first gully trestle had three pile bents, the second had four, and the third, five pile bents. Occasionally large rock would fall from the cliff and damage or remove a pile bent. Should a train try to cross following such a rockfall, the trestle could collapse. White Pass had developed a simple, yet effective monitoring system to warn the locomotive engineers if any bent was missing.

On each trestle, red flags were mounted on each bent. If a bent was missing, a flag was missing. Simply counting the flags provided a warning to the locomotive crew to stop the train and inspect the trestle.

Near Mile 13, a major avalanche chute extended to the track. All vegetation had been denuded by periodic snowslides. For the next 2 miles, steep rock slopes existed to the top of the mountain. In the late fall and late spring, when the daily temperatures rose above freezing during midday, snow avalanches occurred almost daily along this stretch. As a result, the train from Skagway to Whitehorse had to leave early in the morning, before the melting began. Likewise, trains from Whitehorse on the return had to leave late in the afternoon, when temperatures dropped below freezing and the slides stopped. After about 5 P.M., a D-6 Caterpillar equipped with special tracks so as to not damage the rail ties would move from White Pass summit down to Mile 13 and clear the snowslides off the track to allow the evening train to proceed to Skagway. This had been standard maintenance practice for years. Later I advised them on the use of howitzers and exploding shells to bring the avalanches down.

Near Mile 15, a long and high steel bridge spanned a major canyon over 250 feet deep. The bridge was designed in early 1900 for train loads typical of that time. As the loads and length of the trains increased, the stability of the bridge became a concern. This resulted in train speeds being limited to only 5 miles per hour, and crossings restricted to when the wind speed was less than 10 miles per hour. Later a new design involving a shorter bridge and tunnel was developed. I was involved in the geotechnical investigation for the new design.

Near Mile 17, a lookout had been established. From this location the entire Skagway Valley could be seen, including Skagway at the upper end of the Portland Canal. During good weather the passenger trains would stop to allow the passengers to enjoy this view. Today, during the summer, tour trains use refurbished railway cars to carry tourists from Skagway to Lake Bennett in British Columbia and return. It is truly a breathtaking trip and recommended to all tourists, including those from cruise ships, who travel to Skagway.

Theoretical stability analysis indicated the concrete walls had an unacceptably low safety factor to retain the train and granular backfill load. The fact the wall had remained stable for 55 years suggests that the sandy gravel backfill reduced the rate of decay of the wood piles, allowing them to carry this load for this long a period of time.

In the late afternoon, after my site inspection, I returned to Skagway and contacted Friesen with my recommendation to construct a temporary bypass about 15 feet inside the original grade. This would require excavation into rock, and construction of a small dry rock wall where the old wall had been. He accepted the recommendation and asked if I could stay to supervise the work. I agreed.

Five days were required to relocate the track on a temporary basis. The relocation involved two sharp curves which necessitated a reduction in speed. A permanent solution would be developed later.

At twelve other sites, thin concrete walls with granular backfill still supported old trestles. Friesen wanted each inspected. Several walls had major vertical cracks, and showed some differential movement at the top of the wall. There were signs of the walls starting to rotate.

Thin concrete retaining wall constructed in 1910 was strengthened in 1968 with steel channels tied back to bedrock with 30-foot-long tensioned and grouted rock bolts. White Pass Railway near Skagway. Typical winter conditions.

Close-up of back-to-back steel channels to stabilize retaining wall.

Surveys were performed at each site to determine wall dimensions and the bedrock profile above the grade. Allowing for no strength support from the buried trestle, all walls should have failed. The conservative solution was to increase the wall support at each location. It was decided to provide vertical steel channel beams outside the walls, and tie the beams and walls to the strong granite bedrock behind with high-strength tensioned grouted rock anchor bolts. The rock bolts were corrosion protected. A design was developed for each wall, a contract prepared and tenders called. The contract was awarded to a firm from Seattle. Since this stabilization of the walls, no further instability has occurred.

Golder Brawner was awarded several other contracts by White Pass. One contract was to review the stability of rock slopes along the track and selectively improve stability at several locations. I have made it a practice not to perform rock inspections alone, in case I ever fell from the slope. On one inspection I hired Randy, my oldest son, seventeen years old at the time, to accompany me. He enjoyed the scenery, but soon became bored standing or sitting near the track simply watching my inspections. The fieldwork took three days. In the late afternoon of the third day I arranged to charter a small four-seat airplane from Skagway Air Services to fly us to Juneau. The weather was sunny with little wind.

The pilot remarked, "It's a beautiful day. How would you like to take the shortcut scenic route?"

"Sure, sounds great," we said.

The pilot climbed to about 5,000 feet above Skagway, and then headed south toward Juneau through alpine valleys and near snow-capped peaks. The scenery was spectacular, with the bright sun turning the hoar-frosted snow surface into a field of shiny diamonds. The trip lasted only about forty minutes. Halfway through, Randy had run through two rolls of film. Unfortunately, he missed filming the final spectacle of the trip — flying at low level down the Mendenhall Glacier to the Juneau Airport. Randy still vividly remembers this trip.

My next involvement in Alaska was for the Alaska Department of Transportation (ADOT). In early 1970, ADOT were constructing a multilane highway from downtown Anchorage to their International Airport. The new alignment crossed a valley where the foundation materials were comprised of up to 20 feet of peat or muskeg. This material is primarily organic, has low bearing capacity and high settlement under load, characteristics normally not considered suitable as the base for a highway fill.

The historical solution generally involved excavation of the peat and its replacement with stable sand or gravel. This would have been very expensive. In addition, the disposal of the peat posed a problem.

In the past B.C. Highways had used preconsolidation to compress peat and soft organic soils, thereby increasing its foundation strength and reducing settlement. The concept involves applying a fill load in excess of the final load to be applied, then allowing time for near total settlement to occur. The fill above grade is then removed, such that the long-term load is moderately less than the original load (see Chapter 2).

The construction of the new access to the Anchorage International Airport was to be partly funded by the U.S. federal government. Hence, the Federal Highway Agency would be involved, not only for partial funding, but also to be available for technical assistance, if requested.

Bill Slater, Alaska State Geologist, worked on the highway design in the peat area. Many options were considered, including the construction of a concrete viaduct, and replacing the peat with granular backfill. All these options were considered to be excessively expensive. Slater contacted the Federal Highways for their input. The regional geotechnical engineer, Ron Chassie out of the Portland office, answered Slater's call. Chassie advised him of the research we had done in Vancouver, and suggested I be contacted. As a result, I met with Slater in

Anchorage. Following my update on the success of the Vancouver freeway, we reviewed the soils investigation details and evaluated the potential for the use of preconsolidation. Alaska developed a preliminary cost estimate for the program. The cost was considerably less than other alternatives, but one difficulty existed — the supply of sawdust.

This forced the question — was a weightless spacer required? The depth of peat was not as deep as that in Vancouver. Hence, I advised them that I believed that, as long as the peat surface was not disturbed, they could create the fill successfully by developing it in incremental lifts, without the sawdust spacer. This would involve placing 4 feet of clean sandy gravel first, then allowing about one month for settlement and gain in peat strength, followed incrementally by placing additional 4-foot lifts. Each lift would be monitored. The fill would be constructed to grade plus about 6 feet, then allowed to settle until settlement almost ceased. The final phase would be to remove the excess fill. Following this removal, the base gravel and pavement could be placed.

The fill was constructed successfully and is fully functional today. As a result of the success of this design and construction, both Slater and Chassie invited me to assist them on a number of additional projects. I regard them both as being among my very closest engineering colleagues and friends — true professionals.

Slater was not only an outstanding geologist but a championship bridge player. In addition, he was a master fisherman, particularly for Alaska salmon. He recounted to me many fascinating northern projects, some of which he worked on in the field in temperatures below –40°F. To illustrate how cold it could get, Alaskans would say, "When you go outside to pee, the pee freezes so fast you have to back up."

Ron Chassie was the senior geotechnical engineer for the U.S. Pacific Northwest states for the Federal Highways Administration (FHWA), with his office in Portland, Oregon. As a specialist in transportation technology Chassie's major contributions included the development of training seminars and the preparation of design manuals. We worked together on many projects. Both Slater and Chassie are now retired.

Several years later, Don Morefield, a construction engineer in Valdez, contacted me. Atlantic Richfield (ARCO) had recently located the major Prudhoe oil field in northern and offshore Alaska, and received approval from the state and federal governments to develop it. In order to move the oil from the north to the mainland United States, they decided to develop a pipeline south to Valdez, a small port at the head of Prince William Sound. To construct the pipeline, they would need an access road. The southern portion of this would follow the existing Highway 4. About 14 miles north of Valdez, this highway passed through Keystone Canyon, a beautiful but deeply incised winding river canyon.

At one bend in the river, the existing highway passed through a sharply curved tunnel. ARCO had decided to ship the steel pipe in 24-foot lengths on freighters to Valdez. At that point they planned to weld the pipe into 48-foot lengths and haul it on special trucks with back bogies. They soon became aware that the loaded trucks could not pass through the tunnel due to the sharp curvature. However, the economic benefit of transporting the longer pipe length was enormous. Following discussions with the Alaska Department of Transportation, an agreement was reached whereby ADOT would realign that portion of highway to bypass the tunnel.

The bypass involved constructing two bridges to cross the Keystone River and back again. Between the two bridges, a high rock promontory existed which the new highway had to cut through. Morefield phoned and asked me, "Can you come up to advise us on the design of the 320-foot-high rock cut?" I agreed to go.

I flew to Anchorage, met Steve Lowell, a young geologist, and then flew to Valdez on Airlift Alaska, a regional airline. We met with Morefield, who described the proposed project and its alignment. In the field, we had to be helicoptered to the other side of the river since there was no nearby cross-river access.

The most important factor which influences rock slope stability and design is the structural geology; that is the orientation of the rock discontinuities (joints, bedding, shears, faults) relative to the proposed slope angle. In this case, the rock face involved a dip slope of 65 degrees toward the river. If the excavation was to undercut the rock above, that rock could be expected to slide into the excavation, since the angle of friction along the joints was much less than 65 degrees.

Two alternate designs were developed. Alternative one involved excavating the rock slope to follow the dip angle of the joints (65 degrees), scaling any loose rock, and installing rock bolts to stabilize selected slabs. Alternative two involved excavating the slope to 80 degrees, then installing long anchor cables, tensioned to 250 tons and grouted, to tie the oversteepened slope together. The final contract was based on alternative one, which was more practical and less costly.

ADOT required contractors to bid on alternative one, but also allowed any of them to submit another bid based on a design that they considered would be satisfactory, but more economical. This procedure is called "value-added engineering." If the value-added design is accepted by ADOT, and the project costs are less than the lowest standard bid, the contractor receives one-half of the money saved. It's a practice, also used by the U.S. Corps of Engineers, designed to take advantage of contractor ingenuity on unusual projects.

I was not aware that ADOT had allowed value-added bidding on this project, and was not approached to review the bids. About one year later, however, I received an urgent phone call from Morefield advising me that the project had commenced and a very serious accident had occurred. Two of the contractor's employees had been killed by a rockfall. Could I come to Valdez to evaluate the accident and assist in deciding what to do next? He briefly described what had happened. The contractor had submitted a bid to develop the slopes at the steep angle of 80 degrees, and stated that, in his experience, the anchor cables were not required! The estimated saving was about $500,000.

The project geologist, Steve Lowell, sent a letter to his supervisor advising that this proposal was unsafe and should be rejected. A senior ADOT officer was asked to make the final decision based on a discussion with the contractor. He instructed that the bid be accepted! I was not aware of any of this.

On my arrival at Valdez the next day, I was driven to the site. I determined that

Highway rock cut being developed in Keystone Canyon near Valdez. The excavation undercut the bedding. The design called for the cut face to follow the bedding, a recommendation which was not followed.

the contractor had drilled the upper slope with the 80-degree backslope, and that this was subsequently blasted. A Mr. Molusso and his son began working to remove the blasted rock. The son was operating a D-8 Caterpillar and the father was hand scaling. Rock that had been undercut by the blasting slid from above along a steep joint and killed both of them. Steve Lowell had been standing across the river observing the operation. He had a camera and was taking a picture of the construction. As soon as he saw the failure occurring, he took several more photos. They clearly showed the rockfall and the Caterpillar being demolished.

I asked Morefield, "Why wasn't I advised of the change in contract and design?"

"I don't know!" He replied. "What we'd like you to do is help us decide what to do now."

After some thought I advised, "Develop documentation of the tendering and contracting procedure and prepare for potential litigation; cancel the existing contract and re-tender the original design for construction."

The litigation did develop, involving three court cases. In the end the Molusso family was awarded approximately $2.8 million.

The contract was re-tendered and the project successfully completed according to the original design, although it was difficult to pioneer because of no land access — service was primarily by helicopter.

This was one of the few times that a design in which I was involved was changed

The undercut rock failed and killed the Caterpillar operator (A) and his father, who was scaling loose rock. The court awarded the family substantial damages.

without my knowledge. I felt extremely bad about the accident. As a result, on future projects, even when my terms of reference had been fulfilled, I later inquired about the status of the design and tendering.

When Eisenhower was president, the state of Alaska and Canada had reached an agreement that it would be desirable to build a road linking Skagway to Whitehorse, Yukon, and the Alaska Highway. The route would cross the northwestern corner of British Columbia. About 20 miles of road construction were required in Alaska. The route chosen was on the opposite side of the Skagway River from the railway, and crossed the U.S. – B.C. border through Chilkoot Pass, the famed route of the Klondike gold rush.

Most of the route was alongside hill slopes comprised of granite rock. It was recognized that construction would involve much rock excavation at a high cost. Bill Slater, State Geologist, contacted me and asked if I could do a reconnaissance of the route with him and provide advice on rock slope design and excavation procedures. I had come to enjoy working in Alaska and with Alaskans, who are a tremendously friendly and interesting people. Of course I accepted.

We met in Skagway in the late afternoon, had dinner together, then spent the evening reviewing state geologic plans, air photos and the preliminary alignment design.

The next day we drove to the proposed beginning of the project, located the staked field line, and walked to the first rock cut area.

Bedrock was exposed over most of the surface, so the geologic structure could be observed and mapped. The granite rock was very hard, and instability of the rock slopes would only occur if discontinuities in the rock dipped out of the slope enough to result in planar or wedge failures, dipped steeply into the slope enough to allow toppling to occur, or if the contractor were to use excessive blasting during construction. Inspection of the slope indicated the rock contained exfoliation joints parallel to the rock surface orientation, a condition which occurs as the result of a combination of aligned rock grains and stress relief. As the glaciers melted during past glaciations, a great reduction in load occurred near the rock surface. This unloading created rebound and high-tension stresses in the rock. Rock is much weaker in tension than compression. As a result, fractures developed that were generally parallel to the rock surface. The fracture surface developed is quite rough, so the friction along that surface is higher than normal.

At the first rock excavation area, the dip angle of the exfoliation was 65 degrees out of the slope. I advised Slater that the back slope should be drilled, blasted and excavated to that angle. In the past, highway engineers have generally designed rock slopes at 1 horizontal to 4 vertical, or 76 degrees. Excavation of the first slope to 76 degrees would have resulted in failure of the slope to 65 degrees unless expensive rock bolting was performed.

The second rock area had exfoliation at 80 degrees. I instructed Slater, "Design this slope to that angle, 80 degrees, but allow for some rock bolts near the crest to prevent toppling."

Slater said, "How will the contractor react? Won't he challenge the variable slope design, since it'll be more difficult to develop the different angles?"

I answered, "He may. However, the state has the prerogative to develop whatever design they consider appropriate. It then becomes the contractor's responsibility to construct it as designed, and bid the work to provide the profit he seeks. It's also the state's responsibility to develop specifications that are fair to the contractor, such that he can develop that design and make a reasonable profit."

My philosophy regarding rock slope design is to let the rock tell me the slope angle that combines adequate stability with reasonable economy. This requires the determination of the likely type of failure that may occur, and determination of the slope angle needed to ensure that failure does not happen. This may necessitate special stabilization, such as dowels or rock bolts, or the installation of drain holes to reduce water pressure.

For the next two days we inspected rock exposures along the alignment, determined the dip of the exfoliation, and designed the slope angle for each location. Design slope angles varied from 55 degrees to vertical along the length of the project. Where the geology is favourable, I frequently call for controlled blasting and vertical slopes up to 100 feet high with no benches. A major advantage is that any loose rock will fall into and stay in the ditch.

In order for the contractor to develop these slope angles during construction, it was necessary to specify the drill angle for the final line holes in which to place the explosives. It was also recommended that a pre-bid site meeting be held, which potential tenderers must attend. When the contract documents were complete and the contract was advertised, Slater advised me of the site pre-bid inspection date and asked me to attend.

We both arrived at Skagway in the late afternoon and had dinner together. We sat at a table beside several men who obviously were contractors who had come to attend the site inspection. During dinner I overheard one contractor say, "Did you see the different slope angles shown on the plans? Almost all the cuts are different. The guy who designed these slopes was either crazy or drunk. I'd like to meet the stupid son of a bitch!"

Slater and I looked at each other. We both smiled.

I got up slowly, took two steps over to the other table and said quietly, "My name is Brawner, I'm the stupid 'son of a bitch' who designed the rock slopes. We'll take you out to the project tomorrow and show you the reason for the different angles. I'd like to hear your reaction on our return."

He looked at me sheepishly and muttered, "Okay, I'll see you then."

The next day about twenty contractor reps accompanied us along a portion of the project. Some firms had sent several representatives. We stopped at the location of about eight proposed large cuts. I gave them an introduction to structural geology and brought their attention to the through-going exfoliation joints that existed in the rock. Back at Skagway, we met in a meeting room and Slater and I reviewed the design concept, the influence of the structural geology, and described the special blasting program we had specified. At the end of the session we asked for any questions. There were none!

It has taken me several such pre-bid conferences to learn that contractors usually do not ask many questions. They fear that their questions may alert the competition to some special construction procedure, which may give them an advantage. I asked the contractor who had questioned the design the evening before, "Now that you've seen the conditions in the field, how do you feel about the design?"

He answered, "This is the first time I've seen a design based on the conditions of the rock. It looks like a great idea. It should provide more stable slopes with less rockfall. The angle drilling will cause us some problems, particularly the flatter angles. However, you'll save money and accidents over the long run. Well done!" This is one of the few times I have been on the receiving end of a compliment from a contractor. The ensuing contract won a national award for the state.

A major highway extends from Anchorage to Seward, south on the coast. The northern portion runs along Turnagain Arm, an ocean inlet, and parallels the Alaska Railroad. About 30 miles of the highway was paved, but the road was narrow, with many curves. Highway departments regularly liase with police departments to identify locations where the accident frequency becomes excessive, and this stretch was one where Alaskans, who are known to party, frequently had automobile accidents.

ADOT concluded that the highway should be upgraded in the late 1970s. After several years of alignment, traffic and geotechnical studies, including rock mechanics investigations, the first contract was called on the northerly portion. The design proposed was at the base of the rock slopes, along the inlet and beside the Alaska Railway. Straightening of the alignment to a 70 mph standard required rock excavation to remove numerous rock promontories. The excavation involved thin cuts, often 200 to 300 feet high. Starting the cut at the top of the excavation required very difficult drilling and blasting. The contractor described it as "pioneer drilling." Often the initial setup required air lifting men and equipment to the top of the cut using specially outfitted helicopters.

Much thought went into the development of the design and specifications of the first project. I worked closely with Bill Slater of ADOT and his staff, and Ron Chassie of Federal Highways. One major concern was to protect the adjacent railway from flyrock created by blasting. Long-term rock slope stability, after construction, was another key concern. It was recommended that "controlled blasting" near the final rock face be required. I had previously learned that heavy blasting using large amounts of explosives opened joints and developed new cracks in the rock behind the face, which reduces long-term stability. The result is a lifetime source of dangerous rockfall.

I therefore developed the blasting specifications for the project, calling for controlled blasting. Specific requirements included the following:

1. The blasting foreman must have had extensive experience with "controlled

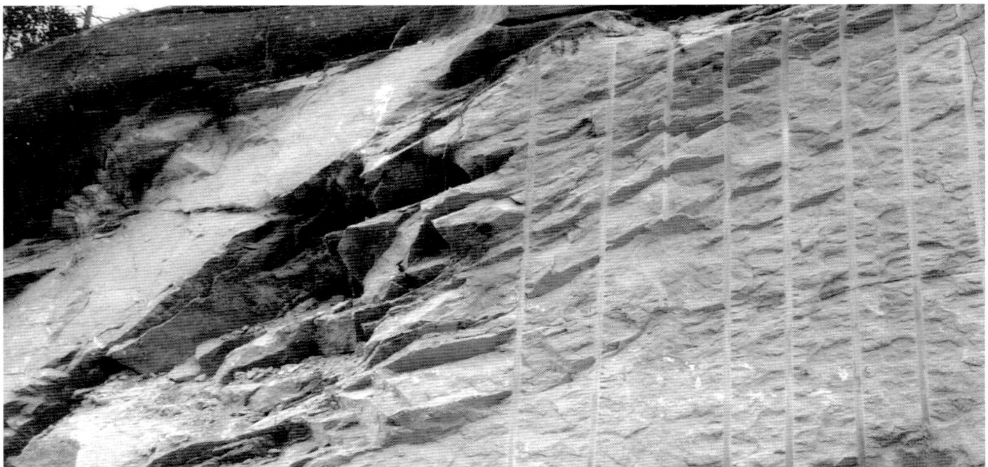

Excellent pre-shear drilling to develop a stable rock slope. Unfortunately, excess explosive was used in the bottom of the hole. The end of the cut did not use controlled blasting. Note the different result. Near Skagway, Alaska, on the highway to Whitehorse.

blasting."

2. Drill holes along the final face were to be drilled at the angle specified for each cut and spaced 2 feet apart down the face. Cut face angles were to be determined from an assessment of structural geology.
3. Hole diameter was to be 3 inches.
4. The explosive charge was to be distributed uniformly down the hole. Light charges were to be used.
5. Where the rock was competent, the final line holes were to be detonated ten holes at a time (simultaneously per delay) along the face, prior to all other blasting in the cut. (This is called "pre-splitting "or "pre-shearing.") Where the rock was fractured or weak the final line was to be detonated last. (Referred to as "cushion" blasting.)
6. The first row out from the final line was to be spaced out 4 feet and 4 feet apart parallel to the final line. In some cases two such rows were called for. (These are called "buffer" rows.)
7. In order to limit the explosive force per hole, "production holes" — those beyond the buffer rows — were to be no more than 4 inches in diameter. Most of these holes were delayed from each other to reduce the charge detonated at any one time.

Prior to ten to fifteen years ago, many contractors treated rock blasting as a "bombing" operation. Such action usually resulted in long-term instability. As a result of this experience, further specifications were developed.

1. The contractor was required to submit his blasting plans for each cut one week in advance for review by the resident project engineer.
2. All line holes were to be drilled within 6 inches of the design location.
3. The contractor was to bid separately for the line drilling at a rate per lineal foot of drill hole. This included the cost of drilling, explosives and blasting.
4. Any hole that did not meet the alignment criteria would not be paid for.
5. Excavation lifts were limited to 30 to 40 feet deep, dependent on the contractor's ability to develop good results with the controlled blasting.

Where controlled blasting is properly performed, the resulting rock face will be uniform, with a portion of the drill holes showing on the surface. Long-term stability will be much better than where uncontrolled blasting is performed. In addition, provided that the structural geology is favourable, steeper slopes, requiring reduced rock excavation, will be more economical.

Peter Kiewit Ltd., one of North America's most experienced contractors, won the first contract. They had a group of skilled pioneer drillers based in mainland U.S.A. who they proposed to employ for the initial rock cut development. However, the local Alaska unions would not allow the out-of-state drillers to work in Alaska. They wanted the contractor to employ only Alaskans! Alaskan unions had become very strong and dictatorial as a result of their success in controlling contracts for the Alaska pipeline.

The Alaskan drillers were not skilled in pioneer drilling, and so a large percentage

Poorly aligned drill holes along the pre-shear line. The contract called for the contractor to be penalized for the misalignment. Anchorage — Alyeska Highway upgrade project.

of the line holes, more than 400 of them, were seriously out of design alignment. These were rejected for payment. Kiewit had bid $2.10 per lineal foot. The average depth of hole was 32 feet. This meant they were losing $67.20 per hole, a total loss of nearly $27,000. Kiewit was furious, and submitted a claim to recover at least a portion of this.

During my association with construction contracts, it has become apparent that many owners and representatives of owners place most of the risk on the contractor. I have been vocal in voicing the opinion that this is unfair. I believe that the owner and contractor should share the risk. If this is done, the contractor will submit a lower bid, saving the owner money, and owner–contractor relations will be better.

I was asked by ADOT for my comments on the claim. I advised, "The contractor bid on the reasonable assumption that he could use his own skilled drillers. He was not allowed to do so because of a condition beyond his control." I recommended ADOT pay half of Kiewit's claim.

On the second project, the contractor outsmarted us! The same specifications were used, but instead of submitting a normal bid, the contractor bid $0.01 per lineal foot for the line drilling. To compensate, he added into the unit rate for rock excavation the amount he deleted from the line-drilling bid. His intention was to ensure he wasn't penalized for inaccurate line drilling. This procedure is described as unbalanced bidding, and most contracts contain a clause that states, "Unbalanced items in a bid may be cause for rejection of the bid." ADOT could have rejected the bid on this basis. However, the contractor's total bid was considerably under the second low bidder, and so the bid was accepted.

On the third contract I advised ADOT to provide a bid value of $2.20 per lineal foot in the contract. This resulted in reasonable payment to the contractor for acceptable drill alignment, and yet also provided an incentive for him to meet the specifications.

One serious problem did develop. Several very large cuts were required on the project. The contractor brought drills onto the project that drilled 6-inch-diameter holes. These were spaced farther apart than normal 4-inch production holes, required much more explosive per hole, and created greater energy release per hole. As a result, the final wall suffered excessive damage and one rockslide involving several thousand cubic yards occurred. A nearby fibre-optics line was endangered.

I recommended ADOT shut the project down until the contractor changed to the smaller drill hole. He was adamant he would not do so. One large cut near the north end of the project remained to be done. ADOT backed off on shutting the project down, but, based on my recommendation, it was decided to delete this cut from the contract and add it to the following contract. The contractor was again furious and submitted a claim. However, the rock quantities on the rest of the project exceeded original estimates, and the claim was rejected.

As a result of experiences such as these, where contracts involve difficult geotechnical conditions, I have made it a practice to consult with a very experienced retired contractor and ask for his comments on constructability, construction methods and appropriate specifications. This has been invaluable. Along this same vein, I recommend the same consulting contractor be retained to review any claims that a contractor submits during or following completion of the contract. This contractor will be aware of typical claim approaches that the owner may be unfamiliar with, such that the owner can be taken advantage of. It is best to retain a contractor who is retired so he is not in competition with the bidders.

In the spring of 1986, on a Thursday afternoon, Doug MacKenzie, a director of the Valdez Gold Mining Co. in south-central Alaska, contacted me. "We've had a very serious slope stability failure in our open pit. We almost had a man killed, and a D-8 Caterpillar's been damaged. I'm concerned about other failures. Can you possibly come up in the next few days to assist us in resolving this problem? It's very urgent. We've stopped mining until the problem is resolved."

I quickly looked at my date-book and noted that I was committed for the next three weeks. I advised him of my full schedule.

He urged, "Is there any way you can adjust your schedule? It's extremely urgent."

I thought quickly and said, "Doug, I can catch a plane to Seattle late tomorrow afternoon and connect to Anchorage to arrive by midnight. If you can arrange a charter flight early Saturday morning, and arrange a return charter for Sunday noon that will get me back to Vancouver Sunday evening, I can come this weekend."

"Excellent! " he replied. "We'll make the arrangements. Is there anything special you need?"

I asked, "Could you prepare a history of the slide area, its past behaviour, and also prepare a geologic cross section?" Then an idea struck me and I also asked, "Can I bring my son, Greg. He's eighteen years old."

Mackenzie answered quickly, "Sure, bring him along."

Greg was hesitant, but agreed to join me.

Valdez had developed a large-scale placer gold mine in the Denali Range of south-central Alaska. Placer gold with occasional nuggets up to several ounces had been deposited along a buried channel of Valdez Creek. Glacial drift had been deposited to thicknesses of over 200 feet over the channel and had been heavily over-consolidated. This soil was excavated to that depth to expose the creek channel. The majority of the gold was deposited in the creek sands and gravels, on the bedrock surface, and in joints below the rock surface to several feet. The glacial drift was sufficiently dense that it was more economical to drill and blast it rather than try to excavate it with mechanical shovels. The drift was excavated and deposited in large waste piles outside the pit area. The pay gravel at the base of the excavation, and the ripped shallow bedrock were excavated with a shovel or backhoe, then loaded into ore trucks, hauled and dumped onto a gravel pad. In the winter, the underpad was heated with hot air pipes beneath the pile to minimize freezing of the ore stockpile. Material from the stockpile was then dumped onto conveyors feeding three large sluice boxes, where the gold was separated from the sand and gravel. About 30,000 ounces of gold was recovered annually.

Airlift Alaska operated a small charter service from Merrill Field in Anchorage. We met the pilot, Jim Archer, who took us to a four-seat Cessna 180. I noticed it had dual controls, so I asked Jim if he would give Greg a flying lesson. He said, "Sure, be glad to."

After about twenty minutes in the air, Jim nudged Greg and pointed to the control yoke. He said, "Greg, this is the control column, you pull it back and the airplane goes up; if you push it in, it goes down. Okay?"

"Yaaa," said Greg, a little hesitantly.

"If you turn it to the right the plane turns right, and if you turn it to the left the plane turns left." As he explained these manœuvres he performed them to illustrate the response of the plane. Again he asked, "Do you understand?"

"Yaaa," Greg answered again, appearing to be more interested.

Jim pointed to the altimeter. "This tells our altitude. We're at 6,000 feet, which we will stay at until we have to cross the mountains. Then we'll go up to 8,000 feet. Understand?"

"Yaaa," Greg returned.

Jim pointed to the compass. "We're on a heading of North 30 degrees West [N30W]. We'll stay on that course for about one and one-half hours. Understand?"

Greg nodded, staring at the compass near the top of the windshield.

Jim looked at him stoically and challenged Greg, "Okay, fly it kid!"

Greg's eyes widened, his mouth dropped and he looked at Jim, totally taken by surprise. "Me, flllyyy it?" he stammered.

"Sure, put your hands on the yoke and get the feel as I do some slow manœuvres," Jim instructed. Greg gingerly put his hands on the control column. Jim took it down 100 feet slowly, up 200 feet, down 100 feet, turned left 5 degrees, right 10 degrees and left 5 degrees. He said, "See, we're back on our flight plan."

Greg smiled a bit. Jim continued through slow manœuvres for about an hour. Then he casually removed his hands from the yoke and said, "Go ahead, do what I did." Greg flew for nearly half an hour and did reasonably well. He generally overshot or undershot each manœuvre, but brought it back to flight line each time. It had been a long time since I had seen him with such an intense level of concentration. Finally Jim said, "We're nearing the mine; I'll take over." He circled the camp, gunning the engine as a signal for a vehicle to come to the airstrip, then, at my request, he circled the open pit so I could look at the slide area and photograph it. I took two very quick overlapping shots so I could later view the area in stereo three-dimension. I have found this to be a very useful tool in assessing site geotechnical conditions.

We landed on the dirt airstrip. Doug MacKenzie met us and took us through security. Doug and the engineering staff briefed me for about one hour, going over the mine history, site geology, mine planning and operations, site hydrology and details of the slide. Following this we went to the mine. As a standard precaution we put on a hard hat, safety glasses and steel-toed boots. Greg was fascinated, not having seen large mining equipment in action before.

I have been involved in many landslide investigations previously, and the conditions I observed were not unusual, except for the need to drill and blast the hard glacial-loaded soil. At several locations on the slope, seepage was actively flowing. This meant there was a water table that would reduce stability. In addition, the blasting would generate a water pressure pulse (hydrodynamic shock), which would also impact stability. The most critical factor, however, was the steep angle the slope had been excavated to — 56 degrees.

We went back to the mine office and I asked for a surveyed cross section down the centre of the slope, both before and after the slide. They said they would have it for me after lunch.

Accommodation and lunch at mining camps are generally very adequate, and this camp was no exception. The buffet-style lunch included a selection of fresh vegetables for a salad, various meats and breads for sandwiches, hot meat of some sort — fish, chicken or beef — cooked vegetables and soup. Pickles, mustard and other condiments were also available. For dessert there was fresh fruit — apples, oranges, bananas — and pudding, ice cream or pie. An abundance of good food!

After lunch I reviewed the new cross sections, which showed the topography, the geology, and zones of seepage. A plan of the open pit also showed surface drainage ditches along the west crest, and a diverted creek along the east side. There was no doubt that seepage from these sources was reducing slope stability. However, modification of the present program would be very expensive and difficult. It was concluded that drilling selected horizontal drains into the slopes was more reasonable. A program for these was decided upon in the afternoon, concentrating on the seepage areas.

Back analysis of the slope, assuming a safety factor of 1.0 at the time of failure, was used to estimate the shear strength at failure. Using this data, it was determined that the slope should be flattened by 8 degrees, increasing stability by 20 percent. This was considered reasonable for short-term stability. The mine life was estimated to be another four years. The dense glacial till was very brittle, so that failure occurred with little initial movement.

The next morning, the blasting program and in-pit drainage were reviewed. It was recommended that each blast hole be delayed to reduce hydrodynamic shock due to a sudden surge in water pressure from the blasting. Also, the in-pit sump was moved away from the toe of the slope. A set of summary notes was left at the site. This would be followed up with a formal report as soon as I returned to Vancouver.

Greg took a keen interest in the mining when we were in the field. I finished my inspection about 11 A.M., and Doug MacKenzie took Greg and I over to a separate building to inspect the sluicing used to remove the gold. There were three sluice boxes operating. Generally one was down to collect the gold, while the other two continued running. There was also a mechanical separator to collect the finer gold. Occasionally nuggets over 1 ounce in size were collected. The entire operation was performed under tight security. An armed guard was inside the building at all times.

After this tour we had a quick lunch and waited for the airplane to arrive. It was the same aircraft and pilot that flew us in. I again asked if he would give Greg a lesson, and he agreed. On the way back we were flying near a mountain that was higher than the plane. Jim said to Greg, "I'd like you to circle the mountain and stay at our present elevation." Greg smiled and slowly circled the mountain. No problem. As we neared Anchorage, Jim took over. He flew us to the main terminal and dropped us off at an entrance near the check-in area.

On board the United Airlines flight to Seattle we talked mostly about the flight to and from the mine. We had checked our bags through to Vancouver, so on arrival in Seattle I said to Greg, "Wait until everyone deplanes and then stick your head into the cockpit and say you're interested in becoming a pilot."

He did that. The pilots kept him there for about twenty minutes, explaining the instruments throughout the cockpit. Greg was ecstatic. On the Vancouver leg he declared in definite terms, "Dad, I want to be a pilot!"

The next morning, Monday, I kept him out of school, drove out to Skyways Aviation in Langley, B.C., and signed him up for flying lessons. He had his first lesson with them that morning.

Greg continued on with his flight training, attended Embry Riddle Aeronautical University in Prescott, Arizona, and obtained a bachelor's degree in aviation science. This was followed by a job flight instructing at Sierra Flight Academy in Oakland to build up his hours. At the same time, via evening courses he completed a master's degree in aviation business. Approximately one year later, Continental Express Airlines hired Greg. Now, seven years later, he is a jet captain. Greg loves flying and owes his career to the Alaska trip.

The cost of flight training is high, since it includes rental of the aircraft for hundreds of hours. After six months with Continental, Greg phoned one evening to say, "Dad, it's payback time! I've learned that I can obtain airline passes for you and Mom, to anywhere that Continental flies." What a nice payback. We use the passes several times a year.

Milt Schumacher, superintendent of the maximum-security prison in Ketchikan,

Alaska, phoned me in February 1987 and advised that a rockfall had occurred which damaged their external security fence, and that they were concerned there would be more rockfall. He stated, "Bill Slater of the Alaska Department of Transportation has given me your name. Could you come up within the next few days to assess our problem?"

He had phoned on a Wednesday afternoon. Mentally adjusting my schedule, I said, "I can fly to Ketchikan this Sunday and be at your office early Monday. When do you arrive in the morning?"

He answered, "I'm here by eight. We'll look forward to seeing you then. Come to the main parking lot and go to the security entrance at the left."

I said, "I'll likely require a survey crew to help me develop a site plan. Could ADOT assist you and have one ready Monday?"

He replied, "I'll make those arrangements."

The best flight schedule resulted in me arriving Sunday about 4 P.M. This was the first time I had been to Ketchikan, and I couldn't help noting the uniqueness of the main runway. It was located along a rocky side hill. To construct it, the rock on the uphill side had to be excavated to a depth of 15 to 35 feet, for a length of about 7,000 feet. Much of this rock was then used to build up the inlet side of the runway, the off-ramp, and a site for the terminal building. Indian carvings inside the terminal created an atmosphere suggesting the original native population. As the airport is constructed on an island, all passengers had to take a short ferry ride to reach Ketchikan.

Ketchikan is constructed along the inlet, with most waterside structures erected on timber piles. This included the cruise-ship terminal, which in the summer was very busy, with up to three ships docking per day. Several streets parallel Main Street, running up and down rocky hills. Ketchikan had been a major fishing port around the turn of the century, and many houses date back to that time. More recently, a major pulp mill has been in operation.

Disembarking the ferry, I caught a taxi, which took me up a winding road to the entrance of the Westmark Hotel, a picturesque building with natural wood panelling and beams, which extends over the brow of a hill overlooking the harbour. A shortcut to the town is via an inclined cable car, an interesting novelty.

The next morning I arrived at the prison by taxi. After I got out, I observed the surroundings. There were rock slopes up to about 50 feet high along the west and south sides. It looked like an old rock quarry. I could see where rock had fallen on the outer fence about halfway along the west face.

I found the security entrance and entered a doorway. There was a telephone with a sign above to pick up the phone and identify yourself and your business. I spoke into the phone, "Chuck Brawner, rock engineer. I have an appointment to see the Superintendent." After about ten seconds the inner door clicked and a voice asked me to enter. Inside, a receptionist told me that Schumacher was on the phone, and offered me a cup of coffee.

Shortly, Schumacher came out, introduced himself and invited me into his office. He began:

"To introduce you to our facility and problem, we operate the primary maximum-security prison in Alaska. Most of our inmates are long term. We've never had an escape, and I want to keep it that way. This prison is built in an old rock quarry with rock slopes on two sides. In that area we have an outdoor compound with double fences twelve feet high. Our inmates are classed into several categories — the very dangerous spend much of their time in solitary cells, and are allowed into the compound separately. Medium-risk inmates are allowed outside six at a time for exercise, and lower-risk inmates are allowed to play basketball or volleyball outdoors, and have access to the prison library. Visitors are only allowed to sit in separate rooms with a window between. Every door is controlled from a central control room and only opened upon receiving a message in code.

"I would like you to inspect the rock slopes with my assistant and advise us of the magnitude of the problem and remediation that you would recommend," he added.

"Okay," I answered, "it will take about an hour to inspect the slopes to decide how much detail I consider is required. Then I'll get back to you with the program I propose."

We went out to the parking lot first. I inspected the rock slopes, particularly noting the structural geology — the joints and shear zones and their dip angle, dip orientation, frequency and surface roughness. I looked for open joints around the

Wedge rockfall (left) that damaged outer prison fence (right) at the high-security prison in Ketchikan. Stabilization included scaling loose rock, rock bolting and rock excavation using Bristar chemical expander rather than explosives to eliminate flyrock.

blocks where little external force could result in a fall. Ketchikan has very high precipitation. Water pressure in cracks and joints can cause ravelling. During the winter, freeze-thaw cycles can cause ice jacking.

I sketched the structural geology in my notebook, pacing off approximate distances. Near the southwest corner of the parking lot we came to the outer fence. The toe of the slope was about 10 feet beyond the fence, and the length of the rock slope near it was about 120 feet. Near the middle of the west-side fence, about 150 cubic yards of rock had slid from the face on two intersecting joint planes. These planes formed a wedge that dipped out of the slope and appeared to daylight at the ditch line. The rock had slid into the fence, torn the buried portion out of the ground and pushed it about 8 feet into the neutral zone between the fences. The fence was not broken. Two large trees had fallen down along the slope face. One had hit the outer fence and bent it. During remediation, the fallen rock would have to be removed but, due to the short distance between the fence and rock face, only small construction equipment could be used.

While inspecting the south face, a major shear zone was observed which dipped out of the slope. The volume of rock behind the shear zone was estimated to be about 300 cubic yards. This volume would likely break the outer fence and reach the inner fence if it failed. It would have to be removed or bolted in place with tensioned, grouted rock bolts.

I returned to meet with Schumacher and said, "There are a number of small unstable areas, one major problem at the south end, and the rock from the failure must be removed. My proposal is to photograph the entire slope in sections and show the stabilization requirements on the photographs. I'll need a surveyor to establish the point from which each photograph is taken, to measure the length along the two rock slopes, and to install chainage stations along the rock face. Can you arrange for him to come this morning?"

"I'll try," he said.

In the meantime I suggested, "I'll go out and photograph the slope. Could you have someone on your staff phone and locate a photo store that can develop one roll of film with all photos enlarged to about eight by twelve inches by this afternoon? I'd like to work on them this evening. I'd also like to meet with a local contract estimator later this afternoon to discuss my preliminary stabilization options. I want to discuss the practical aspects of the construction, the equipment he would suggest be used, unit cost estimates, and potential time to complete the work. Can you suggest a reputable contractor?"

Schumacher was very helpful. He arranged for the surveyor to be there within an hour, located a photo lab that could develop the photos in two hours (this was before one-hour service became available), and arranged for South West Construction to have a contract estimator arrive at 3 P.M.

It was a busy day. I made a second inspection around the slopes and completed taking photos by noon. My proposal for stabilization was as follows:

1. Scale all the slopes of any loose rock that could be removed by hand using steel crowbars. The scalers would have to work down from the top of the slope hanging on ropes tied to trees beyond the crest. Prior to scaling, heavy plywood sheets would be placed against the fence to protect it from falling rock scaled

from the slope. The plywood could be moved from place to place as required.

2. Equipment access could only be from the southeast corner. One rock outcrop near that would have to be removed for equipment to reach the west rock slope. This rock would have to be broken up. I expected Schumacher would not want blasting because of the potential for flyrock. Hence, I decided the rock should be broken using chemical expanders placed in drill holes. A typical expander is called Bristar, a Japanese product. Hole spacing is similar to that used in blasting. The expander is poured into the hole, it hardens, and the chemical reaction causes a strong, slow lateral expansion. After about 18 hours, cracks develop in the rock and it starts to break apart. Where slope removal is required, the rock will continue to fall for nearly a day, but there is no flyrock. The cost is several times that of blasting, but where safety and flyrock control is essential, the extra cost is justified.
3. The large rock volume on the south side should also be broken using Bristar, then removed. After 48 hours the slope would be scaled of all broken rock back to the shear zone.
4. Adjacent to the parking lot there was more room to work. As a result, a number of rock bolts were to be installed to stabilize large blocks, and, to reduce weathering, shotcrete (sprayed on concrete) was to be applied to zones where numerous fractures and joints occurred.

After developing the proposed program, I checked in with Schumacher, inviting him to lunch and asking him to suggest a restaurant. He picked one on the waterfront. During lunch I explained the general condition of the rock slope stability and outlined my proposed program. He listened attentively. He was very agreeable to my suggestion of using a chemical expander rather than blasting. He was surprised that an alternative to blasting was available.

I asked why the prison was located in Ketchikan, rather than Anchorage, the largest city in the state. He answered simply, "Politics." That said it all. I asked if there had been any attempts at breakouts. He said, "Nothing serious. The security is so tight I can't imagine anyone escaping without major outside help, such as destroying an entire wall with a large blast of some sort. We have a good personality profile on all inmates and know who might be involved in such an attempt. They all occupy inside security cells. Our biggest concerns are drugs getting into the prison."

I asked, "How would they do that?"

"Friends on the outside come to the prison as visitors and determine the outdoor schedule for sports activities," he explained. "They then sneak into the woods beyond the rock slope and with a slingshot shoot small packages of drugs into the outdoor area. They have a private code on the package so any inmate knows who it's intended for. This process is random so we don't know the timing. And the adjacent property is privately owned so we don't have access to police the area. I'd like to see the state purchase the property. Other than that, work here is generally routine. Occasionally we have a hunger strike if the inmates want something special, such as seeing or hearing the *World Series*. Good behaviour will result in such rewards."

After lunch he took me through the prison. I was shown a reading room, television room, lunchroom and kitchen, then allowed to look into the high-security cell area. The cell doors there had small barred windows. While we were looking in, several inmates came to their windows and stared at us. It was scary! I was also shown the security control room from which all prison doors were controlled. I wouldn't wish incarceration in the Ketchikan prison on anyone. Over half the inmates had been sentenced to life imprisonment.

I met with the contractor at 3 P.M., and described my approach to the stabilization. He recommended the equipment type and size for the excavation and hauling. In addition, he advised against rock bolting in the fenced area, because the bolts required would be 10 to 14 feet long and there wouldn't be enough room between the rock and the fence for the drill to develop the length of hole required. He also was skeptical about the use of chemical expanders, because he had never used that rock-breakage method before. I assured him that we would provide complete instructions. He finally agreed. We discussed costs and I was surprised at his estimates. They were about double what we encountered in Vancouver. However, I was glad of his input because the project estimate I ultimately developed for budget purposes was within 2 percent of the low bid. This pleased Schumacher.

At four o'clock I picked up the photographs. That afternoon and the next morning I did a final inspection and indicated the stabilization details and location on the photos. These were used in the contract documents. This is a procedure I have found to be very successful for small- to moderate-scale rock stabilization projects. Photographs show it like it is, better than plans do.

I left Ketchikan the next midday. Two weeks later I had completed all design details, plans and specifications so that tenders could be called. Four bids were received. The contractor with the low bid met all pre-qualifications and proceeded to commence the work. I inspected the project when it was about half completed, and later upon completion. The contractor did a good job and was only required to complete some minor work. I enjoyed the project, as it was different than anything I had done before.

During my return trip to Seattle, I sat beside a senior official in the Alaska Fish and Game Branch. I asked him, "If you were to fish for king salmon for your personal enjoyment with a good chance for success, where would you go in Alaska?"

He answered immediately, "Sitka. It has everything; excellent fishing, good facilities, reasonable weather, top-rated guides and magnificent scenery."

I took note of that, and the next spring made contact with a private fishing guide, Paul Yanek, to take my sons and me fishing. It was a great experience. Yanek had an 18-foot Boston Whaler style craft with a small cabin, for use if the weather got bad. It was about an hour run out to the open Pacific where Paul said the fishing would be best that time of year. It was a rough trip out, fighting 4- to 5-foot waves and trying to make reasonable speed.

Out in open water, we stopped and Paul got out the rods and reels. Each was outfitted with 40-pound test nylon line and a jigging lure. The lure looks like a small herring and is slightly bent. A single hook is attached. With jigging, you lower the line to hit the bottom, then pull it up quickly, about 8 to 10 feet, then let the line go slack again so the lure falls in the water, imitating a wounded herring. As soon as the line goes taut you jig it up again.

We caught two 20-pound salmon within about thirty minutes. Paul told us to keep jigging. He cleaned both fish and cut off the heads. He then rolled the innards around a fish head and tied it together in a ball with some string. He did this for both fish. Then he picked up another rod that was shorter, stiffer and stronger, and had a heavier test line. At the end of the line there were two 6-foot-long steel leaders with a large halibut hook on each. He put a hook through each ball and dropped them overboard, letting out line until the bait hit bottom. He put the rod in a stirrup on the edge of the boat and said, "Keep an eye on that one."

My son Randy and myself with our first day's catch. Near Sitka, Alaska. The chinook salmon weighed from 20 to 32 pounds.

The benefit with jigging is that you are busy, unlike trolling where you drag a line at 3 to 5 miles an hour and sit, eyes glued to the rod, waiting for a strike. If the reel screams or the rod bends rapidly, you grab it and give one strong pull to set the hook. Then the fun can begin. The idea is to keep pressure continuously on the fish to tire him out. If you reef the fish in, you will lose more fish than you catch. If the fish tries to run out rapidly, you let him go and the reel sings. Gradually you put pressure on the reel by hand to slow the fish down. He will slowly turn back. With large salmon, each run may go as far as two to three hundred feet. Slowly you reel him closer to the boat, always keeping your rod up and some tension on the line. If you don't and the fish turns sharply, he may spit the hook.

The boat was bobbing in all directions due to the high 8- to 10-foot swells. Even though Randy had taken a Gravol, he started to become queasy. He asked, "Is there somewhere we can go where the water is calmer?"

Paul answered, "Sure, bring in your line, and then the halibut line."

We followed instructions. Randy had just started to reel in the heavier halibut line when he yelled, "It's very heavy!"

Paul put two fingers on the rod. He felt it quiver. "You've got a fish. Work it up slowly with uniform pressure." Randy continued to work the rod slowly — pull up then reel down — with uniform pressure. He had to pull up about 200 feet of line. After about five minutes he shouted, "I see it. Hey, there are two!"

"A little more," Paul said, grabbing a gaff and giving it to me. Both halibut started to splash, seeing the boat. I gaffed the nearest fish and Paul grabbed the other leader.

Suddenly both were in the boat. Paul exclaimed, "Wow. A double-header! I've never had this happen, or even heard of it before." Both halibut weighed twenty-eight pounds. Paul started the boat and we headed to inside waters. After about twenty minutes he pulled into a small bay and stopped near a floating buoy. A line dangled below it. He used a hook, caught the line and pulled up a crab trap. There were three in it. Two were too small, so they were returned to the water. He kept one.

Paul next turned into a channel between two islands and cut the motor. He said, "Okay, drop your lines. This is often a good spot." In about ten minutes I got a strike and slowly worked the fish toward the boat. Suddenly Randy's reel screamed. He didn't need to set the hook; the fish had hit hard. My fish circled the boat and I had to sneak under Randy's rod. I got the fish, headed in and I yelled at Paul, "Get ready!" The fish swam right into the net he had dipped in the water. It too weighed in at twenty-eight pounds. Nice fish.

My son Greg and I — salmon fishing at Nootka Lodge on the west coast of Vancouver Island. Greg was unavailable for one Alaska trip.

Meanwhile, Randy was working hard. He brought his fish near the boat three times, but each time it turned and ran out again. The fourth time he got it to about 20 feet from us. It surfaced, and Paul yelled, "It's a big one! Looks about fifty pounds!" Suddenly the fish dove. Randy's rod bent to the water. Again the rod went slack, and Randy reeled madly to take up slack. The rod again bent sharply again under heavy pressure, and then went slack. It was gone! We could feel the despair as he reeled in.

Paul asked to see the line. He studied the end. "The knot slipped," he said. "I've only had that happen once before. I'm sorry. My fault. It was a big one." Randy caught one more, and we both caught two bottom fish. Paul said, "We'll keep them and throw them to the eagles on the way back."

The sun was setting as we headed in. Sitka has a major eagle population. It was common to count a dozen white heads sitting near the treetops along the shoreline. A beautiful sight. Paul slowed the boat and said, "Throw one of the bottom fish out behind us. Throw it high." Randy threw it a good 50 feet. Immediately an eagle left his perch, swooped over the fish then turned like a dive-bomber and grabbed the fish on the fly with one foot! When he got near the trees, a second eagle gave chase and they both disappeared. The other fish met a similar end.

Over the next three years I went fishing off Sitka twice, once with Randy and once with Randy and Greg. It was always a great thrill — a family thrill.

Ben Pollard, Regional Engineer of the Alaska Fish and Game Branch, Alaska, phoned in June of 1987 to ask if I could help with a problem at the Snettisham fish hatchery, south of Juneau. The hatchery was beside the Snettisham power generating facilities, which provided electricity for the Juneau area. Two pressure tunnels tapped Crater Lake, perched about 800 feet above an underground powerhouse developed in bedrock. The tailrace tunnel was adjacent to the tailrace pond, and 80 to 100 feet from the fish raceways, which contained fingerlings and spawning fish. The main fish incubators were nearby. A new penstock, turbine and tailrace tunnel were required, with the design and specifications developed by the U.S. Corps of Engineers.

Snettisham fish hatchery adjacent to an underground powerhouse. A new tailrace tunnel (5) was required for a new turbine to expand power generation. An air bubbler system was installed along the wall of the pond (3) where fingerlings were kept to minimize blast shock to the fish.

Upon reviewing the plans and specifications, Pollard noticed that no consideration had been given to seismic shock forces due to blasting on the tailrace structures, the hatchery foundations, fish in the raceways or fish eggs in the incubators. He asked, "Are you available in the near future to inspect the site, review the design documents, and assess the impact of blasting on the facilities, fish eggs and fingerlings?"

What a fascinating and challenging project. Without hesitation I said, "Ben, I can adjust my schedule to come up next week, say Thursday and Friday — and Saturday, if necessary. Can you send me a copy of the plans and specifications by rush air mail?"

"Okay, I'll get them out this afternoon," he agreed.

I arrived in Juneau the following Wednesday evening and stayed overnight at the historic Baronoff Hotel. The next morning we flew to the site in a chartered helicopter. I had reviewed the plans and specifications before leaving, and on the plane from Vancouver. As a result, I knew the general layout of the hatchery, ponds and tailrace tunnels prior to meeting with Pollard.

Pollard showed me around the site and I took a number of photographs. The fish incubators were very interesting. All eggs were being carefully incubated under controlled conditions. Also, thousands of fingerlings were being held in the holding ponds until the appropriate time for release. Numerous salmon, some in excess of thirty pounds, worked their way up the entrance fish ladder to a large receiving tank where they would later have their eggs taken from them.

The site geotechnical investigation had been performed by Dames and Moore. The site soils were comprised of soft- to medium-stiff dark grey organic silts with standard penetration values ranging from 3 to 9 blows per foot. The tailrace pond walls were founded on bedrock, and the gate structure was founded on timber piles. Sheet steel piles provided side channel support. The main hatchery building site had been surcharged for about six months with 5 feet of granular fill. After removal of the surcharge, the building was constructed on spread footings and a thick reinforced concrete floor slab.

I discussed the potential impact of blasting, particularly the first few rounds, for the tailrace tunnel. It was necessary to maintain the stability of the hatchery foundation soils and the surcharged loose silts under the switchyard. Other requirements included:

1. Limit water pressure changes in the raceways and holding ponds to ensure no fish mortality
2. Limit water pressure changes in the incubators to ensure no damage to the salmonoid eggs
3. Control sudden water pressure increase on the water elevation control gates
4. Control air blast damage to building components, particularly windows
5. Control the potential for flyrock

A number of steps were proposed to shockproof the facilities. It was recommended that shock-damping material such as neoprene and soft rubber be inserted under all the incubator legs. An air bubbler system was proposed to be installed along the walls of raceways and ponds nearest the blasts so air bubbles would dampen "over pressures" which could develop in the water. The perforated bubble lines had to be weighted to the floor. Air would be pumped through the lines continuously for several minutes before, during and after blasting.

A series of three small test blasts detonating 5, 10 and 15 pounds of explosives was recommended after the damping and bubblers had been installed. Multiple geophones were to be installed around the site at critical locations to monitor each blast. The peak particle velocity, air pressure and water over pressure were to be measured. Based on the results of these blasts, a blasting design and program was to be developed with the contractor. Monitoring and evaluating the blasts was to be performed by a specialist in that field.

The test program was successful, and the tailrace tunnel blasting was performed with no damage to any structures, or the loss of life of any fish.

During construction of the lake tap penstock, in late 1987, the contractor, Pacific Ventures Inc., submitted a major claim for change of conditions. Captain Doug Kennedy of the U.S. Corps. of Engineers contacted me and advised of the following claims:

1. Change of geotechnical conditions within the tunnel and in the vicinity of the lake tap plug
2. Changes in the drill hole locations for the removal of the final rock plug at the lake tap
3. Change in the tunnel alignment
4. Changes to the grouting procedures

The geotechnical information described a major fault zone about 35 feet wide that crossed the tunnel at an acute angle of 20 to 30 degrees. Shotcrete or steel sets were to be used to stabilize the tunnel section. The length of the poor conditions was expected to continue for about 110 feet. Actually support was only required for 75 feet. The contractor overblasted through the fault zone and damaged the rock more than necessary. It was concluded the conditions were in fact more favourable than the contract predicted, and the claim was rejected.

In May 1995, R & M Consultants of Anchorage, Alaska, asked me to attend the inspection of a site for development of a "run of river" hydroelectric project on the Tazimina River, near Iliamna, about 190 air miles southwest of Anchorage. The project was to provide electrical power for the largely Native population in the area, as an alternative to the existing and expensive oil-fired generation system.

The idea was to take advantage of an 85-foot water elevation differential above and below Tazimina Falls. A portion of the river flow could be diverted into a channel and penstock leading to generating turbines below the falls. No water storage dam was required.

The field inspection was made with Charlie Riddle and Peter Hardcastle of R & M Consultants, Duane Hippe of HDR Engineering Inc. and me. We flew by charter to an airstrip near Iliamna and then took a floatplane to a small lake about 3 miles from the falls. We walked, often stumbled, our way to the site. Over half the distance was muskeg terrain with a high water table. About every tenth step one foot would sink through the soft surface of the muskeg. On numerous occasions, my high-top boots sunk below the water, which then poured into my boots. After about fifteen minutes, every step was accompanied by a squish sound. Fortunately, my body heat warmed my saturated wool socks. It took nearly two hours to reach the area of the falls. On arrival at the site, I took my boots off and squeezed the water out of my socks. That felt better.

Site of the proposed Tazimina run of river hydroelectric project. The inlet channel, underground powerhouse and tailrace channel were located based on a field geologic survey. Photo by R & M Consultants, Anchorage.

We inspected the river channel above the falls and selected a preferred river diversion location. The bedrock was composed of several volcanic pyroclastic flows from the adjacent Roadhouse Mountain area. The rock was generally massive and medium hard to hard. A few shear zones were observed, as well as a few areas of weathered rock in the riverbank. A program of site structural geologic mapping and drill hole locations was developed, and the area to be surveyed and topographically mapped was sketched on aerial photographs we had brought with us. The intent was to develop an underground powerhouse, thereby protecting the generating facility from the harsh winter weather. The location for the diversion and intake structure was generally established, as was the location of the inlet and penstock, which was to be constructed in a cut and cover trench. This penstock would be about 170 feet long. A vertical shaft was proposed to contain the penstock and elevator access to the underground powerhouse, which was to hold two turbines. A tailrace tunnel was proposed from the turbines to the exit below the falls.

After the surface inspection, we all clambered down the steep rock face, assisted by a rope tied to a tree at the crest. The rope turned out to be essential for climbing down at several locations on the slope. Below the waterfall, the lower canyon slope was explored for a potential tailrace outlet portal. One requirement for the location was to select a site along the river where the rock excavation could be wasted without detriment to the river environment.

The construction sequence envisioned was initially to develop an access road from Iliamna to the site, and then to construct a site office, parking, construction lay-down area, and a disposal area for the trench and shaft excavation materials. That was to be followed by development of access to the tailrace portal area to allow construction of the tailrace tunnel.

After about four hours at the site, the return walk to the lake commenced. We were tired, and so this walk was even slower than the walk to the site. More wary of the presence of grizzly or brown bears, we were very relieved to reach the floatplane for the return to Iliamna. Reservations had been made to return to Anchorage about 5 P.M. on a scheduled flight. Unfortunately, the pilot calculated that the weight of the passengers, luggage and fuel exceeded his takeoff limit. One of us would have to stay behind. Jim Hardcastle offered to stay. He had to overnight at Iliamna and return the next morning. I was able to connect with a flight to Seattle and Vancouver that evening.

After completing a review of site conditions and a geotechnical investigation, R & M and HDR jointly prepared the preliminary location and design for the project. Based on this information, suggested design parameters for the penstock excavation, the shaft powerhouse and tailrace excavations, and the rock stabilization were developed. These included recommendations for blasting to minimize rock damage, as well as the installation of dowels, rock bolts and shotcrete for support of rock faces in the shaft, powerhouse and tailrace. The program was developed successfully.

This concept of "run of river" hydroelectric development is environmentally friendly. Numerous other similar sites are available in North America and should be considered for such development. They can be developed to serve a specific area, or the power can be directed into the regional power grid.

Alaska — wonderful scenery, friendly people, challenging engineering and great fishing. I recommend it, preferably during the warmer summer season.

6

United States

United States (1968-2003)

Two major mine geotechnical projects are described, the first being the design and development of a 600-foot-high 80° open pit slope at Cleveland Cliffs Republic Mine in Michigan, believed to be the steepest excavated open pit slope in the world. The second was a serious pit wall failure at the Mega Pit of Sante Fe Mining (now Newmont) in Nevada, a failure caused by excess blasting, excavation of the slope toe, and unacceptable delay of a permit to allow open pit dewatering. The delay by government agencies to issue a permit costs mining companies millions of dollars per year.

Other stabilization projects the author consulted on included a massive landslide at Malibu, California, where many expensive homes were badly damaged, and a multibillion-cubic-yard slide to allow a large lock to be constructed at the Bonneville Dam on the Columbia River. Both projects used very long vacuum-assist horizontal drains, a technique developed by the author with dramatically successful results. Where landslides develop and subsurface water is believed to contribute to instability, the installation of such drains is shown to effectively improve stability.

Three highway projects are described, including the design of foundations in volcanic basalt rock for a beautiful concrete arch bridge over the Crooked River Gorge in Oregon. One litigation resolution involving the Bureau of Reclamation is reviewed, emphasizing the need for the designer to familiarize himself with the site conditions.

An interesting historic project undertaken by the author involved development of a program to reopen original tunnels along the historic Columbia River Highway east of Portland, Oregon.

Chapter 6

United States

In 1968, Golder Brawner had been involved on several geotechnical projects with the Kennecott Minerals Company and Anaconda Mining. The mining community in the United States is quite close-knit, and word of new expertise or improved technology soon gets around. In addition, I had presented a paper on "The Three Most Important Factors in Pit Slope Stability" to the American Mining Institute's annual meeting in Minneapolis in 1968. In it I made a bold statement: "If you have had no failures in your open pit slopes, then you have over-designed, and the problem is you don't know how much you have over-designed." As a contractor would say, "You left a lot of money on the table!"

Staff from "the Iron Range" (an area in Michigan where a number of iron ore mining companies operate) attended that presentation. They included Cleveland Cliffs, Empire Mining, and the Hannah Mining Company. Within two weeks, Cleveland Cliffs, near Ishpeming, Michigan, called and asked me to provide a review of their open pit stability, and the design of their Republic Pit.

The waste rock surrounding the ore body at Cleveland Cliffs was extremely hard. On one side of the pit, the contact between the iron ore and the host rock dipped at 80 to 85 degrees, nearly vertical. On the other side the pit crest location was limited by a river. Cleveland Cliffs was developing a 45-degree slope in the host rock, and no instability had been observed in the first five benches. They planned to go down another 400 feet. If they could steepen the eastern slope, they could reduce waste and take the pit another 100 to 200 feet deeper. For environmental reasons, no consideration was given to moving the river on the other side of the pit.

Some rock testing had been done on the host rock. Unconfined compressive strength generally exceeded 40,000 psi, with some results much higher. The rock was massive, with few joints and no faults. If the rock quality and massive nature continued at depth, the rock would stand vertical for thousands of feet.

Michigan had no slope angle regulations, only requirements relating to safe operations, so I asked the mine manager, "What do you think the Michigan Department of Mines would think if a pit slope following the ore contact — eighty degrees — was proposed?"

He stared at me and said, "They'd scream, faint or have a fit. I don't see them ever considering it. They'd probably laugh at us."

I persisted, "I'd like to see three drill holes taken to 800 feet about 100 feet beyond the contact, cored full depth, and carefully logged for structure and any weak zones. The core should be photographed full length in the core boxes, with ten unconfined compressive strength tests per hole. With this data, I would propose classical slope stability analysis to develop a theoretical factor of safety. If the high strength of the rock is confirmed, I'm prepared to recommend the steep contact slope angle, and justify it to the state officials. There are millions of dollars to be saved. The rock is so strong that water pressures are not a problem. The critical factor would be to develop a blasting program that would not damage the final wall."

I had dinner with the mine officials that evening and we discussed the mining, safety and regulatory implications. I also proposed slope movement monitoring using slope inclinometers. They remained very skeptical. A further thought occurred to me. I said, "Would it be possible to invite the state mining regulators to come to the mine in the next day or two to sound them out. If they are adamantly against the proposal, assuming rock conditions are favourable, then a test program would be a waste of time. However, if they give any encouragement, then the program could proceed." I added, "There are many places in the world where near-vertical rock faces higher than that proposed exist." I cited Yosemite Park in California, the Chieftain Cliff near Squamish, B.C., the Matterhorn in the Alps, adding that I could develop a list of many others for the report.

The Chief Engineer said, "I'll contact head office tomorrow and discuss it with them." By nine o'clock the next morning he advised me, "I've got the go ahead."

The following morning officials arrived from the state. The existing mine plan was first presented to them. The favourable stability conditions were then described. Next the cost benefits of steepening the east side slopes were explained. Preliminary cost benefit analysis indicated each degree of steepening could save $2 million in waste removal, and allow an extra million tons of ore. Emphasis was placed on the fact that increased profit would result in an increase in state taxes. The state officials nodded and smiled at that benefit. The key question was then asked, "If Cleveland Cliffs provides a detailed rock mechanics and stability analysis based on site conditions, and develops a slope monitoring program, would the state allow us to considerably steepen that side of the pit?"

They were quiet for a minute, then the senior official asked, "Can we confer amongst ourselves before answering that?"

The Chief Engineer said, "Certainly, you can use my office."

They left and returned in about thirty minutes. The government official stated, "We've considered your request and phoned our head office. The Department is prepared to give consideration to your request provided you can justify the redesign with hard data and valid analysis. The onus will be on the rock stability evaluation and mining method you propose. Safety will be the key issue."

Cleveland Cliffs authorized Golder Brawner to proceed with the study. The investigation confirmed that mining to the 80-degree contact without any benches would be feasible, and could be done safely provided blasting near the final wall was performed carefully, using smaller than normal blasts, and all loose rock was scaled from the walls of each new face that was exposed.

The report was submitted to the state. They requested a further meeting at the mine and a formal presentation of the data. The meeting was cordial but the state officials were cautious. After all, such a steep open-pit mining slope angle had never been approved anywhere in the world. However, they did acknowledge that there were higher steep rock slopes existing naturally at many places in the world.

Following the presentation, state officials asked a number of questions. They wanted more information on the monitoring of the slopes to assess ongoing stability, and more detail was provided. Finally, the spokesman for the group said, "Based on your submission, we are prepared to authorize your proposal to proceed to a depth equal to three benches — 120 feet. Authorization to proceed deeper will depend on the experience with those three benches. Ongoing approval will depend on the safety of your program. You must also understand that any accidents or injuries will be your responsibility. The state will accept no responsibility. Should serious problems develop, the state retains the right to rescind the authorization." It is important to note the significant difference between "authorization to proceed" and "approval" in the state's view. The former assumes no approval of the design, or acceptance of responsibility.

Republic open pit iron mine near Ishpeming, Michigan. The upper slope was developed at 45°. Based on a rock mechanics investigation the lower half of the pit was developed at 80° using special blasting procedures. Tens of millions of dollars were saved by reducing excavation of waste rock. The rock was extremely hard and required special drilling techniques.

Mining proceeded with the new design. Ultimately the steep 80-degree slope reached nearly 600 feet in height! One problem did develop, however. The rock slope was so stable that Cleveland Cliffs decided to erect an underground crusher station near the bottom of the steep slope. The design involved entrance and exit portals to the facility. Rockfall from the face did not create a problem, but during the winter minor seepage froze and created icicles on the face. These would fall periodically and endanger personnel and vehicles as equipment entered or exited the tunnels. There is no doubt that a hard hat would be inadequate protection against a large falling icicle!

Cleveland Cliffs phoned me and asked if I had any suggestions. I advised, "I suggest you drape wire mesh over the entire face directly above the tunnel portals and connect a direct current generator, thereby turning the mesh into a 'heat tape' which will keep the seepage from freezing."

This project stands out as one of the major Golder Brawner successes. Within months of Cleveland Cliffs receiving authorization to proceed, we received calls from Empire Mining and Hannah Mining in the Iron Range to perform similar pit design reviews. Contracts were negotiated with each.

This sort of success has a direct benefit in consulting. The new projects were secured without going through an expensive competitive proposal process, where the potential success rate is only about one in six projects. The lesson is to serve clients so well that they become repeat clients. And they did.

A further lesson was learned. Some state mining regulations are specific *i.e.*, the final slope angle must not exceed 45 degrees. Where conditions are favourable for a steeper angle, however, a variance can usually be applied for, provided detailed investigations can justify the new design as being safe.

Rimas Pakalnis, Associate Professor in the Mining Department at the University of British Columbia, had an uncle who was a real estate agent in Malibu, California, north of Los Angeles. With research funded by Gibraltar Mines and the National Research Council of Canada (NRC), Rimas and I had developed the concept and technique of placing horizontal drains under a vacuum to lower the water table and stabilize soil and rock slopes. Experience had shown that water flow from the drains increased about five times when the vacuum was applied, as compared to flow under gravity alone.

A major landslide had developed in Malibu in 1985, extending from the Pacific Palisades Highway, along the ocean beach, and for nearly a mile up the sidehill. About one hundred houses were located in the zone of movement, many valued at over a million dollars. Some houses had been torn apart, such that you could walk right through the torn portion of the house. The volume in the slide was estimated to be 55 million cubic yards!

Rimas' uncle had mentioned the slide to him, and Rimas described the potential application of horizontal drains to bring the slide under control. The information was passed on to Malibu municipal officials, who expressed an interest.

Rimas then discussed the situation with me and I suggested that we offer to give a free presentation to the municipal staff in Malibu. Prior to any presentation, I further suggested that we should inspect the site and obtain background information such as history, geology, precipitation, hydrology, and any past studies.

The area around Malibu is surprisingly dry, with only 10 to 12 inches of precipitation per year. One situation surprised us. There was limited storm drainage in the overall area and all houses were on septic tanks. I found this absolutely startling for such an expensive subdivision in that day and age. As well, most homes had extensive landscaping which required sprinkling. These conditions added an extra equivalent of about 25 inches of precipitation per year, substantially increasing the amount of water infiltration, and raising the water table. Raising the water table would, over time, reduce both the shear strength of the weathered rock, and the stability of the overall slope. We did no studies to evaluate the original water table elevation, but I would not be surprised if it had risen 50 to 60 feet over the years due to the additional water infiltration.

An American consultant had recommended lowering the water table using wells. Several test wells were installed, but the results were only effective over a short distance. Many wells would be required, with ongoing maintenance and power costs, plus surface disturbance. The use of wells was discarded.

Our offer of the presentation was accepted. During our site inspection, the damage to the houses was observed. The uppermost cracks extended over 3,000 feet from the highway. The upper zone of the slide contained multiple tension cracks, totalling over 8 feet of movement. Numerous inter-slide cracks also existed, along with vertical scarps. Some houses were torn apart several feet, others were tipped, and several had vertical shear failures. Numerous swimming pools were badly cracked and leaking. Damage was extensive. Later on, the residents initiated a multimillion-dollar suit against the county. A number of well-known movie stars had homes within the slide area.

Rimas and I made our presentation to the municipal engineering staff and local residents. We first described how subsurface water impacts stability, then described the local hydrology and geology, and the cause of the failure. We next described the vacuum horizontal drainage system. It was emphasized that the drains would all be drilled from the inner shoulder of the highway at the toe of the slide. No work or encroachment would occur on anyone's property. They appreciated this.

The drains were proposed to be 1,000 to 1,200 feet long. This required a special large track-mounted drill rig which could exert enough pressure to overcome the drill friction and still apply adequate pressure to the bit. On average, three days were required to install each drain.

Vacuum drains installed up to 1,200 feet long along Pacific Palisades Highway to halt a major slide near Malibu, California. The movement involved over 50 million cubic yards. Many expensive homes were severely damaged. The movement developed as a result of seepage from septic tanks and landscape irrigation.

As the first drain was being drilled, and the upper saturated zone was first encountered, the return water was green and smelled like rotten eggs — obviously the result of sewage infiltration. It was most uncomfortable for the drillers. Even the passing drivers noticed it; several convertibles stopped and put their tops up. By the time eight drains were installed, the slide movement had slowed from an original 1 inch per day. This occurred with the drainage flow under gravity alone.

The holes were advanced continuously with drill casings. No soil or rock samples were taken. When the required length was reached, the drilling direction was reversed and the drop-off drill bit was left in the hole. In order to prevent the hole from collapsing, perforated plastic drainpipe was then installed inside the full length of the casing, except for the outer (nearer the road) 30 feet, which was unperforated. The drill casing was removed. Cement grout was injected between the outside of the drainpipe and the inside of the drill hole for the outer 30 feet. This allowed a negative pressure vacuum to be developed within the pipe once a vacuum pump was attached. (It is impossible to develop a complete vacuum in soil or rock.)

The drains were allowed to drain by gravity until all twenty-six were installed. Header pipes were connected to all the drains. Vacuum pumps were connected to groups of six drains. Once all pipes were connected, the vacuum system was turned on. Flow dramatically increased. Within 48 hours the slide stopped moving! In addition to lowering the water table, the seepage direction changed from near parallel to the ground surface (a direction where the seepage friction reduced stability), to a near vertical direction, such that the seepage friction effectively increased the vertical force, increasing the shear strength. There are still nonbelievers in this concept, but the vacuum drainage lowered the water table substantially and was successful.

It was strongly recommended to the municipality that storm drains be installed to collect all street, driveway and roof water, and that a collector sewer system be installed throughout the entire subdivision. In addition, it was recommended to the homeowners that they reduce sprinkling and change to dryland-type landscaping. By reducing the ingress of storm drainage, discontinuing the use of septic tanks, and reducing sprinkling, the lower water table developed by the vacuum drainage would be self-sustaining. The vacuum drainage was discontinued after about ten weeks and has not been required since. The water table continues to be monitored.

The next year I received a phone call from the U.S. Corp of Engineers in Portland, who wanted to obtain information on vacuum horizontal drainage. When I asked about the purpose of the potential dewatering, they provided the following information.

The Bonneville Dam was originally constructed on the Columbia River to develop electricity and provide some flood control. The river was a major waterway for shipping, so to allow barges and shallow draft ships to pass, a lock was installed through the dam. Other dams were gradually developed on the river and shipping on the river increased.

The Bonneville lock became a "pinch point" limiting the allowable ship or barge size on the river. Upriver locks could accommodate larger vessels. The Corp had been asked to develop a design for a second larger lock at the south end of the dam that would take bigger barges.

Inspection of the geology, topography, hydrology and geotechnical conditions revealed that a massive, old dormant landslide existed at the site, with the toe encroaching on the riverbank. The slide had been identified from air photos and was estimated to originally involve 2 to 3 billion cubic yards. The old movement extended several miles south into the mountains.

The lock design required the toe area of the landslide to be undercut about 110 feet during construction. This would reduce the stability of the toe, and the Corp was concerned that the slide could be reactivated, possibly destroying the site for the lock. They wished to explore the potential of increasing the stability of the slide by lowering the water table in its lower area. Vacuum horizontal drains were considered to be an option, and so I was approached to provide information. As in many past cases, I offered to give a seminar. The Corp considered this approach and a seminar was set up. Representatives of the Oregon and Washington Highway Departments, the Federal Highways Agency, and the Corp of Engineers attended, about forty people in total!

My presentation included the significance of subsurface water on the stability of soil and rock slopes, geotechnical conditions and their effect on drainage and stability, field investigations recommended, methods of subsurface drainage, a detailed discussion of vacuum horizontal-drain design and installation, and a review of case examples. Shortly after the seminar, I was invited to visit the site and review the preliminary design developed by the Corps. The proposal was to install drains from two different elevations. One set of about twelve drains was proposed from the lower area of the slide at ground surface, entering the slide at a 2 to 3 percent grade. Previous drilling revealed that the depth of the old slide at the lock was about 90 feet. The Corp proposed a vertical shaft down to a drill chamber about 110 feet below the surface, that is below the slide failure zone. Another twelve drains were proposed at this elevation. To monitor the water table lowering, numerous piezometers were installed in the slide area, extending several thousand feet up the slide. Drains ranged from 600 to 1,200 feet long, and all were developed to allow application of a vacuum.

I was not involved in the implementation of the stabilization. However, I was advised that the drain installation, drainage and construction of the lock were successful.

In early 1994, I received a phone call from Glen Thommen, foundations engineer with the Oregon Department of Transportation (ODOT). Glen said, "ODOT wishes to evaluate the feasibility of reopening portions of the old Columbia Gorge Highway as a tourist attraction. The original highway between Hood River and Mosier along the south side of the Columbia River Gorge was opened in 1921, after two years of very difficult construction along the rock cliffs. The rehabilitation would be to allow hikers, bikers and vintage vehicles to use the old route."

The initial construction traversed along about 1,500 feet of precarious volcanic cliffs, and included two tunnels. The construction was a major challenge in its day. The history of the Mosier tunnels in particular is fascinating, and is described in the U.S. National Register of Historic Places. The contract was let to A.D. Kern of Portland in 1919. The two rock tunnels were lined with 12-inch by 12-inch timber supports and lagging. The four concrete portals were protected with a finish of stone masonry. An observation gallery was developed outside the rock face between the tunnels.

Considerable overbreak occurred from the roof of Tunnel No. 1 prior to timbering. Loose ground was also encountered in Tunnel No. 2. After completion, ongoing

The original Columbia River Highway through the Columbia Gorge east of Portland, Oregon.

problems occurred from 1921 until 1953. A review of old records, and interviews with ODOT personnel, revealed the following stability conditions.

Cliffs up to about 500 feet high existed for 700 feet west of Tunnel 1. Serious rockfalls occurred from the portal of Tunnel 1 west as soon as the route opened, and were much more frequent during rainfall and freeze-thaw cycles. Vehicles were often damaged by falling rock, or struck rock on the highway. Reference is made to three deaths from falling rock, one between the two tunnels and two in the tunnels.

Traffic on the road gradually increased in volume and truck size, so that traffic lights were installed, and eventually one-way traffic was instituted. On numerous occasions rocks struck vehicles waiting near the west traffic lights. Reference is made to vehicles hitting and damaging the timber lining. One gas truck caught fire inside the tunnel. The damaged timbers were never replaced.

On several occasions wire mesh was attached to the rock face west of the tunnels in an attempt to control rockfall. This proved unsuccessful, as the mesh filled with rock, tore loose and collapsed on the highway, narrowly missing two vehicles. ODOT attempted to scale loose rock from the face on two occasions. This was only partially completed over the high rock cliffs, and rockfalls were only reduced slightly.

During one period of ongoing precipitation, a flagman was placed at the west end of the cliff to watch for rocks coming down. He would tell the motorists when it appeared to be clear, and instruct them to "go like hell." Maintenance crews fre-

The original Columbia Gorge highway grade and tunnel was closed due to dangerous rock conditions. An investigation and project was developed to rehabilitate the route for hikers, bicyclists and motorcyclists. The west approach experienced serious rockfall.

quently cleared rocks off the highway.

After two severe rockfalls in 1953, one that broke a headlight and another that tore off the running board of the car of the local mayor, the road was closed to traffic, which was then diverted to the new freeway under construction along the river below. Following closure of the road, a number of hikers were injured by rockfall in the tunnels. This resulted in a decision in the early 1960s to fill the tunnels with talus rock and sand. The materials were placed on the grade and pushed up to near the roof. Filling was done from east to west.

In 1981, Tenneson Engineering Corporation was retained to determine the feasibility of reopening this section of old highway. The report concluded it was feasible, but that the impact on the local communities would be insignificant. The procedures to rehabilitate the tunnel area were not developed.

Following a request by the "Friends of the Columbia Gorge" in the early '90s, ODOT was directed to develop a program to preserve and restore the continuity of the Historic Highway for the Columbia Gorge National Scenic Area. The HNTB Corporation was retained by ODOT to evaluate the redevelopment of the Mosier Tunnel and west approaches. On the instruction of ODOT, HNTB subcontracted the rock and tunnel stabilization evaluation to Brawner Engineering Ltd.

The area was inspected by Glen Thommen, Larry Pierson and Randy Davis of

ODOT, along with myself. Nearly a full day was spent walking along the old highway, inspecting the rock cliffs and talus slopes below, and the most easterly and westerly portal areas. Numerous photos were taken. Many old reports were reviewed, as well as ground and air photos. ODOT obtained a copy of the original drawings. Geologic maps were also obtained and reviewed, as were climatic conditions and the earthquake history of the area. Nearing the end of the site investigation, I said to Larry Pierson, "Larry, the project really should be separated into two quite different programs — rehabilitating the two tunnels and developing a safe alignment under the rock cliff at the west end and between the tunnels. What do you think?"

He answered, "That's a reasonable approach for the evaluation and design. Both, however, could be dealt with under one contract."

"Sure, that would be feasible," I concurred.

While we were photographing the west portal, a rock fell from the cliff and crashed onto the pavement behind us. It broke into several chunks, and one rolled over the road and down the slope toward the railway below. The railway had constructed a rock catch berm and a wire-mesh fence to control such rockfall. They were successful and the rock did not reach the tracks.

As we made our way under the rock cliff to the car, Larry laughed a bit uneasily and said, "Keep one eye on the cliff above, one eye on the rock-covered pavement, and be ready to run!" Amen. We made it okay.

Photos taken during the original tunnel construction showed some overbreak, but generally stable rock faces. Roof support was required due to the potential of ongoing rockfall, but these rockfalls were likely be the result of rock damage by excess blasting rather than poor rock conditions. One potential concern, nevertheless, was that weak interbedded layers between subsequent lava flows might exist in the tunnel roof, leading to an unstable roof. The lava flows themselves are very hard basalt rock and provide good roof conditions.

Our recommendations for tunnel rehabilitation included one suggesting that the tunnels be excavated in short increments, leaving the lower third of the tunnel filled to provide a working platform closer to the roof. The timbers were to be removed first. The roof should then be scaled of loose rock and shotcreted with about 3 inches of steel fibre reinforced shotcrete with 5 percent silica fume admixture. This admixture improves cohesion, and increases the strength and durability. For safety during construction, both scaling and shotcreting could be performed with remote-controlled equipment. The lower fill would then be removed and the tunnel walls stabilized.

The development of a safe roadway below the steep rock cliffs was a very different problem, and several options were reviewed:

1. Stabilize the rock slopes above
2. Develop a rock shed
3. Tunnel inside and parallel to the cliff face

A bench existed along the cliff about 200 feet above the road. Option (1) involved developing a catch bench at that location to intercept rock falling from the upper cliff. Vehicle access would be provided to this bench to allow it to be cleaned as required.

The lower cliff face would be completely scaled of loose rock. Being on the south side of the valley, there were many freeze-thaw cycles, and so such scaling would be required about every five to six years. Alternatively, wire mesh could be draped over the rock face. This drape allows the rock to ravel behind the mesh and drop into the inner ditch, where it can be removed. Long-term maintenance is required.

Generally drier conditions prevail from May to September in the area, and rockfall is substantially less during low precipitation periods. If this option was selected, it was recommended the road be closed during winter when wet or freeze-thaw periods were prevalent, and rockfall much more frequent.

A structural engineer would design the rock shed option. The roof had to be strong enough to resist a rock size up to about 3 cubic yards, without deflecting rock onto the railway below. To disperse the impact, the roof would be covered with 3 to 4 feet of sand. Foundation supports were to be developed for the outside shoulder wall footings. Rock removal from the roof would be required every eight to ten years.

Both options 1 and 2 were expensive. The most expensive but safest design, requiring the least maintenance would be a tunnel 40 to 50 feet in from the face. About every 100 feet a lateral tunnel to the face could be developed to ensure ongoing ventilation of the tunnel.

Option 2, the rock shed, was chosen, and the tunnels were rehabilitated under the supervision of ODOT. No difficult problems were encountered so I was not called. This project was, nevertheless, very interesting, particularly due to its historic significance.

I was contracted by the California Department of Transportation in the fall of 1998 to present a seminar on "Rock Slope Stability for Transportation Projects." John Duffy, Caltrans Geologist from San Luis Obispo, contacted me regarding the arrangements. The state had limited funding for training, but Duffy wanted to have about sixty state engineers, geologists, foremen and technicians attend. I suggested they invite a number of contractors and consultants and charge them each $200. With this strategy, he developed the necessary funding.

The seminar was presented twice, to two separate groups. At the end of the second day, Cliff Tillotson, President of TCI Construction Inc., came up and asked if he could speak to me privately about a project. I agreed and said, "Wait a few minutes until I answer the questions of the attendees."

After about fifteen minutes, Cliff and I went outside on the veranda, and he said, "We're a rock stabilization and environmental construction company. We recently completed a project for the Bureau of Reclamation to stabilize a rock slope above the parking lot of Folsom Dam, outside of Sacramento. The rock conditions were much different than what we expected, and we didn't recover all of our costs. Do you ever assist contractors with claims for extra costs?"

I said, "Yes. My approach is to review the circumstances for the potential claim, and then advise the contractor whether I consider there are reasonable grounds for extra payment. Sometimes I don't agree with the contractor, and I advise him accordingly. Can you describe the circumstances briefly?"

Cliff described their problems. "The rock was presented as being competent, and the slope, up to 200 feet high, was to be stabilized by installing tensioned grouted pattern bolts. However, we encountered many very soft zones and voids which made it impossible to develop the required tension in many of the bolts. Many anchors slipped. We had to redrill for many bolts, install double anchors, or change anchor types. We also had to move our crane much more frequently than planned because we were directed to stabilize many rocks where the bolts didn't coincide with the design pattern. We had expected to drill and install bolts in a vertical line perpendicular to the slope, so that only one crane location would be required for each line. Instead up to ten crane moves were required per line."

I said, "It sounds like you have reasonable grounds for a claim. I'd want to see the contract, your summary of costs and payments, and visit the site. I can't do it this trip due to other commitments, but I could be available in about two weeks. Could you in the meantime document the basis for the claim, and send it to me before my site review?" I also described my fees and costs.

Cliff said, "Good, I'll contact you in several days regarding timing, and arrange for the documentation you request."

I added, "We should meet with the Bureau at that time. It's desirable to obtain their initial reaction and attitude toward your basis for claim. They may be rigid, or they may be prepared to entertain a claim. Can you arrange that meeting?"

Cliff said, "Okay, I'll look after that."

Two weeks later I flew to Sacramento, where Cliff met me with his superintendent and their wives. We drove out to the site.

Folsom Dam is a concrete dam that stores water for domestic, agricultural and hydroelectric development. It's about 200 feet high, with a tailrace that directs water into the original narrow canyon-like river channel below. A major rock excavation up to about 220 feet high had been developed to provide storage space and a parking lot. The dam provides tours for interested parties. On one occasion a large rock fell from the cliff and demolished a parked car. Fortunately no one was in the vehicle. That incident resulted in the decision to stabilize the rock slope.

The slope is comprised of granite rock, which generally appeared hard and competent, and scratching the rock with a geology pick seemed to confirm that. Looking visually from below, however, areas of very weathered rock were not noticeable. The contract documents included photographs of the slope and a summary of some dip-and-dip direction data that had been obtained from surface geologic mapping. No subsurface information was provided. The rock was described as "competent granite with occasional weathered zones," a general statement.

That evening, after the site inspection, we had a pleasant dinner together and talked generally about the dam. Cliff's wife, Renee, was the office manager and in charge of accounts. She was very friendly and intelligent, with a great smile that made everyone feel comfortable. She had an excellent grasp of the project's financial problems.

The strategy for the next morning's meeting with the Bureau was simple: indicate that as a result of rock conditions that were unexpected, TCI had suffered a substantial loss on the project. TCI had completed the project to meet the specifications, and had fulfilled the requirements of the project. The items which were to be claimed for were then summarized briefly, and the rationale given for each item.

At the meeting, TCI introduced me and presented my credentials. I gave my confirmation of conditions not described in the contract, and then asked the Bureau, "We're prepared to submit a claim for extra payment, what is your process?"

The Bureau was represented by five individuals, including the project site engineer and the geologist who had inspected the site prior to developing the contract. They both were firm in stating that they believed the problem was due to inefficiency on the job. The senior Bureau representative was more conciliatory, and indicated the Bureau policy was to entertain claims which were submitted with backup details and information.

I asked, "Did TCI meet the project requirements?"

He said, "Yes, the work is satisfactory, but the project over ran our time estimate."

A contract officer who would decide the claim also attended. He was congenial and confirmed they would review the claim, saying, "Please ensure you provide financial details, time sheets, and equipment rentals in detail for all aspects of the claim. It is not the desire of the Bureau to have to settle contracts in court."

On the way to the Sacremento airport to drop me off, we discussed the next steps. These included:

1. Summarize the worksheets for all personnel, equipment and supplies, cost them and develop the project costs. Include all administration and overhead costs. Develop the total project cost and compute the project loss.
2. Prepare a summary of evidence of change of conditions, and related project problems and delays.
3. Prepare the claim with costs — include an allowance for profit.
4. Request Bureau project documents, including the Resident Engineer's diary for our review.
5. Brawner to supply an external technical report to back up the claim.
6. Arrange for a meeting to present the claim.

I reviewed the project documents that TCI sent to me, and concluded that the following problems had occurred:

1. Numerous voids and decomposed granite layers were encountered in the slope. Over one-third of the anchor zones encountered problems.
2. Numerous holes collapsed prior and during installation of the bolts.
3. Oversized holes developed in the decomposed granite requiring extra grout, and creating unacceptable anchorage conditions.
4. Drill rods became stuck due to inability to blow all cuttings from the holes.
5. Special anchorage systems had to be developed; some required double anchors.

6. Many holes were abandoned, which required new holes.

It was concluded that a significant change of conditions occurred which the contractor could not anticipate from the contract documents, or from the site inspection. It was noted that the Bureau refused to adjust the design requirements to accommodate the adverse rock conditions.

I prepared a report outlining the basis for the claim. The conclusions were as follows:

1. Inadequate geotechnical site information was provided by the Bureau. This was a surprise after the Bureau's experience of failure at the Grand Teton Dam and the Corps of Engineers failure at the Libby Dam.
2. The site conditions encountered were not expected, based on the documents and site inspection.
3. Spot bolting replaced pattern bolting, requiring many more equipment moves, with no price adjustment.
4. The Bureau refused to adjust the bolting requirements to match the site conditions.

I concluded by stating, "If the Bureau is not prepared to pay extra based on a change of conditions, I recommend TCI initiate court proceedings."

In early 1998 I attended a further meeting between TCI and the Bureau. On the way to the meeting, I suggested to Cliff and Renee, "You should have a bottom-line dollar amount in mind that you're prepared to accept. If the meeting goes well and the Bureau is responsive, they may offer a settlement today."

Cliff looked surprised, "Really?" Then he continued, "We've never submitted a claim before. This is all new to us, since we've made a profit on all our past projects."

What a refreshing statement. I concluded that in the past they had bid carefully and accurately, provided a job well done, and left a satisfied client. Indeed, the company has an excellent reputation in the western United States.

TCI presented its claim in detail. The original Bureau staff members were present, except that the contract engineer was different. On several occasions the resident engineer and geologist raised objections or criticized the claim. Following the claim presentation by TCI, I reviewed the technical merit of the claim. I then asked the geologist, "Would you please describe the design process the Bureau used for the project?"

He hesitated, but was requested to answer by the contract engineer. "I spent time inspecting the surface of the slope and measuring the orientation of exposed joints. I also took a number of photographs of the rock face. No drilling was done. This information was forwarded to the Denver office to their rock mechanics engineer, who then developed the stabilization design and specifications."

I asked, "When did he inspect the site?"

The geologist answered, "He didn't come to the site."

There was quiet for a few moments. The significance slowly set in. Finally I stated, "I want to confirm what you said. You performed a site surface inspection, which included rock identification, joint mapping and photographing the slope. You sent this information to a rock mechanics engineer in Denver who developed the design and specifications. He did not visit the site to inspect rock conditions. Is that a correct summary?"

He slowly answered, "Yes."

The Bureau staff obviously realized that a "paper" design had been developed. Not to have visited the site prior to developing a design to stabilize a dangerous rockfall site is inexcusable. A judge would crucify the Bureau staff, and they must have realized it.

I said, "Do you realize the legal consequences of a paper design?"

The Bureau engineer said, "Let's break and come back after lunch."

On our return after lunch, I was asked not to attend the meeting. I suspected the Bureau wished to settle, and so only wished TCI present.

About an hour later the meeting broke up. As we met in the hall, Cliff and Renee were all smiles. She said, "They wished to settle and we've agreed on an acceptable price." As is usually the case, the amount is not made public.

It was a happy ride to the airport, and since that event my wife and I have visited the Tillotsons in Santa Barbara. A great couple and a new friendship.

In 1994, a proposal had been developed by the Montana State Department of Highways to upgrade Highway 2 near Kalispell from two lanes to four lanes along the Flathead River, at a location referred to as Badrock Canyon. The alignment was very restricted between the river on the outside, and a major cliff and state park on the inside. In addition, two major power lines existed on the top of the rock cliff.

Robert Peccia & Associates of Helena, Montana, had received a contract to evaluate the highway design options through the canyon. They had retained Joe Armstrong, retired State Materials Engineer, to assist them with the geotechnical considerations. Joe appreciated that the major rock excavation required might seriously impact the rock stability. He contacted me and asked, "Chuck, Peccia Associates is putting together a team to upgrade the highway through Badrock Canyon near Kalispell for Montana Highways. Would you be interested in taking the lead on the rock stability and slope design for the design team?"

I had enjoyed working on several projects with Joe in the past in Montana. This project was in a beautiful area near Glacier National Park. I indicated my interest, "Joe, I would be available and look forward to working with you again."

One of the great benefits of consulting engineering is the opportunity to trade stories and experience with other engineers. It is amazing how much new technical knowledge one gains from others during these professional associations. There are thousands of great experiences and engineering successes that are never recorded in the technical literature. It is not uncommon for engineers to contact others who might assist them to better understand an engineering challenge they are faced with, and when they do, great friendships often develop. This is what has occurred between Joe and me. In a way, it is like developing your own "experience Internet."

A field program was performed to map the structural geology. The rock was comprised of bedded limestone and argillite. The bedding dipped from 18 to 35 degrees to the east, and varied from 0 to 30 degrees from the strike of the proposed centreline. There had been past movement on some of the beds, hence the shear strength would be low, near its residual value. The movement had an angular component toward the proposed highway slope, and a prominent set of near-vertical joints was nearly parallel to the highway.

The rock itself was hard and generally unweathered. The rock face at the highway was near vertical, about 80 feet high, stepping back in several benches to about 200 feet above the highway. From the roadway it looked like a very steep slope could be developed. Such a design would allow negligible infringement on the river. However, widening the highway cross section by about 50 feet would require a rock back slope nearly 200 feet high, and cut slightly into the park boundary.

There is a very visible and active environmental lobby in that part of Montana, and they had succeeded in stopping highway widening at nearby Columbia Falls only a few years before. While Joe and I were inspecting the rock conditions along the highway, an old car stopped and a woman got out. She was about thirty years old, had long braided hair, wore a well-filled plaid work shirt and a pair of faded blue jeans, torn on one knee. She had a pretty face, but one of the coldest pair of eyes I had ever seen.

Joe saw her coming and suggested to me, "Don't talk to her. She's the local activist leader."

I turned my back toward her, but she approached me and demanded, "What are you doing here?"

The way she said it got my back up, so I took the offensive. "I'm here to advise the state on how to widen the highway through the canyon," and then I emphasized, "to improve the safety!"

She wasn't phased at all and stated, "I've already stopped one project and I'll stop this one too. I'll have you know I'm a paralegal and know how to stop destruction of our parkland and beautiful scenery!"

I responded, "Well, you're obviously not up on the law. Several years ago a rockfall in Colorado killed eight passengers in a Grayline Bus. The U.S. government acted very quickly and established the *Title 23 U.S. Highway Code* and the *1991 Intermodal Surface Transportation Act* which states, it shall be the National Policy to bring all Federal Highway Systems up to standards and to increase the safety of these systems to the maximum extent. Highway facilities must be developed to meet existing and probable future traffic needs in a manner conducive to safety, durability and economy of maintenance."

She stared at me, and I continued, "I will advise Montana Highways officials of your threat. You don't seem to be aware that if you stop a project and an accident occurs where loss of life or serious injury results, you can be held personally liable!"

She continued to glare, took a deep breath, then turned and left.

I have never understood why governments are so hesitant to take strong legal action against radical environmentalists. I believe in protecting the environment, but when it endangers lives, or costs many jobs, that's another matter.

Joe and I spent several hours inspecting the rock along the highway. I had learned many times that you also have to look above the highway for signs of potential instability, so I said to Joe, "I want to get up top and get a better overall picture of the site conditions. Where can we rent a helicopter?"

He answered, "There's a sightseeing chopper company near the park entrance."

I said, "I'd like to rent it for about a half an hour to do a detailed aerial inspection and get some good air photos. Can we drive there now and see if we can arrange it? The weather is good."

He agreed, "Sure, let's go."

The helicopter was on a pad near the highway. We went into the adjacent office and asked, "Can we rent the helicopter for half an hour to do an air inspection of Badrock Canyon and take some air photos?"

"Yes, I'm free now," the pilot offered.

"What's your charge?" I queried.

He answered, "Four hundred dollars, plus fuel."

I agreed to that, and after we were airborne the pilot said, "Tell me where you'd like to go and at what elevation."

I was in the left seat and pointed a bit to that side and said, "We'd like three

Major tension cracks along the west side of Badrock Canyon near Kalispell, Montana. The state proposes to four-lane the highway without encroaching on the river. Realignment into the rock requires a stepped slope design to guard against a toppling failure.

passes, low level and slow, high level, and then circle at midlevel. I'll tell you when we're at the right elevations."

As we came up on the site, I was surprised to see four major tension cracks almost parallel to the highway on the top of the cliff. One went beyond the first power line. I asked the pilot to drop down lower and slow to a walking speed. The cracks looked about 5 feet wide. I said to Joe, "The entire cliff has undergone some toppling. We'll need to determine if the movement is active."

We took many pictures out of the side window at three elevations, and several more as we passed parallel to the cliff, flying about 50 feet off the river. I also took several stereo pair photos.

On our return to the helipad, the pilot spotted a grizzly bear about 100 feet behind a barn. The pilot circled and the bear looked up, then scrambled into the bush. It was a golden brown; the pilot guessed about 600 pounds in weight.

Upon landing, we went back to the site and walked up an old trail where we could climb to the cliff top from behind. There we inspected the tension cracks. They were 4 to 6 feet wide, and debris had fallen into them, indicating that toppling had occurred many years ago. No active cracks were observed.

A field structural-mapping program and crack-monitoring program were later developed in the field, which confirmed the toppling was no longer active. I speculated that a glacier had filled the Flathead Valley about 12,000 to 14,000 years prior. This provided high pressure against the cliff. As the glacier melted, stress relief occurred, water pressure developed in the cracks and the joints of the major joint set parallel to the highway opened up.

To develop a wider cut for the highway, a design was developed with three benches, one between each tension crack, and each located at the quarter elevations of the cliff. Special requirements were developed for controlled blasting whereby the contractor would be required to retain a blasting consultant. Winter construction was recommended due to the very heavy summer tourist traffic. Also, construction in rock during cold weather is very practical.

The project is awaiting final permits and funding.

In August 1993, Jerry Peterson of Terracon Consultants in Boise, Idaho, contacted me with a request that I provide them with rock slope engineering assistance on a project in southeast Idaho. The project was to upgrade 14 miles of winding highway from Montpelier to Geneva.

The existing two-lane highway, constructed just prior to 1940, had numerous sharp curves and steep grades. Rock cuts up to 100 feet high were envisioned for the upgrade. The rock types involved limestones, shales and sandstones, ranging from very weak and weathered to hard. Numerous rock exposures existed along the route. In many locations bedding, joints and faults dipped out of the rock faces, which would have a major effect on stability and the rock slope design.

The new design was to provide a 50 mph design speed, and maximum grade of 6 percent with passing lanes. The project extends from the 5,977-foot elevation near Montpelier, to 6,943 feet at the summit. Winter conditions are severe, with up to 8 feet of snowfall.

A site inspection was performed with Dennis Williams, Senior Geologist with Terracon. He had extensive geological experience in Idaho and was a great help. He also had a great sense of humour. We spent three days inspecting and mapping rock-slope faces and outcrops above the highway. Air photos were also inspected.

Two large old rockslides were located. At other locations high rock cuts would be required to improve the alignment. Major rock excavation at the toe area of the old slides could reactivate the movement. Relocation of the highway in these areas warranted serious consideration.

In order to develop the review report many other details were obtained. These included:

1. Climatic data.
2. Further air photos, including colour photos and infrared, if they were available. These are useful in locating shallow water courses and subsurface seepage.
3. Water well drilling records.
4. Seismic history.
5. Regional geology reports.
6. Past rockfall history from highway maintenance staff.

In addition to the surface mapping of the rock faces, a large number of drill holes were required, and so their locations were selected. Rock core was to be obtained and mapped for structural geology, rock type, lithology, hardness and RQD (Rock Quality Designation). In order to evaluate the shear strength along discontinuities and fault zones, instructions were provided to obtain and prepare samples for direct shear tests. This information is required to perform rock slope stability analysis. In addition, piezometers were to be installed to determine the depth to the water table. Many site photographs were taken to include in the final report and contract documents.

In my report to Terracon, the following specific recommendations were provided:

1. Preliminary slope angles for each rock cut. This angle was dictated by the rock conditions. There is no reason for all rock slopes on a project to be the same slope angle.
2. Ditch design to catch random rockfall. This was based on the modified Ritchie criteria, developed for the state of Washington.
3. Specifications for scaling of loose rock.
4. Stabilization procedures and specifications for the installation of rock dowels, rock bolts, water depressurization drains and shotcreting.
5. Specifications to minimize blast damage to the rock slopes for long-term stability.
6. Require that a pre-bid conference be held with all potential contractors in attendance.
7. Require that a geologist be assigned to the project to assist the State Project

Engineer during construction.

Based on this report, as well as aerial and ground topography surveys, the project geometry requirements, realignment potential in the areas of the rock slopes, and base gravel sources were to be determined. Test-hole drilling and soil testing were also to be performed.

Terracon developed a final geotechnical report and submitted it to the state. It was recommended that a site inspection be made with all senior project participants attending, including representatives from Fisheries, Environment, and design and construction engineers from the state.

A major issue developed in the area of the lower old rockslide. The state had designed a relocation alignment that crossed the creek twice in the centre of the valley, and encroached to a major extent on the south valley wall. Nearly one-half million cubic yards of rock excavation was proposed. I had recommended to Terracon and the state that a retired experienced contractor attend the meeting to discuss the actual construction of the highway as to sequencing, materials balance, construction equipment, traffic control and detour, *et cetera*. The contractor retained was David Hilde, the former owner of a major construction firm with extensive experience in Idaho and Montana. His input was excellent. In particular, he pointed out that to balance quantities, the large volume of rock excavated near the lower slide would require hauling uphill for about 6 miles. This would cost 4 to 5 million dollars, and was not warranted economically. The design near the slide was reviewed and modified.

This experience confirmed my conviction that geotechnical and highway projects should always be reviewed for practicability. While in the B.C. Highway Department, I had said on numerous occasions that design engineers should build what they design. More practical and economical designs would result.

The project is presently underway. I have made two field construction inspections to the site, one to evaluate a slide in soil activated by undercutting a slope, and one to evaluate a small rockslide area. In general the project is developing according to plan.

In September, 1994, tension cracks were observed up to about 150 feet behind the west highwall of the Mega Pit of the Twin Creek Gold Mine near Winnemucca, Nevada. Originally two separate open pits had existed at Twin Creek — the Chimney Creek South Pit and the Rabbit Creek Pit. Over time, Sante Fe Mining purchased both properties and gradually developed the pits into one large open pit, hence the name Mega Pit.

Dean Peltonen, Chief Geologist, contacted me and asked, “Could you come to the mine in the near future to look at the cracks and assess stability? We have active mining in progress directly under the cracked area.”

I replied, “I can be there next Monday. Can you fax the latest geology and mining sequence by this Friday?”

He answered, “Yes. And we’ll look forward to seeing you.”

On my arrival, I met with Grant Dye, Chief Engineer, and Bill Faust, Manager of Mining Operations. The following information was reviewed and discussed:

1. Pit plan with a plot of the surface cracks
2. Geologic cross sections
3. Location of movement hubs and mirrors
4. A dewatering program
5. Blasting plan — April to September
6. Precipitation records

The geology of the site comprises hard caliche (cemented sand and gravel) to a depth varying from 200 to 400 feet. This zone overlies weak, clayey siliceous siltstones and shales. The contact between the two zones dipped toward the pit below a portion of the slope. A thrust fault dipped at a shallow angle into the slope near the existing pit bottom. The water table was above the toe of the slope. Several steep dipping shears existed which angled about 30 degrees to the face, and exited the pit slope at the north end of the cracked area.

The open pit slopes were designed by other engineers. The slope face in the upper caliche was 50 degrees, and in the bedrock below 45 degrees. This angle in rock was steeper than the original design. The total pit depth was about 650 feet.

Several deep water wells had been developed in the area to supply mill water and road maintenance requirements, plus lower the water table. One horizontal drain had been installed previously just south of the movement zone. It was flowing at about 50 gallons per minute.

Blast holes were 9 inches in diameter. In the final line they were 9 feet apart. In the caliche and soft rock no explosives were placed in this line. In all other holes 85 to 90 pounds of explosive were used. Blasting involved up to 400 holes. Prior to the cracks developing, the last blast, nearest the toe area of the slope, was a confined blast, totally below the existing pit floor. The seismic shock forces due to such a blast are several times greater than if the blast had been toward a free face, which is recommended practice. This confined blast may have triggered the cracking beyond the crest.

The factors that contributed to the cracking were considered to be the following:

1. The large confined blast at the toe of the slope
2. The sloping dip of the caliche — bedrock contact
3. The low strength of the siltstone — shale bedrock
4. Water and seepage flow pressures in the lower slope
5. The steep shears in the north half of the movement zone

At the time of my inspection, a new blast pattern had been drilled at the toe, but the holes had not yet been loaded with explosives. Based on my discussions with mine staff, and a review of documents and plans, I prepared a report that contained the following recommendations:

1. Establish a pit slope movement-monitoring program using an EDM (electronic distance measurement) instrument to monitor distance to prisms established down

three cross sections of the pit wall. Contact me if movement accelerates. Cease mining if movement reaches 1.5 inches per day and vacate that pit area.

2. Fill in all cracks and construct a surface water interception ditch beyond the cracks to minimize surface water entering the slide area.
3. Do not load the drilled bench area at the toe of the slope, or excavate that bench.
4. Within 150 feet of the slope toe, limit blasts to 50 holes per blast and delay every hole .
5. Install 10 horizontal drain holes 250 feet long below the movement area. The perforated drains should be 60 feet apart. Install perforated plastic 2-inch inside-diameter pipe with no perforations in the outer 30 feet. Place concrete grout in this outer section to fill between the walls of the hole and pipe so that a vacuum can be applied to augment gravity drainage. Experience has shown that an applied vacuum increases flow from the drains about 5 times. This will lower the water table and increase the shear strength in the slope.
6. Call me if the movement begins to accelerate.

I had a heavy work schedule the next several months, but on December 21 I received an urgent phone call from Bill Faust. He stated, "The movement is increasing significantly and now totals over one foot! Can you come in the next day or two?"

I was astonished — more than a foot of movement and I hadn't been called sooner. I answered, "Bill, that's serious. I'll be there tomorrow. I'll fax you later about my arrival time."

I always carry the *OAG Pocket Flight Guide* for North America in my briefcase. It lists all the airline flights between cities in North America. I took this out and looked up the flights to Reno, Nevada, from Vancouver. In 1994 there were no direct flights except charters for gamblers. I had to route via San Francisco. I made up my own flight schedule and phoned my travel agent, Mike Barwell, at English Bay Travel. Flights were arranged to get me to Reno by ten o'clock that evening. I overnighted in Reno, got up early and drove three hours to the mine via Winnemucca, arriving there about 10 A.M.

On my arrival I went immediately with Bill Faust and Don Walton, Mine Engineer, to the pit. Several of the cracks near the crest had opened about a foot. One crack could be followed down the face of the slope. We then drove to the bottom of the pit. Seepage zones were observed at numerous locations along the lower bench face, and water had ponded along the toe of the slope. I didn't see any new horizontal drains.

I asked Bill, "Why haven't the horizontal drains been installed?"

He answered, "We applied for a permit immediately after your last visit, but it hasn't been received yet!"

A three-month delay! I asked, "Were they aware that the slope stability depended on the drainage?"

"We explained the reason in the application. We've found the permitting process often very slow."

This has become a serious dilemma for the mining industry. Between thirty and forty permits from a variety of agencies or government departments are usually required for a mine to go into production, a process that usually takes several years. I am aware of several projects that took four years to permit. The permitting staff don't seem to realize that feasibility studies, environmental studies, economic studies, permit applications and reports are usually paid with borrowed money. Permit delays cost mining companies millions of dollars per year. I'm surprised the mining companies don't resort to court action for a remedy.

Some excuses for delays or refusal are quite unreasonable. An EPA (Environmental Protection Agency) officer once said to me, "Cost considerations are your problem. We don't take cost into consideration." On another occasion, a Fisheries officer threatened in a meeting, "If you kill one fish, your permit will be cancelled!" Often heard from North American regulators is, "In your reclamation program, you must use native species." I wonder if they realize palm trees are not indigenous to Hawaii.

We went back to the mine office, where I asked to see the slope-movement graphs that were available. The movement vs. time charts for prisms in the upper slope showed the same pattern. Reasonably uniform movement occurred from September 23 to about November 15 of approximately 0.08 inch per day. From November 15 to December 12 the rate increased to about 0.13 inch per day. Beyond December 12 the rate increased to 0.8 inch per day.

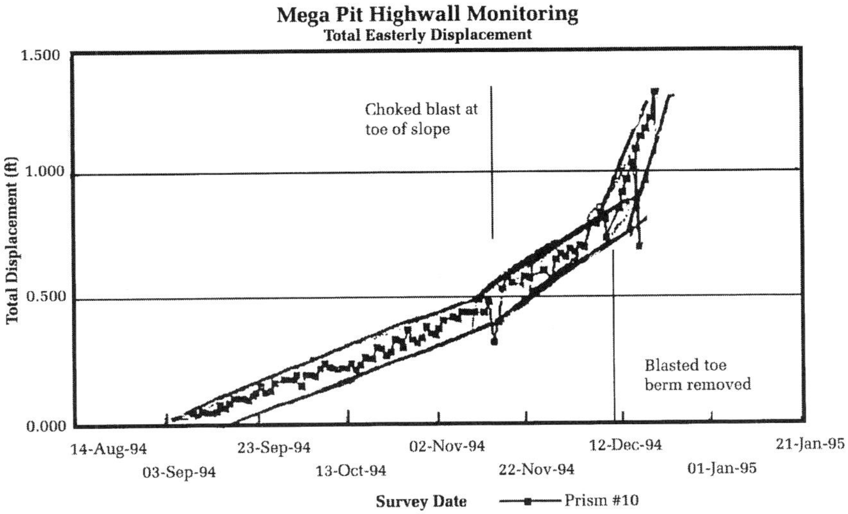

I asked Don Walton, "Was any work performed near the toe about November 15?"

He looked at the mine reports and said, "A blast was detonated at the toe."

I asked him, "Any mining around December 12?"

He again looked at the mine reports and answered, "Yes, the blasted bench was dug out."

Each increase in movement velocity was caused by a mining event at the toe of the slope, first the bench blast, and later excavation of the bench. Both actions I had advised them not to do in September! I looked again at the graph. I said, "Don, the movement is accelerating sufficiently fast that I believe the slope will fail within several days. I strongly recommend you move all equipment out of that portion of the pit as soon as possible," and I emphasized "today!"

By late afternoon all equipment except one pump was removed, and the lower access road was closed. The operation was closed for Christmas. I returned to Vancouver late that evening.

Mega Pit northeast of Winnemucca, Nevada. A major slide developed which was monitored during mining. The failure was predicted and the mine was evacuated three days prior to failure. The failure occurred in weak shale at depth with high water pressures.

On December 27, I received a phone call from Walton. He said, "The slide came down over Christmas closure. No one was here. The automatic slope movement instrumentation indicated major movement late December 23, and total failure December 26." He asked me to visit the mine in mid-January to assist in planning cleanup of the failure.

I made a further visit on January 12 and 13. A strategy was developed to allow the debris to consolidate for another two weeks, after which time removal of the slide debris from the bottom of the pit could be commenced.

The upper scarp from the slide was near vertical for about 200 feet, too steep for safety. It was recommended that this slope should be flattened about 20 degrees. Long-hole angle drilling and blasting was proposed. This blast was designed, drilled and all holes detonated simultaneously. It was very successful, and the new slope developed to the design angle.

A program to "back analyze" the slide was also recommended. With this data, a new slope design incorporating a setback of the crest was developed and implemented.

The Oregon Department of Transportation planned to construct a new four-lane bridge across the Crooked River Canyon, near Bend, Oregon. In 1996 they commissioned me to review the site investigation, rock coring and rock testing done by ODOT, and provide recommendations for foundation design for the concrete arch structure. The structure was designed by DGES Consultants of Olympia, Washington, and featured an arch span 440 feet long. Senior designers were David Goodyear and Norm Smit. Jon Heacock was the project manager for ODOT.

The site geology was comprised of inter-canyon and plateau basalts typical of this region of Oregon. The rock mass is moderately hard to hard (R3 – R4). At some locations open joints, vesicles, near-horizontal interflows and open voids were encountered. Oregon DOT performed an extensive site investigation, including eight boreholes. The core was not oriented (relative to the canyon wall), but a laboratory investigation included unconfined compression tests and point load tests on selected core. Core recovery, RQD, joint conditions and core description were logged. The core was photographed, and retained for inspection by the contractor.

For a 500-year earthquake occurrence at the site, the peak rock acceleration is 0.07g from a magnitude 7.75 subduction-zone event and a magnitude 5.75 crustal event. Based on the drilling results, it was apparent that the north-thrust block rock conditions were less favourable than for the south-thrust block. RQD values at the north ranged from 53 to 93 percent, and for the south-thrust block area, 84 to 100 percent. The recommended allowable design bearing values were 80Ksf for the south-thrust block and 50Ksf for the north block. The north area also had more voids and cracks. As a result, foundation consolidation grouting was considered necessary.

Rigid arch structures are sensitive to lateral differential settlement. Based on the recommended loads, the foundation settlement was estimated to be 13 to 20mm, with differential about one-third the total. The settlement would include excavation load recompression, joint closure and elastic compression. To assess settlement, it was a requirement that settlement readings be taken at each foundation corner during construction. The front faces of the thrust blocks were not to be closer than 16 feet to the cliff face.

At the initial site meeting with representatives from several departments, the Fisheries Department representative said, "You will not be allowed to put any blasted rock in the river."

I interjected, "That's not practically or economically feasible. Extensive rockfall from natural causes continues to fall from the canyon walls. Some remains on the lower talus slopes but some, including occasional very large ones, roll or bounce

North-arch thrust block and abutment under construction. Special blasting was required to minimize blast rock entering the Crooked River in the bottom of the canyon. Near Redmond, Washington.

Crooked River Canyon concrete arch near completion. A beautiful structure at a dramatic site. An environmentally successful project. Design by DGES, Olympia, Washington. Photo by DGES Consultants.

into the river. With controlled blasting, the amount of blast rock falling into the river can be limited. Any large blocks which enter the river can be broken up, or if they are a hazard, they can be removed when the bridge is completed."

Considerable discussion followed. Fisheries gradually modified their position to allow some rock to enter the river, particularly during periods of no spawning or fish runs.

A pre-bid conference was held at the site and all proposed bidders attended. Major attention was paid to the blasting and rock excavation requirements, and special blasting procedures were specified. The contractor was required to retain a blasting specialist. To minimize damage to the rock, specifications included pre-split blasting along the final walls of the excavations for the arch abutments. Maximum lift depth was 15 feet and the total excavation depth 75 feet. Blasting patterns were developed to lift the rock and not throw it toward the river. Blast hole spacing and delays were designed to limit the size of the blast rock. It was estimated that no more than 10 percent of the blasted rock went over the cliff face, so that very little rock reached the river.

Most rock was excavated with a backhoe that was lifted in and out of the excavation as required by a large crane. The backhoe loaded the blasted rock into a large bucket, which was also lifted out with the crane. The rock faces were bolted and meshed for construction safety and long-term stability.

The north foundation imprint was grouted to a depth of 15 feet with various cement grout mixtures. Grout pressures were kept low so as to not "lift" the rock. More grout than expected was required in about 20 percent of the holes. Two core holes were drilled to confirm that grout had filled most cracks and voids. About 27,500 cubic yards were removed for the thrust block foundations, but through skilful blasting and rock removal, the river was successfully protected.

This example emphasizes the need for controlled blasting to reduce rock damage, and the need for foundation grouting where voids or cracks are prevalent.

In 1993, the U.S. Federal Highway Agency (FHWA) sponsored the preparation of a *Rockfall Hazard Mitigation Manual*. The manual was conceived by Ron Chassie, FHWA Geotechnical Supervisor, and contracted for preparation by me. In the U.S., all funding for freeways, and about 90 percent of funding for primary highways, are subsidized by federal funds. However, most of the design, construction and maintenance is performed by the individual states. I prepared a questionnaire to determine state policy and procedures, and sent it all to the state highway departments. Twenty-four replied that they had rock problems on highways in their state. Hence, there were twenty-four different standards. One of the purposes of the manual was to develop standardized practice for rockfall mitigation across the United States. It was to be readable by highway maintenance foremen as well as engineers.

Following the responses to the questionnaire, I arranged a meeting of state geotechnical engineers from several states to discuss the contents of the manual. Those invited were:

- Ron Chassie, Senior Geotechnical Engineer, FHWA, Portland
- Barry Siel, Geotechnical Engineer, FHWA, Denver
- John Duffy, Geologist, California Department of Transportation
- Rich Andrew, Geologist, Colorado Department of Transportation
- Alexander Yatsevitch, Engineering Geologist, New York Department of Transportation
- Richard Cross, Geologist, New York State Thruway Authority
- John Ledbetter, Engineering, North Carolina Department of Transportation
- Steve Lowell, Chief Engineering Geologist, Washington Department of Transportation
- Larry Pierson, Chief Engineering Geologist, Oregon Department of Transportation

Two chapters that the panel wanted to include caught me by surprise. They were "Aesthetics" and "Legal Implications." A big issue for environmentalists was the conviction that aesthetics should be the number-one issue, rather than safety. Fortunately, the U.S. Congress had recently extended the Transportation Act to emphasize that safety was to take precedence over all other issues. This change to the Act resulted from a rockfall that hit a tour bus in Colorado and killed eight tourists.

To prepare the chapter on "Legal Issues," I contacted the Attorney General's departments in six states for a statement on each state's liability policy concerning loss of life or damage to vehicles due to rockfall.

As part of the contract, I was required to present seminars on rockfall mitigation at about thirty locations throughout the U.S. to engineers, supervisors and foremen. It was a busy and challenging year!

7 Australia

Australia (1972-1999)

The Victoria State Electricity Commission has one of the largest combined coal mining and electric power developments in the world, mining over 60 million tons of brown coal each year with bucketwheel excavation at Morwell and Yallorn. Major problems developed at these mines, including pit bottom heave and pit slope failures which damaged mining equipment. Dewatering, using wells and horizontal drains, was designed by the author to improve stability and control floor heave. Specially located wells were recommended to lower the water table and level the adjacent town, which had settled differentially over 5 feet. Extensive movement monitoring was required to evaluate stability and settlement.

Serious problems developed at the Blackwater and Goonyella coal mines in Queensland. Slope stabilization and modification of blasting procedures were introduced. The stabilization involved alteration of the blasting program to reduce pit wall damage, excavation of a weak clayey coal layer prior to placing dragline spoil, and drilling basal depressurization holes to stop pit bottom heave. Rock slope design and stability problems were investigated for the railway from Townsville to the Freeport Nickel Mine in Queensland.

Major instability occurred at the Hammersley Iron Mine where the haul road location undercut structural discontinuities which dipped into the pit at angles steeper than the friction angle along the bedding. The haul road should have been located on the other side of the pit where the geologic structure was favourable.

Deep-sea fishing in New Zealand with the author's wife created an exciting change of pace. He landed a 290-pound marlin after an hour of hectic manual exercise.

Chapter 7
Australia

At Golder Brawner, I made a practice of reviewing the overall consulting engineering workload near the end of each week. This allowed me to better allocate staff to projects, and to determine whether the workload was increasing or reducing. If it was reducing, the plan was to develop a strategy to look for more work.

"Sandy" Sanderson, former bridge engineer of the B.C. Department of Highways, had established his own consulting practice. I provided consulting on several of his bridge projects. During lunch with him after an engineering meeting one day, he made a statement that I have never forgotten: "In the consulting business, you have to look for additional work while you're still busy. If you wait until things get slow, you'll be too late. It generally takes about three to four months to develop contracts."

To review the overall project picture, I decided to meet with my Associate partners twice a year over a weekend. We would go to a resort within two to three hours' driving distance from Vancouver. The group included Bob Wilson, Norman McCammon, David Campbell, David Pentz, Glen Gilchrist and our wives. Harrison Hot Springs was our favourite resort.

The typical program involved a Friday evening dinner followed by dancing. The music suited us — mainly of the 1950s and '60s era. On Saturday morning, the ladies met and arranged their own day program. From about 9 A.M. until 5 P.M. Saturday, my Associates and I discussed a wide range of corporate matters, and developed new policy wherever we considered it necessary. On Saturday evening we again enjoyed dinner and dancing. On Sunday we slept in, then went for a leisurely swim and hot tub. Later we went golfing in the summer, or curling at the local rink in the winter, or went sightseeing locally. We left for home in midafternoon. In those days such corporate events could be charged as business development. Not so today.

In early 1972 my review of the pending workload revealed a downturn, in line with an obvious slowdown in the North American economy. I set up an urgent partners' meeting, with three subjects on the agenda: the sequence of staff layoffs, associated consulting opportunities and expanding our business development to other areas.

Developing a sequence for layoffs was difficult. The majority of our staff had been selected carefully, and our rating of each individual tended to differ. We finally developed a system to rate each individual staff member. This was a painful exercise, one I never want to go through again.

The second item — other consulting opportunities — only took about an hour to deal with. The only specialty that we considered we might expand into was environmental engineering. We listed potential environmental areas to consider, including cleanup of contaminated soil, garbage and landfill technology, ground water contamination, and closure of industrial projects. It didn't take us long to realize we would have to hire one or more senior specialists in those fields. This would pose a problem because we weren't familiar with these fields and would have difficulty in determining terms of reference. Without this familiarity, how could we evaluate performance? We put the idea on hold. (Many other consulting geotechnical engineering companies subsequently did expand into this area, and the Golder group ultimately did also.)

Thirdly, we discussed trying to develop work in other areas — Southeast Asia, Australia, South America or Europe. We had already broken open the United States market, and Golder had other offices in Boston and Atlanta, so that was not considered. After several hours of discussion, we selected Australia. It was decided that I would first contact the Australian Association of Mining Engineers.

The following Monday I prepared a letter to the president of that Association, outlining our experience in geotechnical mining engineering, with emphasis on case examples in rock mechanics and rock slope stability. I offered to give lectures or seminars to their members at a location or locations of their choice. At the time, the Canadian government had a program to pay travel expenses for foreign business development. If a project developed, the company repaid the government upon completion of the project. Accordingly, I stated that we would pay my expenses. I hoped for an answer within a month.

The day after my letter was received the president telephoned. He said, "I have your letter offering lectures on rock mechanics in mining. I've spoken to all our major branches in Australia and there is significant interest. We'd like you to give lectures to our Association branches in Brisbane, Sydney, Melbourne, Adelaide and Perth. We'll arrange for all your expenses in Australia, send out invitations to our members, arrange the lecture location and provide for cocktails afterwards. How soon can you come?"

I thought for a moment and suggested, "I'll phone you tomorrow; what is your number?"

I was flabbergasted! What an opportunity. I called the partners together. It was decided that I would go in four weeks. I called back and advised of my timing.

For the next three weeks I had colour slides and overheads prepared presenting a number of interesting and challenging case examples. I also intended to review the most important concepts of rock mechanics as they applied to surface mining. My intent was to lecture with no notes, but to use slides and overheads as props. There is nothing more boring than a speaker who reads his presentation.

On my arrival in Sydney, I was taken to the Wentworth Hotel — excellent accommodation. I was advised that they had not only arranged for a lecture to the Association in the evening, but for me to speak to the mining students at the local university in the afternoon. Those lectures were fun, as the students were very interested and asked many questions. My other lectures were held on Monday in Sydney, Wednesday in Brisbane, Friday in Melbourne, and the following Monday in Adelaide and Wednesday in Perth. I then spent the weekend in Perth. I consider Perth to be the most beautiful city in Australia, and have enjoyed returning there several times.

The atmosphere in each of the five cities is vastly different. In my opinion, Sydney is wide open, a combination of San Francisco and Las Vegas. Brisbane is cosmopolitan, low key and not unlike Portland, Oregon. Melbourne is artistic, proper, formal, and hosts the Melbourne Cup horse race, with an atmosphere similar to Lexington, Kentucky, although Melbourne is a much bigger city. Adelaide is a desert city, somewhat comparable to Phoenix, Arizona. Perth is the flower city, clean and sedate, but when it's America's Cup sailing time, the excitement builds.

The lectures were a great success, and the exposure to the mining companies was excellent. I stayed another week to meet with mining executives, and I returned home with four major mine stability contracts — Hammersley Iron Mine, Western Mines, Hail Creek Coal Mine and the Goonyella Mine, plus a contract with Gutteridge Haskins & Davey, consultants to the Freeport Nickel Company, on a new railroad from their mine to Townsville, in Queensland. As a result of this work, instead of laying off staff, we had to hire additional geological engineers and technicians.

Several weeks after returning from Australia, I received a telephone call from Doug Stewart, Chief Engineer of the Victoria State Electric Commission in Melbourne, Australia. He informed me, "We're having stability problems at our Morwell and Yallorn open pit coal operations, and we wish to interview a number of consultants to potentially assist us in solving the problems. I'm sending Ken Lamin and John Alexander, senior engineers, to interview consultants. Would you be available to see them in two weeks in Vancouver?"

I replied, "Yes, that would be fine. Please advise us of their travel plans and I'll meet their airplane. In the meantime, could you send information describing the stability problems as soon as possible, so I can study them before they arrive? The reason for this is to familiarize myself with the problems. I may have encountered them elsewhere."

The information arrived in about one week. The problems were threefold. The first was serious heave of the base of the pit which had tipped a large excavator onto its side. The second involved large slabs of coal sliding off the excavated face, hitting and damaging two of their large bucketwheel excavators. The third was large settlement of up to 5 feet occurring in the portion of the town closest to the pit. I had encountered all these problems before, the first at the Iron Ore Company of Canada (IOC), near Schefferville, Labrador, the second at a pit near Estevan, Saskatchewan, owned by Saskatchewan Power Authority, and the third related to settlement caused by dewatering.

When Lamin and Alexander arrived, I spent the entire afternoon with them reviewing photographs, plans and summary reports. They indicated that their Board of Directors had instructed them to interview four or five international geotechnical mining consultants, and then select one to investigate and develop solutions to their stability problems. They did not want to retain a university group.

The author demonstrating a high-pressure triaxial cell used to test rock. The equipment was portable for field testing of core samples of coal.

I asked who the other consultants were. (In preparing proposals, it is very useful to know who your competition is.) Lamin answered, "Bechtel in San Francisco, D'Appolonia in Pittsburgh, the National Coal Board in England, and Rheinbraun in Germany." These were extremely formidable competitors, except for the National Coal Board, which dealt primarily with underground coal mining. They shouldn't have been a competitor.

I reviewed the mine stability experience of Golder Brawner internationally, and introduced them to my senior rock mechanics Associates. I showed them our testing laboratory, which we were very proud of. Later I arranged to have dinner with them at Trader Vic's Polynesian restaurant in the Bayshore Hotel.

The next morning I took them for a tour of Vancouver, including the Grouse Mountain skyride. Fortunately it was a spectacular day. Looking over Vancouver and the harbour from 4,000 feet was fabulous. It is a beautiful city.

Early that afternoon I drove them to catch their plane, feeling confident that we had made a good practical impression. I emphasized our experience in dealing successfully with similar problems, and gave them references to contact. The next day I sent a letter thanking them for their interest in Golder Brawner, enclosing some technical information and a summary of our experience.

I heard nothing for many weeks. I had all but given up on the project when, late one afternoon, some two months after the interview, I received a telephone call from Doug Stewart. He advised me that they had selected us, and asked if I could come to Melbourne in the near future to meet with them, visit and inspect the mine, review the problems and provide a preliminary assessment, to be followed by a further investigation. I was elated. The timing was excellent as I could integrate the trip with work at the Hammersley project in West Australia.

Two weeks later, I arrived at about eleven in the morning after an overnight flight via Honolulu. Lamin and Alexander met me at the Melbourne airport. They drove me to my hotel and said, "We'll pick you up tomorrow morning about eight o'clock."

I had slept reasonably well on the plane, and hadn't yet felt the effects of the six-hour time change, so I said, "If it's okay with you I'd like to have a shower, change my shirt, have lunch with you, and work this afternoon."

They were taken aback. Australians don't do that. After an international flight, they take a day to adjust to the new time and location. However, they reacted favourably to the workaholic approach.

During the early years of my international travels, I had difficulty sleeping on airplanes. Then, by chance, on a return flight from Zambia, Africa, I picked up a book on self-hypnosis in the Montreal airport. I tried to teach myself, via hypnosis, to sleep on airplanes, but it didn't work. I still felt interested in the process, however, so I discussed it with my doctor. He recommended a physician who used it in his practice. After four, one-hour sessions with this doctor, I learned to go to sleep in very noisy conditions. The process involves becoming completely relaxed, at which time you readily drop off to sleep. It has made multiple time-zone travel much easier.

I had lunch with Lamin, Alexander and their general manager, Bob White. After lunch I spent about four hours going over plans and reports with mining engineers, design engineers and geologists. The project was enormous. They were mining about 60 million tons of coal annually and burning it to develop thermal electricity for Melbourne and most of the state of Victoria, a population of 4 million people. In addition to slab failures from the open pit slopes, and the pit bottom heave, the town of Morwell, with a population of about 10,000 adjacent to the pit, had settled about 5 feet nearest the pit, and 1 foot on the far side of town. This had caused many of the sewers throughout the town to flow in the wrong direction. Multiple pumps had to be installed to move the sewage. Fortunately, the differential settlement had spread quite uniformly over about ten blocks, so major structural damage had not occurred.

On Thursday and Friday, I met all of the senior office staff and spent most of the time reviewing and discussing reports and plans. The Morwell-Yallorn coal belt contains over several billion tons of brown low-grade coal with a high moisture content, and the Morwell coal seam itself is about 450 feet thick, near horizontal under some 50 feet of silty overburden. Annual precipitation in the area is about 25 inches, with most falling during the winter, May to August, and originally the water table had been near the ground surface. A dense clay layer 10 to 15 feet thick underlay the coal, followed by 14 to 20 feet of very clean sand. Below the sand, another coal layer existed, some 200 to 250 feet thick. Mining had extended to a depth of about 400 feet below original ground. The boundary of the city of Morwell was within 1,000 feet of the pit crest.

I spent Thursday with the Chief Geologist. He indicated to me that he believed the pit bottom heave, near the centre of the pit, was the result of pressure from the two sides of the pit, causing the coal to buckle upward. The distance across the pit was about 2,000 feet. I considered his theory unlikely, given the distance from the toe of the slope, but said nothing.

On Friday I met with engineers from the Civil Department. One young engineer, Adrian Brown, particularly impressed me, so much so that I later hired him to come to Vancouver. He had only recently completed his master's degree in Geohydrology (the relationship between surface water, groundwater and geology), and said that he believed the sand below the coal was an aquifer (a zone of pervious material filled with water in the voids) under pressure. If this pressure in the sand layer exceeded the weight of the coal remaining above the aquifer, the pit bottom coal would rise up due to the unbalanced pressure. I considered this to be a reasonable possibility, but it would be necessary to drill below the bottom of the pit to determine the precise water pressure in the sand. I was surprised that the mine staff had not already done this.

I had been advised that I would be taken to the mine on Monday, spend two days there, then return to Melbourne. They wanted me to spend Wednesday alone to evaluate the information I had reviewed and the field conditions I had seen. I was requested to prepare a review of my preliminary interpretation of conditions for a presentation to senior management on Thursday. At that presentation, I was to review the cause of instability, present preliminary recommendations, and develop an ongoing program for a definitive stabilization program.

On Friday afternoon the Chief Geologist asked, "Are you a golfer?"

I answered, "Yes, I enjoy golf, but I don't play frequently enough to be very good."

He said, "I belong to the Melbourne Golf Club, and I'd like to invite you to play with me tomorrow afternoon."

I replied, "I'd enjoy that."

Morwell open pit coal mine owned by the State of Victoria Electricity Commission. A 450-foot-thick near-horizontal deposit of brown coal is being mined by bucketwheel excavators.

He picked me up at my hotel at 1 P.M., and we drove to the golf club. On our arrival, he took me into the lounge and ordered two beer! Beer before a golf game? I was the guest so I didn't disagree. And besides, the Australian Foster's beer is good.

After the beer, we went to the clubhouse and he arranged a set of clubs for me. The course was challenging and in very good condition. In spite of the beer, I was only one down coming to the 18th hole. He informed me, "If I win my

A planar failure, on a joint dipping about 50° into the pit, covered the bucketwheel and tipped the machine, which was badly damaged. Structural geologic mapping was implemented to locate similar joints. Where they were encountered, the pit slope was excavated to the joint angle.

friends will be happy."

I asked, "Why will they be happy?"

He answered, "I told them I was playing a good golfer from Canada."

Australians are great gamblers; they will gamble on anything. Suspecting, I asked, "Your friends are betting on our game?"

"Yes," he said.

Up until now it had been a fun, no pressure game. Suddenly it was Canada against Australia!

I had won the 17th hole, so I had to tee off first. Canada against Australia. I teed up with my driver, took a deep breath and swung. I hardly saw the ball. It hooked a bit but stayed in the fairway, 220 yards away. My second shot landed in light rough off the edge of the green. He landed on the edge of the green. I chipped to within five feet of the pin and sank the putt. He two-putted. Both pars, so the game ended with me still down one. Australia won! My consolation — I had been a good guest and enjoyed the beer.

One thing did bother me though. About every third hole the Geologist gave me a replay of his buckling theory for the pit bottom heave at the mine. I wondered, "Was I being set up?"

On Monday morning the Chief Design Engineer, Don Holmes, drove me out to the Morwell operations. The Morwell valley is very attractive, with many small sheep and cattle ranches. Low mountain ranges frame both sides of the valley, and monster eucalyptus trees highlight the landscape.

At the open pit offices, I was met by Ken Lamin, Operations Manager. After being introduced to other mine management staff, I was taken on a tour of the pit. It was imposing, about 1 mile long and 2,000 feet wide at the base elevation, and nearly 400 feet deep. Seven mining benches had been developed. Eight large bucketwheel excavators were digging coal, developing 50- to 70-foot-high mining faces at 70-degree slopes (called batters by the Australians). Each

Pit bottom heave due to water pressure in a regional sand aquifer which exceeded the weight of coal in the pit bottom. One similar heave tipped a bucketwheel excavator. Morwell Coal Project.

A typical dewatering well installed to reduce subsurface water pressures to control pit bottom heave. The water was corrosive and warm which required special well and pump design.

bucketwheel excavated coal and deposited it on a conveyor, which carried the coal to a central storage pile. From there it went to a briquette plant to be squeezed to reduce the water content of the coal, such that it would burn more readily. Fortunately the coal had a very low sulphur content; hence acid rain was not a problem. Farmland, cattle, sheep and the town exist up to the edge of the mine property without noticeable environmental problems, except for settlement.

I was taken to a location where heave of the pit bottom had occurred. The heaved area was about 15 feet high, with numerous cracks around the heave. Water was flowing from some of the cracks. I asked Lamin if a drill was available to drill down into the sand to monitor water pressure.

He answered, "Yes. We can have it set up by noon. Tell us where you want the hole drilled."

We got out of the vehicle; I walked around the heave and selected a location near the centre and set a stake. He radioed the request to the mine foreman.

There was fine coal dust on much of the pit floor. During the summer, the "in pit" temperatures could reach 120°F, meaning that spontaneous combustion and fire were a constant concern. To reduce the risk of fire, the slopes (batters) were sprayed periodically with water. In addition, all equipment used in the pit, including personnel vehicles, was undercoated to prevent sparks.

In one instance, several years previous, two fires started about the same time. The mine had only one fire truck. One fire was extinguished, but the second quickly developed into a major fire engulfing about one-third of the pit. Water and fire trucks from the entire region were commandeered to fight the fire. In addition, the Australian air force was called in to drop water and fire retardant. At one stage they were preparing to direct the adjacent Morwell River into the pit. Suddenly, and thankfully, the wind direction changed about 180 degrees and the fire was blown back on itself. Had this not occurred, they would have lost the pit. Had they lost the pit, that region of Australia could have been without electricity for at least several months — a major disaster.

Adrian Brown reviewed the geotechnical and hydrological conditions at the mine for me and presented his interpretation of the cause of the pit bottom heave. I found him to be very knowledgeable and thorough.

At three o'clock we received word from the drill crew that they had reached the clay zone and were nearing the sand layer. We left the office for the pit.

At the drill site, we looked at samples of the coal and clay. Both were partially disturbed by the heaving. The driller said, "With the next run [5 feet], we should hit the sand." After about another ten minutes of drilling, water suddenly gushed up through the drill rods. Within several minutes it spurted about 15 to 20 feet above the drill. I asked the driller to case the hole and install a shut-off valve and pressure gauge. An 80-foot pressure head was measured!

We had seen enough. There was no doubt that aquifer water pressure exceeded the weight of the coal and clay above the aquifer and had caused the heaving. The coal only weighed about 72 pounds per cubic foot — pcf. The next question was to determine whether the aquifer was confined, with little recharge, or widespread and recharging. For the former, pressure relief wells would reduce the pressure rapidly and no pumping would be required. On the other hand, if the aquifer was extensive, pumps would be required in the wells.

Tuesday afternoon we drove back to Melbourne. On Wednesday I was given a room to myself to prepare my presentation for management the next day. I developed my summary notes on overheads for projection on a screen.

The presentation to management took place in a conference room on the Executive floor. Five members of the Board of Directors, the Manager, Chief Engineer, and representatives from the Civil and Geology Departments attended. John Alexander chaired the meeting. After I was introduced to those present, using the overheads, I described the mine's geology and hydrogeology, its history of instability, my interpretation of the causes of the failures, my recommendations for stabilization, plus some further investigations that I considered were required. I concluded that the high aquifer water pressure had caused the heaving. This all took about two hours, after which I sat down.

Alexander asked the Chief Geologist if he had any comments. He was a big man, about sixty years old, and very experienced in coal geology. He stood up, looked at me, looked around the room, looked at me again and said, "I disagree with Mr. Brawner's theory." Actually he said it in stronger words than that, but I won't repeat them.

You could have heard a pin drop! I don't know how long there was quiet but eventually the chairman said, "Mr. Brawner, have you anything else to say?"

I didn't think but simply said, "Mr. Chairman, you gave me two hours to present my interpretation. I suggest you give the Chief Geologist the same time to present his case."

Everyone looked at the Geologist. Instead of asking for time to prepare his interpretation, he stood up and began to present his assessment of the pit bottom heave. He obviously wasn't prepared, and had no information with him. He rambled for about ten minutes, then sat down. Sweat had beaded on his forehead and his complexion had gone grey. Alexander closed the meeting.

I felt sorry for the Geologist, but it taught me a lesson. Never present an opinion without facts to base it on, and be very careful about stating you are 100 percent certain. Murphy has an uncanny way of creeping into any picture.

After my return to Vancouver, a pump test program was developed to determine whether the aquifer was confined or being continually recharged. The program involved two pumping wells and six water-pressure monitoring holes. It revealed a very extensive aquifer with its source in the adjacent hills. As a result, twelve large-volume, long-term dewatering wells were dug. Special pumps and pipes had to be used since the water was highly corrosive.

Shortly after developing the dewatering program, a slide occurred at the Morwell pit along a joint that dipped 55 degrees into the pit from one of the operating faces. The planar-type failure occurred rapidly and buried the bucketwheel, bent the wheel arm and tipped the machine. Reduced loss of production occurred for many months. I was called back to Morwell to evaluate this failure.

The mine had not been effectively mapping the pit slope geology for joints, so it was recommended to them that an ongoing geological mapping program be developed. The manager instructed the Geology Department to institute the necessary program. Where joints were located that would dip out of the slope in excess of the angle of friction on the joint, about 30 degrees, that face would have to be excavated to the joint dip angle. This unfortunately reduced productivity of the bucketheels, but was necessary for safety.

Concern was also expressed that the toe of the overall batters was getting close to the weaker underlying clay, and excavation to that level could destabilize the entire slope. Piezometers indicated a relatively high water table, partly due to the slope sprinkling. Some tension cracks had already appeared beyond the pit crest, between the town and pit limit. If instability developed, the cracks could extend into the town and toward the administration buildings.

It was recommended that horizontal drains about 300 feet long, spaced 50 feet apart, be drilled into the toe of the operating slopes, with perforated plastic pipes installed in the holes to drain and lower the water table. The Australians have a tendency to select a more conservative approach than designed. They subsequently installed many of the drains 500 to 600 feet long. Being a public agency, cost did not seem to be a major parameter.

To evaluate the ongoing stability of the slopes, slope indicator casings were installed in vertical drill holes and cemented into the holes, which extended into the lower coal seam. The casing has four grooves at 90 degrees, down which a sensitive instrument with an internal pendulum is lowered. The angularity of the borehole is measured at 1- to 2-foot intervals and integrated the full length of the hole. The initial readings are assumed to be vertical, and all subsequent readings are related to this datum. The instrument is very accurate: relative lateral movements in excess of about 0.05 inch can be picked up at depth. Prior to initial readings, the casing must be checked to determine if it has spiralled or twisted. In one hole at Morwell, the casing twisted about 120 degrees.

The slope indicator instrument had been developed by the Slope Indicator Company in Seattle many years previously and was fully tested and proven. I recommended the Australians purchase it. However, they had a mechanically minded geotechnical engineer who decided they would design and build their own.

Back in Vancouver, the first set of instrument readings I received showed the bottom block in the slope moving more than the upper portion. One failure mechanism at Morwell involved blocks moving near horizontally on coal-clay contacts, but I considered this type of movement to be unlikely. If you ride a horse, you must move as the horse moves. I phoned Morwell and asked them to check their instrument. Instead they retained a university professor to show mathematically that their data could be correct. Upon receiving his analysis, I disagreed with it, and again phoned and asked them to check the instrument. I received no answer.

About three weeks later I was in Melbourne and asked about the instrument. They had not checked it. I spoke to the Chief Engineer and General Manager, stating strongly that the readings were wrong and they must check the instrument. The engineer was ordered to do so.

The next day I was advised that the accelerometer in the instrument had been installed backwards! When it was corrected, the readings showed the upper portion of the slope was moving and the lower portion was not — as should be expected. Australians can be stubborn.

The dewatering due to mining had lowered the water table about 350 feet under the edge of the town nearest the pit. Under the far side of town, however, about a mile away, the water table had only been lowered about 100 feet. The variable water table represented a typical drawdown profile toward the open pit.

The coal at Morwell is slightly compressible. By lowering the water table, the weight of the coal increased by the buoyancy reduction of the water (62.4 pcf). Hence the drawdown of 350 feet had increased the total pressure in the lower coal by 350 x 62.4 pcf, or about 12 tons per square foot. The 100-foot lowering at the other end of town increased the load by about 3.5 tons per square foot. This differential loading caused settlement of about 5 and 1.5 feet respectively. The entire town surface topography had tipped by about 3.5 feet!

I recommended that the mine install about six dewatering wells under the centre and far edge of town to lower the water table more uniformly. This would cause the town to slowly level out! The final elevation would be about 5 feet lower than original, but the sewers would drain as designed.

The mine dewatering resulted in a lowering of the water table as much as 20 miles away, to the edge of the aquifer. Geodetic benchmarks within this area had to be relocated.

About one month after returning to Vancouver, I received a phone call from Charles MacArthur, President of Utah International Mining Company in San Francisco. He advised that they had some major cracks which had developed beyond the highwall of their Blackwater open pit coal operation in Queensland, Australia. He said, "We don't know their impact on stability in areas where we plan to move our big 60R drill and 60-cubic-yard dragline. Would you be able, in the near future, to visit the mine and advise us?"

I said, "Yes, I would; I'll phone you tomorrow regarding timing."

One of the great problems in specialty international geotechnical engineering is the frequent long distance travel and efficient use of the available time. Because of my extensive presentation of technical papers, articles, lectures and experience around the world, the client generally demanded that I come personally. This played havoc on my available time. Gradually I developed the approach of saying to the client, "Yes, I will come and I will be bringing Mr. X, my Associate, along with me." I did not ask to bring him; I made it a condition of my coming. If they were going to demand my time, I was

Excess blasting in the opening cut developed serious cracking beyond the highwall. Multiple delays were incorporated in a revised blasting program. Blackwater Mine, Queensland. Brian Stimpson, Associate partner of Golder Brawner and Ian Sutherland, Chief Engineer.

going to set some of the conditions! After that first trip, in which we met and they gained confidence in my Associate, I would then arrange for him to play a major role in the project, and thus reduce some of my travel.

I also learned that when you bring an Associate you must ensure that he contributes effectively to the discussions. You can't sit him in the corner just to listen; it is important that he contribute. You also have the responsibility to pass on as much knowledge to him as possible. You should be training him to be a partner, or even your replacement some day.

This time I chose Dr. Brian Stimpson, a graduate from Imperial College, London, England, to come with me. We arrived in Brisbane and were flown on a chartered aircraft to the airport nearest the mine site. Ian Sutherland, Chief Engineer, met us and drove us to the mine.

We spent some time reviewing the mine site conditions, geology and mining procedures. Following this we inspected the mining area. An irregular crack about 1 foot in width had developed back from the mine crest for about 100 feet. The coal deposit, at a depth of about 100 feet, was covered with horizontally bedded siltstone and sandstone. No major vertical joints were observed in this material, indicating that the cracks were not related to the geology. I considered that they were likely due to excess blasting. I asked Sutherland to arrange for me to meet with the blasting engineer.

He said, "We don't have a blasting engineer. We have a blasting foreman."

I said, "May I speak to him?"

He said, "Yes," and arranged for the foreman to come to his office.

After introductions I asked the foreman, "Would you describe your drilling and blasting procedures." He advised as to the depth of boreholes, spacing of boreholes, pounds of explosives per hole, *et cetera*.

I then asked, "How many pounds of explosive did you detonate for the last three blasts?"

"About 350,000 pounds, 400,000 pounds and 450,000 pounds respectively," he replied.

I asked, "How many delays did you use?"

He looked at me with amazement and sputtered, "Delays?" There had been none! This meant all blast holes detonated near simultaneously. You didn't have to be a rocket scientist to imagine the massive explosion and energy release. To top it off, the majority of the blast was below the water table, increasing its impact.

Blast holes can be delayed, such that a specific number of holes are detonated at one time, followed by other holes. This reduces the seismic energy (shock force) at any one time, minimizing damage to the adjacent mining wall. No more than 5 holes should have been initiated per delay at the mine during each blast. As many as 200 holes per blast could be detonated at this site, provided only 5 holes were detonated at a time, about 25 to 50 milliseconds apart. Near the final walls the energy must be further reduced, so the final wall is not damaged. It is also important to take into account that seismic shock is greater below the water table, and where the blast is fully confined, as for a drop cut.

I concluded that excess blasting energy was at fault, and advised them accordingly. I then worked with the blasting foreman to modify the design, which included allowing only 5 blast holes per delay. Again they reacted very conservatively, delaying every hole.

Upon my return to Brisbane, I contacted the local seismic office and asked if, during the past two weeks, they had any record of earthquakes in the area of the town near the mine. The seismologist returned in about ten minutes and exclaimed with surprise that there had been three earthquakes with Richter ratings of 2.8, 3.0 and 3.1. Brisbane was 500 miles from the mine! The heavy blasting was creating small earthquakes!

To this day I have found that most mines use excess blasting energy, which damages the final mine walls. This results in the ultimate use of flatter mine slopes than if controlled blasting was used. The mining industry is penalizing itself hundreds of millions of dollars per year by using poor blasting design and practices near the final walls. Typically, at these mines, more emphasis is placed on production than on pit wall stability.

On my second trip to the Blackwater Mine, we were flying half an hour out of Brisbane on a charter. Our pilot had the cabin radio receiver on so we could hear the airway communications. An unidentified pilot radioed in, "Calling Brisbane tower. Can you give me the time?"

The air traffic controller, who did not ask for any identification, answered, "If you are QANTAS, it's ten hundred hours, if you are ANSETT, it's ten A.M. and if you are a bush pilot the big hand is on the twelve and the small hand is on the ten." Australians can also be informal.

While I was at Blackwater, an adjacent mine, Goonyella, contacted me and asked if I could look at their stability problems — not problem, but problems. When I had completed my work at Blackwater, Goonyella staff picked me up.

Their mine was a mess! The highwall, excavated by a large dragline, had many slope failures, some of them covering up coal that had been exposed by the dragline. The waste pile placed on the opposite side of the exposed coal was sliding into the excavation, and when the 30-foot-thick coal seam was excavated, the pit bottom heaved. This further disturbed the waste pile and the highwall stability.

My field inspection, review of the geology and hydrogeology, geotechnical data and blasting procedures took one day. The next morning I reviewed my interpretation of the conditions with the Goonyella staff and provided my recommendations. That afternoon I prepared a written summary.

The highwall instability was caused by excessive blasting forces. I developed a modified blasting pattern and delay design to minimize instability. The waste piles were sliding on a 1-foot-thick layer of weak clayey carbonaceous coal which underlay the sought-after coal. I recommended that, as soon as the coal was mined, this layer be removed to expose sound sandstone below. The waste pile would be stable on this stronger material due to higher friction.

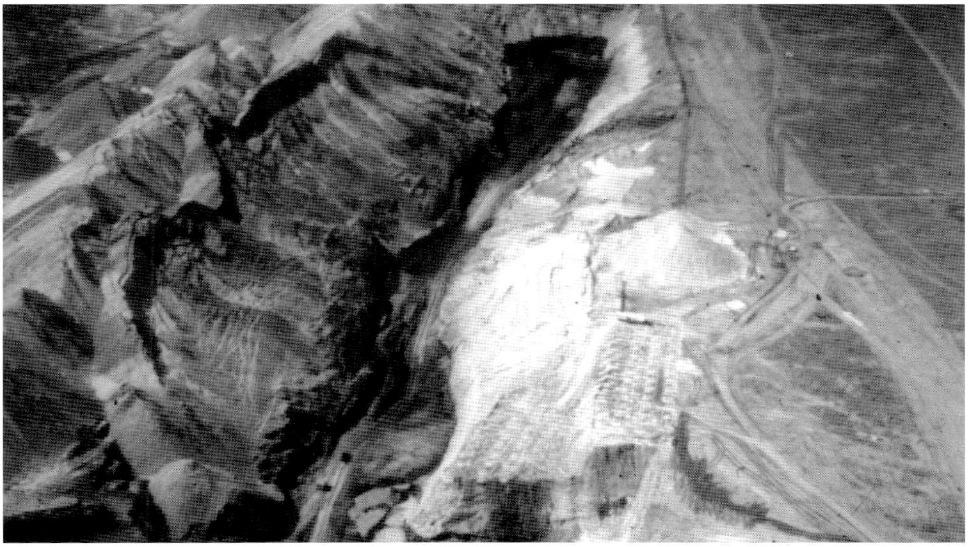

Goonyella strip coal mine in Queensland. The waste pile was failing on a weak coal–sandstone interbed. The highwall was badly damaged due to heavy blasting and the pit bottom was heaving due to artesian pressure. All problems were evaluated and addressed.

A review of the hydrogeology revealed high water pressure in the sandstone beneath the coal. The pressure was sufficient to cause the pit bottom to heave. This heaving in turn removed toe support for the highwall and waste pile. The coal was being blasted prior to excavation. In order to reduce the water pressure, it was recommended they drill every tenth blast hole in the coal an additional 12 feet — into the sandstone — and fill the lower 10 feet of the hole with clean pervious sand. The subsurface water would flow into the sand-filled hole. When the coal was removed, water would flow up from the holes in the sandstone, reducing the pressure. This procedure, described as the use of pressure relief wells, stopped the heaving.

The staff were surprised that I was able to evaluate their problems and recommend simple solutions in such a short time. It's called experience, and such experience saves substantial money on engineering projects. Great satisfaction is gained from providing the successful solution to a difficult problem, and this success justifies the higher hourly rates charged by experienced engineers.

The Chief Engineer of the Hammersley Iron Mine in Western Australia contacted me first in 1972 and again in 1990 to evaluate their pit-slope stability and design. The area is very dry, with only a few high-intensity rainstorms per year. The water table is very low and has not affected stability. Surface erosion, however, can be serious.

Hammersley open pit iron mine in West Australia. A study was made in 1970 to evaluate pit slope stability and potential deepening of the pit.

Hammersley open pit slope where benches were failing along the bedding. The haul road should have been developed on the opposite side of the pit where the geologic structure dipped into the slope. Today this slope would be developed without benches to the dip of the bedding. Brugg-type fences could be installed periodically to catch ravelling rock.

In 1972, the west pit slope was ravelling severely. The iron formation is bedded with periodic shale layers, which, on the west side of the mine, dip about 40 degrees into the pit. The design was to develop a 45-degree slope with benches every 40 feet, but the benches were ravelling back to the bedding. One haul road was on this side, and it had to be frequently cleaned of ravel. It was recommended that the slope be redesigned to the bedding angle. This could be developed using variable depth boreholes. It was also recommended that the slope be well scaled as each lift was

Blasting in the pit used inadequate delays with the result that excessive seismic forces were damaging the pit walls. Modifications were made to the blast program.

excavated. In addition, it was recommended that the haul road be built on the opposite side of the pit, where the bedding dipped into the slope and a 55-degree average slope could be used — 10 degrees steeper than the existing slope.

In 1990 I was asked to provide recommendations to allow deepening of the pit up to 400 feet. Based partly on my first visit, the structural geology was well understood. Shortly after my arrival, I was advised that a blast in the pit bottom was to be detonated in about one hour. I asked to see it. About fifteen minutes before the blast, we drove to a lookout at one end of the pit. Iron ore is much heavier than typical rock and requires more explosive per ton of rock for breakage.

The blast was in the central section of the pit bottom, about 800 feet away from the steepest highwall. When the blast was detonated, dust from the highwall was almost as great as that which developed in the immediate blast area. There was no question in my mind that too much explosive had been used, and with too few delays.

The following recommendations were made:

1. Detonate all blasts to a free face
2. Detonate no more than five rows per blast so the blast does not become choked
3. Within 200 feet of the final walls, the blast rows to be perpendicular to the face, except for the line holes
4. Use controlled blasting near the final wall, with two buffer rows between the final line and the production holes
5. No more than two holes per delay for the production holes

Bob Rivett, President of Gutteridge, Haskins & Davey (GHD) in Brisbane, contacted me regarding my assistance with rock slope and fill stability for a new railway, the Greenvale Railway, from Townsville west to a proposed nickel mine for the Freeport Nickel Company. The railway would extend from a deep-sea dock in Townsville, up through the Hervey Mountain range to the mine area. Negotiating the mountains required many side hill cuts, high fills, and two major tunnels. I was requested to review the designs in the rock cuts, including those which led to the tunnels.

The rock was generally a weathered granite. Some zones were very hard; others were weaker, with soft weathered layers along shear zones and joints. The area was subject to a wet season, during which tropical storms would drop 5 to 10 inches of rain in only a few hours. This created major slope erosion and surface water drainage control problems. It also created a slope design problem. If typical slopes of 1 horizontal to 4 vertical (1H:4V) could be excavated in the rock, the excavation quantities would be moderate. However, the railway had retained a professor of civil engineering from the University of Queensland to assess the slope design, and he had recommended a much more conservative approach — 1H:1V. This increased the excavation quantities by over 2 million cubic yards, as well as the right-of-way that had to be purchased.

I spent two full days out on the project inspecting conditions with GHD engineers. Access to many locations had to be made over private property. At one access point a sign placed by the local farmer said, "Trespassers will be shot." I was startled to see it but was advised that they had obtained special permission to trespass from the owner.

The inspection involved climbing numerous rock slopes and mapping geologic structure — joints, bedding planes and shear zones — to determine the dip angle and dip direction of the discontinuities in the rocks. The hardness of the rock was also qualitatively assessed by hitting it with a geologist's hammer. "Ping" meant hard and "thud" meant soft.

Following the fieldwork, we spent a day in the office developing the details for the ongoing investigation. The railway company had established the design grades, plus the curvature and width of the grade which would support the railway ties. The investigation included field surveys to establish the alignment and approximately locate the tunnels, bridges and culverts. Because of the high-intensity rains, all drainage ditches and culverts had to be designed for high flood conditions and to control erosion. Drill site locations were established to assess soil, rock and groundwater conditions at proposed bridge sites, large culverts, high fills and deep cuts. Drilling was performed to obtain samples of the soil and rock, and to install piezometers.

I stayed at the Travelodge Hotel, across the street from the ocean beach in Townsville. On returning from the field the first day, I was hot and tired, so I went up to my room, put on my bathing suit and grabbed a towel to go for a swim in the ocean. As I walked through the lobby, near the check-in desk and on my way to the front door, a female voice yelled, "Where are you going?"

I stopped, turned around, concluded she had yelled at me and answered, "I'm going for a swim."

"Not in the ocean, you're not! It's jellyfish season."

I walked over to the desk and asked, "How bad are they?"

She answered, "If you get bit you could be very sick for a few days, and if they bite your doodie you might lose your manhood!" Australians tell it like it is! I went for a swim in the hotel pool instead.

Australians are very good hosts. I was invited to several of their homes for meals and taken out to several very good restaurants. They are also very aggressive drivers. To make matters worse, I wasn't used to driving on the "other" side of the road. It didn't take me long to try for the backseat wherever we went. I felt safer there.

In the outback, the dry interior of the country, most of the access roads were dirt, so that driving was very dusty and rough. GHD used a Toyota 4x4 with very hard springs. Sheep and cattle, and of course kangaroos, were everywhere. On several occasions we had near misses. To protect the front of the outback vehicles, solid metal protection frames, called "roo guards," are mounted on the front bumper. Road builders don't usually install bridges or culverts in the small creeks. Instead they pour a concrete road surface down through the creek bed to drive on — provided the water level is below the floorboards.

I made three trips to the railway project during various stages of investigation and design before the final review. After assessing all the geotechnical data, I developed a range of slope angles for the rock cuts. Where the rock quality was good, I recommended 1H:4V slopes; where the rock quality was moderately hard, 1H:2V; and where it was weathered and erodible, 1½H:1V. The latter angle is typical for soil. I also recommended wider catch ditches below the steeper cut slopes. Below high cuts, the ditches were to be shotcreted to control erosion.

The university professor retained by GHD had recommended all rock slopes to be 1H:1V. I expressed concern that this did not adequately deal with the variable rock conditions, potential serious erosion, or extra cost. He was adamant that his design was satisfactory. GHD set up a meeting for us both to attend, along with the railway engineers.

Rather than get into an argument, I arranged to speak first. I reviewed my rock mechanics background and experience, stated the reasons for my design recommendations, and then said to the railway company's Chief Engineer, "It's your money. I have provided what I consider to be a reasonable, safe and economical design. If you accept the professor's design, I accept no technical or financial responsibility."

The Chief Engineer said, "OK, I guess we'll have to make a choice."

Some months later I learned that the major cuts were designed using my recommendations, and the small cuts, with limited excavation, using the professor's design. Not an unreasonable compromise.

The workload continued to increase in Australia to the point where I made eighteen trips there in one year. It became obvious that we should open an office in Australia, but this posed a problem. Australians are nationalistic. We believed that sending a Canadian there to start a business would run into the "you are not Australian" attitude. We decided to try to find a local firm that we could associate with.

The firm that Golder Associates developed an association with was Brickell Moss, who had an office in Sydney and another in Auckland, New Zealand. Our strategy was to get established, then buy Brickell out. This approach worked well and within one year the local firm was purchased. Within a year offices were also opened in Melbourne and Brisbane. The Melbourne office started with a bang. They obtained the geotechnical consulting work for the Melbourne city centre underground railway. The Vancouver office supplied much of the rock mechanics design, headed up by Dr. John Sharp.

On most of my trips to Australia, I would be tired upon completion of my assignment there. As a result, I would usually stop for several days in Fiji or Honolulu. On one trip I decided instead to spend several days in New Zealand. I flew directly from Melbourne to Auckland, arriving about ten in the evening, and was I bushed.

After going through immigration and customs, I picked up my bag and looked for a taxi. I intended to flake out in the back seat, but the cabbie, a jolly, rotund individual, said, "Jump in the front governor; you'll see more." Actually it was already dark.

Nevertheless I obliged, and during the trip to the Intercontinental Hotel he asked, "Why did you come here?"

I answered, "For a little R and R."

He chuckled, "Then I'm your man. I used to be a tour guide. I'll show you the city tomorrow! You sleep in, have breakfast in bed and I'll come for you at 10 A.M. I'll show you the interesting parts of Auckland until about 5 P.M., charge twenty-five dollars, and you buy the lunch. How's that for a deal?"

I thought about his offer — sounded good. "Okay, you've got a deal. Pick me up at 10 A.M."

I had a great time. He was articulate, appeared to know his history, and had a good sense of humour. I saw the upscale housing areas, a few slums (there weren't many), art galleries, museums, the Maori Cultural Centre, the many volcano cones, the harbour and the Auckland harbour bridge.

As Auckland expanded some people wanted more privacy, so they moved to the south shore of the harbour. Initially they were served by ferries. As the south-shore population grew, they began to demand a bridge. Finally a four-lane structure was constructed. This led to such a south-shore population explosion that the bridge traffic became a snarl. Not wanting to bear the cost of a new bridge, the country called for proposals from international consulting bridge engineers to evaluate the possibility of developing two more lanes on the bridge. Japanese consultants came up with the idea of adding a lightweight aluminum lane hung on each side of the existing bridge, solely for the use of automobiles. The cabbie was telling me this story as we approached the bridge. As we reached the north end he laughed, with his stomach bouncing, and said emphatically, "We call it the "Nippon Clip-on!" I enjoyed the New Zealand humour and laughed along with him.

As a result of this exposure to New Zealand, I arranged to spend another week there on two more occasions, again after completing Australian assignments. Pat accompanied me on both trips. On the first trip we flew from Melbourne to Wellington and rented a car to drive north to Auckland and the Bay of Islands. Midway between the two cities we overnighted at Lake Taupo and enjoyed the hot springs. There were pools to bathe and soak in, but the highlight was separate cabañas with jacuzzi-like tubs. Each cabaña was private, and at the entrance the temperature of the water was displayed, ranging from 90 to 120°F. As the tubs were private, you could remove your bathing suit if you wished.

In Auckland we arranged a flight on an amphibian plane to Russell on the Bay of Islands. With Pat occupying the co-pilot's seat, we took off from Auckland harbour and landed on the water at the Bay of Islands. The pilot guided the plane toward the beach, lowered the wheels from the side of the plane and taxied up onto the beach, where we disembarked in style. We stayed at a motel for three days.

The second day I arranged to go deep-sea fishing in a 35-foot fishing cruiser, skippered by an old sea captain called Jack Britain. We had to first catch bait — small fish about 2 to 3 pounds — then we headed for the open sea. This was about 8:30 in the morning.

After catching the bait, Britain had instructed me how to handle the rod and reel. He said, "You must keep your left arm stiff and use your back to bring the fish in. If you use only your arm, you'll be played out in fifteen minutes. Your back is much stronger."

He placed my feet on a circular metal ring step, saying, "Brace yourself and raise the rod with your stiff left arm until it is vertical and your back is tipped well back. The back of the seat moves. Now reel in and lower the rod, keeping strain on the line. You must never let the line go slack. If the fish suddenly turns, the twist may throw the hook. Develop a rhythm of rod up, reel down — keeping tension."

We trolled with two 80-lb. test lines. He had threaded a large, very sharp hook through the side of the bait fish and tied the steel leader so the fish flopped along the surface of the water. He said, "This makes it look like a wounded fish."

At one location we slowly circled a small island. As soon as we reached more open water I noticed a fin about 50 feet away coming toward the bait. I said to Britain, "There's a fin," and pointed toward it.

He said, "It's a marlin; don't move."

Suddenly the marlin sped up toward the bait, came partly out of the water and struck! Zing! The reel screamed. Britain yelled, "Get in the chair!" He slowed the boat, turned and strapped me into the swivel chair. He lifted the rod and reel out of its cradle, braced himself, put the end of the rod into the chair holster and said, "Put your hand on the cork band and the other on the reel handle." The fish ran out with the reel screaming. Britain added, "Let him go; the brake will gradually stop him."

But the marlin kept going! Britain reached over my shoulder and tightened the reel brake. Suddenly the fish jumped clear of the water. It must have been 1,000 feet away! Britain yelled, "It's a striped marlin and a big one! Keep the line tight. Try reeling him in. He's taken out a lot of line. There's not much left on the reel."

I reeled, rod up, reel down, rod up, reel down. The marlin was really resisting. Ten minutes, twenty minutes; sweat was pouring off my forehead. The fish hadn't come in more than 100 feet. Suddenly he started jumping. Once, twice, three times! Pat was taking pictures and yelling encouragement.

Thirty minutes, forty minutes. He was still well out. I would reel him a bit but he would run again — even against the brake. Then the line started to dip. Britain said, "He's diving, reel hard. We don't want him to do that." I did a furious round of reeling and the marlin came up.

I suddenly felt about sixty years old. My arm was almost dead and my back ached. Then, without warning, the marlin broke the surface — on his side. Britain said, "He's finished; reel him in."

I got him beside the boat. Britain gaffed him and asked me to hold the gaff. He took a rope and draped a lasso knot around the neck. He said, "Move forward ten feet." He also lassoed the tail. He said, "He's too big to get into the boat. We'll tie him onto the step at the back of the boat." After that was done he said, "Beautiful fish. The striped marlin is the smallest of the marlin family but the fiercest fighter, pound for pound. It looks about three hundred pounds." He added, "Do you want to take it home or have it stuffed?"

Successful marlin fishing with my wife at the Bay of Islands near Russell, New Zealand. She accompanied me on many consulting engineering trips.

The fish was 12 feet long — take it home? At what cost? I thought for a bit and said, "No, but I'd like pictures at the dock."

"Okay," he said. "I'll radio in and have a photographer at the dock. At about what time?"

I had rented the boat for the day. However, I was tired out. I'm sure I couldn't have landed another one, so I said, "This has been a fabulous experience. I'm ready to go in now."

He turned the boat and we headed back to the dock. There were about twelve people there.

The photographer had already arrived. Britain manœuvred the boat up to the dock where a pulley and scale were located. He wrapped the pulley rope around the tail and lifted it slowly until it was high in the air. The scale read 290 pounds. He filled out a record board and stood it beside the fish. The photographer took several photos of Pat, Britain and myself — with the fish. Britain said the fish would be given to the local hospital.

While standing on the dock I heard a local New Zealander say, "Those bloody Americans; they come and catch all our fish!" With my brush cut, there is no doubt I was easily taken for an American, but I turned to him and said softly, so only he could hear it, "We're Canadians." He suddenly raised his hand and said loudly to everyone on the dock, "They're Canadians; they're Commonwealth." At that point many came over and offered congratulations.

Several months later I received a trophy from the Russell Fishing Club, which was for the third-largest striped marlin caught in New Zealand water that year.

With that introduction to deep-sea fishing, I made fishing an annual event, fitting it into my business or holiday travels. For larger fish, locations have included Mexico, Costa Rica, Fiji, the Cook Islands, Tahiti and Bermuda. A further annual event has been fishing for salmon off Vancouver Island, Douglas Arm near Prince Rupert, and off Sitka, Alaska, with my sons Randy and Greg. We have also enjoyed fishing for walleye (pickerel) and northern pike at Lloyd Lake in northern Saskatchewan. In return, Randy and Greg have taken me golfing each year, including one trip to the British Open at St. Andrews in Scotland, the Firestone Open at Akron, Ohio, and Pebble Beach in California.

8

Philippines

Philippines (1968-1998)

Placer Mining Company of Vancouver developed a major open pit copper mine on the Island of Marinduque. Periodic typhoons flooded the mine, closing it until the water could be pumped out. The author designed a drainage tunnel under the pit to dewater it by gravity. An inspection of the tailings dam at the mine indicated it could fail during an earthquake. The author changed the design of the dam from upstream to downstream construction.

Underground block-cave mining beneath an infilled tailings pit caused the tailings to liquefy, then flow into the underground copper mine at the Atlas Consolidated Mine, killing six miners. The author describes the ensuing investigation and stabilization program, where the tailings were drained using wells into the underground, where water was pumped to the surface. In the adjacent Biga Pit, a 55 million-cubic-yard slide was stabilized using drainage tunnels, a procedure which the low cost of Philippine labour made less expensive compared to the installation of horizontal drains by a North American contractor.

An unusual tailings disposal procedure at Black Mountain Mine near Baggio involved "controlled flushing." Annual typhoons were relied upon to wash the tailings away. Environmentalists lobbied to stop the procedure, but local farmers protested, pointing out that the mine tailings flooded the rice paddies downstream and increased productivity.

Chapter 8
Philippines

Typhoons typically strike the Philippines several times a year, dumping as much as 25 inches of rain in one day. In 1968, such a storm hit the island of Marinduque and the Marcopper open pit copper mine, an operation owned by the Placer Mining Company, out of Vancouver. The mine flooded and took many days to pump out. Along with the flooding, the mountainside pit slope suffered severe erosion. Tony Triggs, General Manager, contacted me and asked if Golder Brawner could visit the site and provide geotechnical advice.

I made arrangements to travel on United Airlines to San Francisco, and from there to Manila on Philippine Airlines. I had learned by then that overseas flights were generally emptier midweek. Sure enough, my flight overnight from San Francisco was only half full, and, as a result, I was able to find three seats together in the economy section, lie down and get a reasonable sleep. The flight was supposed to be non-stop on a DC-8 jet. However, we encountered severe head winds and had to land at Wake Island to refuel. The stop at night was unique but uneventful. The airport was old and looked like it hadn't been maintained since the war.

Triggs met me at Manila and we drove to the Intercontinental Hotel in a suburb called Makati, a new business district midway between the airport and the city centre.

The next morning we flew in a twin Aero Commander plane from Manila to the private airstrip at the mine. The short airstrip was situated in the mountains between two steep gullies, making for an exciting landing. Triggs briefed me on the mining operation and then turned me over to Geology to inform me of the rock types, stratigraphy and geologic structure.

Generally the rock conditions involved about 100 feet of weathered rock overlying competent rock. With the high annual rainfall, a high water table existed. This resulted in adverse subsurface water conditions and many wet holes for blasting. If the slopes and pit bottom could be effectively drained, the overall slope could be steepened some 5 to 6 degrees, and less expensive ANFO explosives (ammonium nitrate and fuel oil) used, as compared to the much more costly emulsions or slurries required in wet holes.

After considering the pit dewatering required after typhoons, and the loss of production during that period, I proposed an under-pit drainage tunnel. A tunnel about 1.8 miles long from the Makulapnit valley could be constructed to extend under the pit. Drill holes about 8 inches in diameter could then be drilled vertically from the existing pit bottom to intersect the tunnel and drain the open pit by gravity. Several years later this idea was put into effect, and a major additional benefit developed. The tunnel lowered the water table, so that many of the blast holes were dry. The less expensive explosives required saved nearly $1 million per year.

In the late afternoon of the third day, Triggs asked, "Would you like to see the tailings dam before dinner?"

I accepted.

Starter dykes had been developed at each end of a mountain saddle using local laterite (completely weathered rock), compacted to a controlled density and moisture content. The dykes were a maximum of about 60 feet high. A 12-inch-diameter tailings line was constructed along the top of each dyke. Small-diameter tailings outflow pipes (spigots) were located about every 30 feet along the main pipe. Deposition of the tailings upstream of the dykes was performed continuously. The coarser sand sizes settled out quickly and the finer gradation (slimes) flowed out farther. Deposition continued along both dams until the tailings beaches were near the elevation of the starter dykes. The main pipe was moved to the outer edge of the starter dyke and the coarse sand pile was raised. A small Caterpillar tractor then spread and compacted the sand and shaped a sand bench. When this was raised about 10 feet, the main distribution pipe was raised and spigotting began again at the new elevation, building a new sand berm. At the top of the starter dykes, to neutralize any acidic seepage that occurred, coarse crushed limestone was placed along the toe of the first sand bench. At the time of my inspection, four sand benches had been developed. The height of the dams was about 100 feet. This method of construction is described as the "upstream method."

Tailings dam at the Marcopper mine on Marinduque Island, Philippines. The dam was being constructed using the "upstream" design. The Philippines are subject to frequent earthquakes. As a result, liquefaction could occur. The dam was redesigned.

On average, one major earthquake, Richter 7 or greater, occurs each year in the Philippines. Uncompacted saturated sand has the potential to liquefy and flow due to vibration or high stress. The south tailings facility was located at the head of a valley that continued to the city of Santa Cruz, several miles downstream.

Failure of many tailings dams in Chile, only a few years previously, had occurred due to the El Cobre earthquake. Twenty-six people were killed and liquefied tailings had flowed over twenty miles. I brought this to the attention of Triggs with the comment,

"The dams being constructed here will not resist earthquakes." He wasn't convinced.

I said, "Tony, do you have a drill rig on site? I recommend the sand density be tested. If the standard penetration test (N) values are less than 10 to 15 blows per foot, the saturated sand will liquefy under a moderate earthquake." (The N value is the number of blows of a 140-pound hammer dropped 30 inches that are required to drive a 2-inch-diameter sampler 1 foot into the sand.)

He agreed to the test. At several locations the N values were less than 5 blows per foot, and in one location, at a depth of 20 feet, the sampler sank under its own weight. There was no question that the dams should be redesigned to resist earthquake stresses. Triggs asked, "Would you develop the redesign?" I agreed.

I was also concerned about the adequacy of the construction of the starter dam. Sandy Laird, the Chief Engineer, provided me with the original plans. They called for compaction of the laterite in 8-inch lifts to 100 percent standard proctor density. No density records could be located. Sandy said I should speak to the Philippine engineer who had supervised the construction.

I asked the engineer how the compaction had been performed. He looked at me blankly and didn't answer. I asked, "Was a rubber-tired, sheepsfoot or a vibratory roller used for the compaction?" He answered, "A vibratory sheepsfoot roller."

I didn't ask another question. There was no such type of roller at that time. I concluded he was going to give me the answers he thought I wanted to hear, and so any other questioning would be a waste of time. This is still a problem in many Third World countries. Any engineer working internationally must deal with this concern. All designs must be rechecked.

I arranged for sampling of four test holes and testing of the soils in the starter dykes. The next day I departed for Vancouver. It was agreed that Golder Brawner would develop a revised tailings disposal design, and that I would return with it in three weeks.

On my arrival in Manila three weeks later, a customs officer opened my briefcase. I had a small soil vane tester in it, which I used to test the strength of clay. It was in a small leather case, and had cost me thirty US dollars. The officer took it out of its case, looked at it and said, "The duty on this is fifty dollars."

I looked at him and said, "It's not worth that much; you keep it."

He stared at me, and then said, in a questioning manner, "Well, twenty dollars?"

I said quickly and a bit sternly, "I'll give you five U.S. dollars. It's not worth more."

He hesitated, looked around and said softly, "Okay." I took five dollars from my wallet and waited for him to put the vane tester back in the briefcase before I gave him the money. Triggs said afterward, "It was good you gave him something, because they are very poorly paid and need the extra income to live on. If you hadn't done that, it could have cost you several days in jail."

I have encountered similar situations in numerous countries. As a result, I keep a small item — such as a compass — in my briefcase. A compass isn't worth time in jail and it's wise to be prepared for such a situation.

The original dam design was changed to one based on the "downstream method." The new design required that coarse tailings or laterite fill be placed downstream in compacted lifts to develop a new top elevation. A materials balance study indicated there would not be enough coarse tailings to develop the new dam sections each year. As a result, the dam had to incorporate compacted zones of laterite. To ensure the dam would be drained, the upstream slope required a low-permeability compacted laterite membrane to reduce seepage through the dam. In addition, a free-draining sandy gravel underdrain with drainpipes was constructed under the entire new downstream dam section. Such a dam would resist the typical Philippine earthquake. A new downstream dam raise was planned for each year.

Several years later, Marcopper developed an exploration program to try to locate a new ore body. The drilling was carried out on a grid spacing of about 1,000 feet, with one of the holes located in the centre of the tailings storage area. This hole struck copper ore! Further drilling proved over 70 million tons of commercial ore, all covered by millions of tons of tailings.

A feasibility study revealed it would be economical to remove the tailings and develop a new open pit. Marcopper obtained approval to deposit the tailings in an ocean bay about 14 miles from the original tailings area. The plan was to dredge the tailings from the existing pond and move them by gravity in a combination of flumes, tunnels and pipes to the bay. Development of this new system was initially very expensive, but operating the ocean disposal program proved to be reasonably economical. The secret of the design was to achieve an adequate gradient, such that the tailings continued to flow.

Finding an ore body under the tailings pond was certainly an embarrassment to Marcopper. Examples such as this have since resulted in the mining industry developing a procedure called "condemnation drilling" prior to design and construction. On potential mining projects, drilling is now performed in all areas where buildings, waste piles, leach pads or tailings disposal areas might be developed. It's a costly procedure, but regulatory agencies now require it.

There was a branch of the American Society of Mining Engineers in Manila, and Tony Triggs asked if I would speak to them on my next trip to Marcopper. I agreed, and proposed a topic of "Case Examples of Stability on International Surface Mining Projects."

The lecture was presented about six weeks later. One hundred and forty mining personnel turned out, representing about twenty Philippine mining companies, government Mines Department staff, government safety inspectors, and international consultants resident in the Philippines. My lecture was comprised of a review of eight projects where stability had been a serious problem, including projects in Canada, United States, Mexico, Chile, Australia, Zambia, Spain and Ireland. The examples included failures of open pit mine slopes in soil and rock, tailings dams and waste piles. Colour slides were used extensively and I did not use notes. The presentation was a success, with questions afterward lasting for about forty minutes.

Before leaving the Philippines, I was invited to provide stability reviews at the Atlas Consolidated Mine in Cebu, Marinduque Mining Co. on Mindinao, and at Lepanto and Black Mountain on Luzon. The success of the lecture in securing these contracts alerted me to the effective use of such presentations for business development. The entire thrust of the lecture is to use case examples to illustrate, in a very low-key and professional manner, experience and capability. No high pressure or competitor bashing is involved. The greatest benefit was the direct development of contracts without having to prepare a proposal in competition with other consulting firms. The majority of Golder Brawner international contracts up until 1978 were sole sourced in this way.

Over the years, I found it very frustrating to be asked to provide a proposal on a project and learn that as many as ten consulting firms had been approached. Owners either didn't understand the cost of preparing proposals or didn't care. On quite a number of such projects I refused to submit proposals. Some companies simply shop for the lowest cost and do not give recognition to experience.

Charles Clark, Vice President and General Manager of the Atlas Mining Company on Cebu phoned me shortly after I had completed the Marcopper tailings design change. He said, "Chuck, we've had a serious tailings inflow into our underground mine and six miners have been killed. How soon can you come to help us resolve the problem?"

I answered, "I can be in Manila next Sunday evening. I'd like to stay at the Intercontinental Hotel."

He replied, "Excellent. I'll arrange for a driver to pick you up from the hotel at 8 A.M. He'll take you to our corporate hangar at the airport and we'll fly you down in our plane, a Kingair. You'll arrive at the mine about 10 A.M. and be met by Harry Tolle, our Chief Engineer."

The flight to Manila was uneventful. I find long-distance travel westerly much easier than easterly. Arriving in Manila in time to get a good night's sleep, I adjusted easily to the six-hour time change.

The weather in Cebu, nearly a two-hour flight from Manila, was unsettled. As a result the plane ride was very rough. Upon landing, I was met by Harry Tolle, a big man with a great sense of humour. He had worked in the tropics much of his life, and so had adjusted to the warm humid weather. I found it uncomfortable, partly because my clothes always felt damp. Moving frequently from the air-conditioned office to the warm outside, I caught a cold on the second day.

Tolle drove me to the office to meet Mr. Clark. He was a quick-speaking and decisive American mining engineer who had spent a lot of time prior to Atlas in Peru. He expressed regret for the failure and loss of life, saying, "We closed that part of the mine after we removed enough sand and ore to recover the bodies. I'd like you to try to determine the cause of failure and advise us what we need to do to get the mine back in operation. We've instructed all mining staff to provide you with any information and records that you want. As soon as you can provide me with conclusions and recommendations, please contact me."

Atlas operated two mines in close proximity: an underground mine developed as a block caving operation, and a large open pit mine. The accident occurred in the underground mine. Originally, ore at the mine was recovered from an open pit which extended to a depth of 500 feet. The resulting pit crest eventually came within 100 feet of the mine buildings. At that point the surface mining changed to underground. The ore grade was moderately low, generally ranging from about 0.6 to 1.0 percent. It was crushed to a gradation of silty sand and then processed to produce copper concentrate. The coarse sand in the tailings was split off using cyclones and deposited in the open pit. Rather than allow the pit to fill with water, they reasoned that the sand would be more desirable and safer. However, the sand in the pit was now saturated. The fines, primarily silt sizes, were dumped into the nearby river and carried to the ocean. (At the time, 1970, environmental issues had very low priority in the Philippines.)

The block caving method uses a series of parallel drifts (tunnels) about 50 feet apart, developed about 200 feet below the bottom of the pit. These drifts were concrete lined for stability. Small raises about 3 feet square were developed 20 feet apart on each side of the drift. These extended upward for about 15 feet. A metal frame was bolted to the concrete face to hold 4-inch by 8-inch timbers across the openings or draw-points. Ring drilling above the drift was performed through the concrete for about 75 feet. Blasting the upper 60 feet fractured the rock above the raises. The broken rock flowed into the raise, filling it against the timbers. Following removal of ore through the draw-points, the opening above the raise would collapse and extend the volume of broken rock upward.

Small mining (muck) cars that would hold about 10 tons of rock, 12 to 15 connected together, were moved along under the draw-points. The timbers at the draw-point were raised to allow the broken ore to flow into the muck car until it was filled. The timbers were then replaced over the draw-point. In order to remove the ore evenly, it was usual to draw uniform volumes of ore from adjacent draw-points.

The mine was developed in this way in blocks about 200 feet by 200 feet horizontally. As the ore continued to be drawn, the roof of the draw zone would collapse and more broken ore would collect over the draw-points. The weight of the collapsed rock would create pressure, pushing more ore down through the draw-points.

Occasionally a draw-point would plug. The local practice was to tie a half stick of dynamite to the end of a bamboo pole and push it up the draw-point to refusal. The detonating cord would be carried about 20 feet down the drift and then triggered. This started the flow of ore again. Rudimentary, but effective.

The old open pit mine was nearly full of tailings. These were saturated from rainfall, with the water level near the top of the tailings. As the rock continued to cave in, the block, a funnel-shaped depression, occurred in the surface of the tailings. Mine management paid no attention. Suddenly saturated tailings flowed out of many draw-points into a drift, filling the drift and trapping six miners, who were drowned in the tailings.

In trying to reconstruct the accident, I asked for and studied the draw-point records. Instead of uniform withdrawal of the ore from all draw-points, it became apparent that drawing of ore was concentrated from draw-points where the ore flowed more readily. The day before the inflow, nearly 600 tons of ore was pulled from only 6 adjacent draw-points, with no ore withdrawn from draw-points on either side. The active draw-points were directly under the surface depression in the in-pit tailings. I concluded that the concentrated withdrawal of ore had caused excess removal of ore, leading to an upward collapse of rock until it encountered the saturated tailings, which then liquefied and flowed down into the drifts. The foreman's notes referred to considerable water flow from the draw-points involved. I also concluded that the availability of water within the tailings was necessary for the liquefaction.

I spent considerable time with Atlas staff studying the mine geology, mining plan and layout, mining operations and draw sequencing, as well as the tailings disposal. I learned that there was considerable excess underground water pumping capacity. After dinner that evening at the guesthouse with Clark and Tolle, I went to my room and reviewed my notes, reports and plans. By midnight I had reached conclusions on the reason for the inflow, and developed a plan for redeveloping stability.

At breakfast next morning, I advised them that I considered that concentrated removal of ore from only 6 draw-points likely precipitated the inflow, and that the high water pressure and liquefaction of the saturated tailings led to sufficient volume of tailings inflow to trap the miners. I further advised, "It is recommended that the permeable sand tailings in the pit be drained before that area of the mine is reopened, since unsaturated tailings are much less susceptible to flow." I recommended that four, 8-inch-diameter holes be drilled and cased through the tailings to the bottom of the pit, arriving at a point above underground stations where pumping to the dewatering system could be developed. Diamond drill holes would be drilled below the pit bottom through the well casings to intersect the underground drifts near a pump station. At that time, a 6-inch well screen would be lowered full length into each hole to the bottom of the tailings, and the casing pulled. This system would drain the tailings by gravity to the underground pump stations, where it would then be pumped to the surface.

Dewatering well being developed in the old tailings pond area. Several wells were drilled, screened and developed. Each well was located above an underground pumping station. Water from the tailings drained by gravity to the pumping stations to be pumped to the surface.

At the same time, in order to minimize seepage down into the tailings, the entire surface of the tailings was to be slightly sloped and covered with 2 feet of local compacted laterite. Four piezometer installations were recommended to monitor the progress of the dewatering.

At this point the Philippine Mines Branch advised Atlas that, as soon as the mine was reopened, the tailings could no longer be dumped into the river. Atlas therefore obtained approval to pipe the tailings to the ocean for disposal in deep water. A steel-pipe tailings line was constructed to the ocean and carried out on anchored pontoons to a vertical drop pipe about 100 feet long. The disposal point was in about 300 feet of water at the crest of a deep-water drop off, descending to over 20,000 feet.

Clark asked Golder Brawner to supervise the tailings dewatering using a Philippine contractor. I selected Bill Turner, a hydrological engineer on my staff, to do the supervision. He did a good job. The Philippine contractor was competent and followed directions explicitly.

The first two gravity wells were developed with no problem. Surveys of the third well indicated the diamond drill hole would be about 8 feet away from the underground opening. Miners were brought in to enlarge the opening. This was successful and the "hit" was made. The fourth hole was also okay.

After establishing the wells, the water table dropped rapidly — about 2 feet per day. Nearing completion of the wells, when it appeared they would be successful, I flew from Vancouver to Manila and on to Atlas. Turner was ready to come home. The tailings line to the ocean had just been completed and tailings were flowing.

Clark, Tolle, Turner and I were having dinner at the guesthouse when Clark remarked, "I wonder what the tailings are doing to the fish at the bottom of the drop pipe?"

Bill Turner offered, "I'm a scuba diver. If you can obtain an underwater camera and the necessary scuba gear, I'll go down and get some photographs for you."

Clark queried, "You would do that?"

"Sure," returned Turner.

Clark said, "Prepare a list of exactly what you need and I'll get it as soon as I can."

The equipment arrived two days later and Turner made the dive. I was still there. After returning to the surface he said, "The fish were so plentiful around the outlet of the downspout pipe that, along with the tailings plume and air bubbles, photos really wouldn't show anything. I've never seen so many fish outside an aquarium. They really seemed to enjoy the turbulence, and none seemed to be in trouble." The issue was resolved.

While I was there, Clark asked me to look at their tailings pond with a view to establishing an overflow spillway. I reviewed the drawings of the dam and the original geotechnical report. The north side of the dam abutted a slope with shallow rock. I recommended a rock cut be made through the abutment, such that the spillway could be developed in rock. The outcrop at that location was hard andesite. I stayed on to assist with the design.

The next morning a very heavy rainfall developed. (Later I learned that 8 inches fell that day.) The water level in the tailings pond was soon within 4 feet of the crest, as the inflow volume was in excess of the capacity of the outlet culverts. The concern was that the water would overflow the low point of the dam crest. I went with Tolle to the dam and saw the potential urgency. I asked if they had any spare tailings line pipe or 2-foot-diameter culvert pipe. He replied, "Yes, near the concentrator."

I said, "I recommend you bring four sections down to the dam as soon as possible and install them through the abutment area, just below the crest elevation, as overflow pipes."

Within two hours, as a temporary measure, four shallow cross trenches were dug with a backhoe and the culverts installed and backfilled. Within six hours they were flowing half full. The dam was saved! This example shows the benefit of having repair materials on site.

At the same time, the surface water had eroded the hillside slopes and undercut numerous banana plants, which have very large leaves. One of the culverts became plugged with several plants, so two of the workers went to that culvert and attempted to pull the plants out. As one of the plants was extracted, the accompanying sudden surge of water partially sucked one worker into the culvert. Fortunately, several workers managed to quickly grab both arms and pull him out. Everyone had a good scare and learned something new about water flow and pressure.

Atlas had decided to increase production as soon as the underground stability problem was resolved. To do this they required additional electrical energy. Arrangements were made to purchase a large barge-mounted diesel-powered generation plant from U.S. Navy surplus. After setting up the plant in the closest bay on the coast, power poles were installed along the proposed line to the mine. The next phase was to string power lines along the route. After several days of stringing line, the wire began to disappear, obviously taken down at night by locals. A decision was made by Atlas to string just one line continuously over the course of one day. That evening the line was energized. The next day, despite the fact that numerous danger signs were installed along the route, the line was found cut, and one Filipino was found severely burned. No further problems developed with the power line.

About a year later, I received another call from Charles Clark at Atlas. A major slide had occurred along the west side of the Biga open pit. About 40,000 tons of ore per day were being mined from this pit, which ranged from about 400 to 700 feet deep. The slide extended down the pit face for 300 feet, and then up to the top of an adjacent hill about 3,000 feet away. The top scarp had dropped about 20 feet, and major tension and shear cracks bounded most of the slide. Atlas had established several movement hubs within the slide, and by triangulating from a baseline across the pit once per week, they established that the slide was still moving, about an inch per day. (This was before electronic distance measurement [EDM] had become established.)

A geotechnical investigation was developed which involved four diamond drill holes down the approximate centreline of the slide. Standpipe piezometers were established in each hole. Because the slide was moving, the holes were drilled through

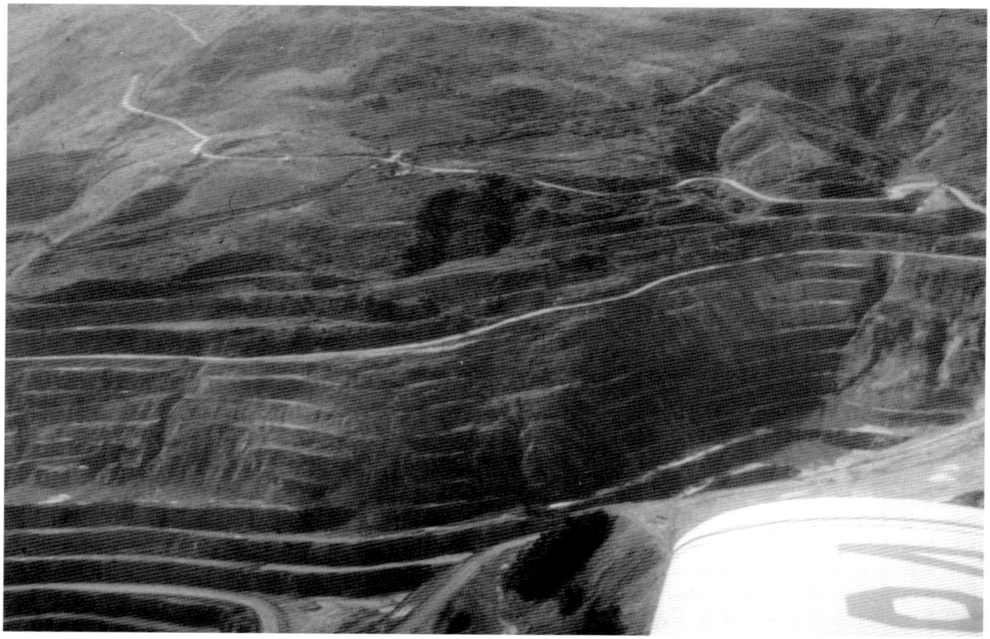

A major slide involving about 55 million cubic yards developed. Note the head scarp at the top of the mountain. Movement occurred about 300 feet deep in the contact zone between saprolite and unweathered rock. Atlas Consolidated Mine on the island of Cebu.

about 300 feet of weathered rock, until they reached into about 10 feet of hard rock. The piezometers were installed rapidly, so that it would be several days before they would be sheared off at the failure zone. The shear zone was found at the pit slope face, near the contact of the weathered and hard rock.

The piezometers revealed that the water table averaged about 100 feet below the surface, and the failure zone averaged about 240 feet below ground level. The volume within the slide was estimated at 55 million cubic yards! Based on experience, it was obvious that the only economical procedure to stop the slide would be to lower the water table.

A back analysis was performed on the central cross section. The unit weight of the rock was determined from drill core, and the piezometric head and depth of failure from the drilling. The safety factor was assumed to be 1.0. Cohesion was considered to be zero due to the movement. The back analysis revealed an effective residual friction angle of 19 degrees. Lowering the water table was calculated to increase the safety by 10 to 16 percent, depending on how effective the reduction was.

Because of my experience with horizontal drains, I discussed this option with Clark. I suggested the program should be performed by an experienced contractor in this field, that is using equipment, drill crew and drainpipe brought in from North America. We then discussed the cost. I estimated about $20 per lineal foot of installed drain: for thirty 500-foot-long drains, a cost of about $300,000, plus supervision.

Clark said, "Our labour costs here are about one U.S. dollar per day. We could put five or six drainage tunnels into the slope for less than the cost of the drains you suggest. Are tunnels a possibility?"

I had been scooped! I should have taken the substantially different labour costs into consideration. In all future international work in Third World countries, this consideration was high on my priority list. I said, "Mr. Clark, drainage tunnels are a definite option. I have designed several that have been successful. Give me several hours and I'll come up with a preliminary program. I can then meet with your engineers and arrive at a cost estimate and potential schedule."

The program I proposed comprised two tunnels 1,000 feet long commencing just below the failure zone, with two 200-foot-long laterals extending from each, a total length of 2,800 feet. Every 100 feet along the tunnels, I proposed a drill station from which five percussion drill holes would be developed, fanned out upward and perpendicular from the adits. These fanned holes would increase the effective diameter of the tunnels, improving the speed and volume of water interception, and the lowering of the water table.

The Filipino people are shorter in stature than most North Americans and generally use smaller mining equipment. As a result, instead of developing the 8-foot by 8-foot tunnels typical in North America, Atlas decided on 6-foot by 6-foot dimensions, except at the drill stations, which were 10 feet by 10 feet, and 8 feet long. The smaller size required less support for roof stability. Instead of shotcrete support (sprayed on concrete), they used timbers obtained locally. Both tunnels were to be constructed at the same time.

All miners were locally hired; no expensive expatriates or equipment were required. Working with Manuel Reyes, Senior Engineer, a cost estimate of $180,000 was developed, much less than that for the horizontal drains. The time

Drainage adit developed under the slide area shown in the previous photograph. Two adits 1,000 feet long with four laterals were developed. The resulting drainage stopped the movement. Had it been necessary, bulkheads could have been installed near the entrance and the adits placed under a vacuum.

required, however, was a little greater; three months compared to two. The tunnel design was approved.

As soon as the tunnels had progressed to 300 feet, the water table began to drop, as measured by the piezometers. At the same time, the rate of movement of the slide started to slow. By the time the drainage tunnels had reached 500 feet, the movement had stopped. And lowering of the water table had a dramatic effect on stability. The weathered rock had a surprisingly high hydraulic conductivity.

At the same time as the tunnels were being dug, to prevent surface water flowing into the slide mass, the cracks on the surface were being filled in with small Caterpillar tractors. This was important, since the annual rainfall at the mine averaged about 90 inches.

After the slide stopped, Clark asked if the pit could be deepened in the area of the slide.

I answered, "Yes, provided the slide area was not undercut, and provided the blasting forces to break the rock within 300 feet of the slide were reduced by delaying every blast hole." He agreed to this.

On another occasion, I planned a trip to Australia to visit the Morwell project, followed by a trip to the Philippines to review the Marcopper tailings dam. I arranged for Pat to come with me.

After meeting with the State Electricity people in Australia, we had dinner with John Alexander, the Chief Engineer at Morwell, and his family. John had a teenage son who was interested in old railway clocks, which were becoming collectors' items. He was just finishing the refurbishing of one from a small station in Yarrow. It was dated 1880, and had an internal spring mechanism that had to be wound once a week. I asked him what it was worth. He said he would sell it for $150 dollars (Australian). I bought it partly out of personal interest and partly as a public relations gesture with Alexander. The boy was pleased and arranged to pack it well for our return travel.

I hadn't thought about what I would encounter with Philippine customs. Upon our arrival in Manila, the customs officer asked what was in the package.

I answered, "An old clock from an Australian railway station."

He hesitated for a minute. It was obviously the first antique Australian railway clock he had been confronted with. Finally he said, "What did you pay for it?"

"One hundred and fifty dollars Australian," I answered. This was about 1,000 Philippine pesos. "We'll only be in Manila for five days, then we return to Canada. There should be no duty."

He countered, "We have no assurance that you won't sell it here!"

I hadn't thought of that. "But what would a Filipino want with an antique Australian clock?" I asked.

Suddenly he took on a more rigid attitude. "You'll have to pay one thousand pesos duty!" he asserted.

I returned, "Would you please check with your supervisor?"

He stared at me for several seconds and then went to a back office. Several minutes later the supervisor approached us and said, "You can check it with us here. We'll give you a claim slip, and when you leave for Canada come to this office, show us your boarding pass and claim check, and we will return it to you."

That seemed fair enough, so I agreed.

Tony Triggs of Marcopper and his wife Nora picked us up outside immigration, and we had dinner with them at an exclusive restaurant near the Intercontinental Hotel. It had a lively floor show after 9 P.M. specializing in Filipino songs and dances. One dance was a pole dance. Two Filipino couples, two men and two women, squatted at ninety degrees facing each other. They held onto two thin poles about 8 feet long, one in each hand, such that the poles formed a cross. Each couple held the poles about 3 feet apart, and initially a very athletic dancer stood in the square space in the middle of the poles. As the music began, each couple would bring the poles together quickly. The dancer jumped in the air to avoid the poles hitting his feet. It looked a bit like skipping rope. The dancer would turn, hop on one foot then on the other and perform various dance manœuvres, at all times missing being struck by the poles. Then he would suddenly jump away and a female dancer would take over. The music was extremely lively and the clickety-clack of the poles kept the beat. The entire audience clapped.

We were enjoying it until an attractive female dancer came to our table, took my hand and said, "Come, I will teach you." It happened so fast that I didn't have time to object. Suddenly I was on the dance floor. She stood me in the centre and said, "We will start slowly. Just jump up and down. We'll match your beat." They did. I started slowly and gradually sped up the tempo. All was going well until they began moving the poles farther apart. I almost fell. They stopped, stood up and clapped. It was over so quickly that I wasn't embarrassed, until I went to sit down and noticed that most people were laughing. Following this experience, whenever we have been at local country folk shows, I sit in an area where I think the dancers won't easily find me in the crowd.

The next day we flew on the company charter aircraft to the airstrip at the mine, and Pat came with me. On our arrival at the airstrip, Trigg met us. As we got into the car to take us to the main guesthouse, he said, "We had real excitement here yesterday. Another Aero Commander came in with a group of Japanese engineers. The pilot forgot to lower his landing gear and the plane landed on its belly, sliding along the runway. Sparks were everywhere but no one was hurt. As soon as the plane stopped, the Japanese rushed out and ran from the plane yelling in Japanese. They weren't happy campers!"

The guest accommodation was excellent. Marcopper had developed a large group of houses for expatriate senior staff, along with one main five-bedroom guesthouse. Much of the inside finishing was Philippine mahogany or walnut. A cook and housekeeper attended to the house chores. We had several very enjoyable dinners there with the Triggs, Sandy and Betty Laird, and John Dodge and his wife.

It was nearing Christmas and the local school principal had asked the mine to provide something for the students. Betty Laird took on the job of finding something for nearly 200 children. Pat went shopping with her. Betty was very resourceful and decided to buy a pencil for every student. This may not seem like much, but in an impoverished country the pencils were greatly appreciated by the students.

One aspect of the tropics that is disconcerting to the ladies is the presence of geckos, tiny lizards that can be found anywhere. One evening Pat found one in her bed! By the time we were leaving, we had given several of them names. They are actually quite harmless.

Another problem is the continual high humidity. Clothes never seem dry when you put them on. In an effort to strive for dry clothes, a light bulb is generally left on in all clothes closets to provide some heat and evaporation.

Sam Black, President of Black Mountain Mining, contacted me after my lecture in Manila to ask if I could provide an inspection of the tailings disposal facilities at their mine near Baggio, north of Manila. I agreed to go.

His son, James (Jim) Black, picked me up at my hotel and we drove to the domestic air terminal. Black Mountain owned a twin Cessna airplane which he then flew to Baggio. I learned that James had flown for the U.S. Air Force in the Pacific during the last war, and I found him very interesting to talk to. His hobby was looking for World War II aircraft that had crashed in the island jungles, and his goal was to completely reassemble a Japanese Zero fighter aircraft. He said he had enough components for about half a plane.

The tailings system at the mine was unique. The Philippines have a very wet season, typically with one or more typhoons per year, which can dump up to 25 inches of rain in one day. What are normally only creeks during the "dry" season — seven to eight months of the year — become raging torrents about 25 feet deep during and immediately after a tropical storm.

The tailings were stored along one side of the river behind a coarse tailings dyke pushed up about 20 feet high. After one pond about 100 feet long and 70 feet wide was completed against the adjacent hill, a second pond was developed downstream, and then a third, and so on. At the time of my inspection, six such ponds had been constructed.

Jim said, "The rainy season will start in about one month. During the first major storm, the river floodwaters will scour all the tailings and carry them downstream. Next dry season we'll start again and develop the ponds as you see them now, and rely on the following year's flood to remove them again. We call this method "controlled flushing."

I asked, "What happens to the tailings when they hit the floodplain downstream?" I knew this area to be generally developed as rice paddies and market gardens.

He laughed, "The government wasn't too concerned until the international environmental crowd became very critical and brought in the media. The government advised us we would have to find another procedure. Then an interesting reaction developed. The farmers downstream banded together and did not want the flooding to stop. After the first year's flooding, which deposited several inches of tailings over their land, their crop productivity had increased by about one-third! The tailings had a major fertilizer benefit, similar to the rejuvenation of the farmed flood plains along the Nile River in Egypt. When the Aswan Dam was constructed, farm productivity dropped considerably as the flooding stopped. The government order was rescinded."

There was nothing I could propose to be of benefit. They had it well under control. It is interesting to note the procedure could never be licensed in North America.

Flying back to Manila, Jim asked if I would be in Manila the coming weekend.

I said, "Yes, I don't plan to leave until next Tuesday. I have several meetings Monday."

He extended the following invitation, "A group of us own a small island about forty minutes from Manila. There's an old emergency airstrip on the island and several supply buildings. We've developed one as a clubhouse. My wife and daughter and a flying friend are going over for a picnic Sunday. Would you be available to join us?"

I accepted. It would be a nice break.

His driver picked me up at the hotel and drove to their home in a very fashionable area of Manila. All properties were gated and surrounded with metal fences or rock walls, with wire mesh or cut glass at the top to discourage intruders. Jim introduced me to his wife, a very attractive blonde American woman, and to Sharon, his ten-year-old daughter. They had three servants: a driver/handyman, a gardener, and a cook/housekeeper. The house was beautiful, with an abundance of very expensive mahogany and teak furniture. Attractive paintings, hand-carved tables and murals by Philippine artists adorned the rooms.

We flew to the island in his private plane. The beach was beautiful coral sand with very green-blue surf lapping on the shore. The ocean was calm, and we went swimming for half an hour, then lay in the shade on the beach.

Black spoke openly about his time in the Air Force and his encounters with Japanese planes. Near the end of WWII he had been stationed at Clark Air Base, outside of Manila. He found that he liked the easygoing life in the Philippines, and that he got along well with the locals. They are a very friendly and compassionate people, very family oriented. He decided to stay.

It was hard to leave the peacefulness of the island, and we got away a little late. Jim had filed a "visual flight plan" on leaving Manila, but by the time we neared Manila International Airport, it was almost dark. He had to radio in and obtain an emergency instrument clearance to land. Fortunately, the airport was not too busy and clearance was approved. The outing was a very pleasant break in a busy foreign schedule.

Prior to one of my trips to the Philippines in 1988, I received an invitation to present a seminar on tunnelling to the Japanese Tunnelling Society. Since the entire seminar would require a translator, the actual presentation time, about seven hours, would take two days. Each sentence was to be translated by a Japanese geotechnical engineer who had obtained a doctor's degree from the University of Missouri. He spoke and understood English very well, including the necessary tunnelling technology. They offered a generous honorarium.

As was my practice, I contacted the Canadian Embassy for advice on protocol. Shortly after my arrival at the Imperial Hotel, in the early afternoon, I heard a soft knock on my door. A short Japanese man introduced himself as the Protocol Officer for the Canadian Embassy. I invited him in. His English was adequate.

He said, "You will be invited this evening to a special dinner here in the hotel with the Japanese Tunnelling Society President and Information Officer, and your translator. It will be a private room. They will discuss your program and method of presentation. There is a guest protocol for the dinner. The first course will be sushi. Many foreigners do not like it, but it will show honour to your hosts if you eat some of the sushi. There will be sauces to assist you to eat at least some."

I said with a definite manner, "I will eat all of the sushi." I was not fond of it, but I understood the implication and would use the sauces.

He smiled, then added, "There is one other protocol. Are you prepared to answer questions at the end of your presentation?"

"Yes, of course," I replied.

He continued, "In the midafternoon of the second day there will be a break. This is not specifically for tea. During this time, those attending will try to decide who will ask the first question. It is a professional honour for that person. If they cannot agree, there will be no questions! Please understand this will not be a personal reflection on you."

Appreciative, I said, "Thank you for that notice and advice." We then talked about my presentation and I advised, "I have arranged it all on slides and will require a carousel projector, large screen and pointer."

"That will be no problem," he said.

The dinner was "semi-formal." They were excellent hosts and had placed a vase in the centre of the table with both Canadian and Japanese flags, a nice touch. They approved of the fact that I ate all the sushi. The two tunnelling engineers tried to speak English, and I respected their attempt. Fortunately, the translator was available to assist. One protocol I had forgotten to ask about was the procedure for bowing. Luckily, they all seemed to be comfortable shaking hands.

Sixty engineers and contractors attended the seminar, and it initially went well. By midafternoon, I found it a bit nerve-wracking wondering whether there would be questions. However, when I finished the presentation and thanked them for the invitation, a grey-haired executive-looking gentleman stood up and asked that first question. Questions continued for about two hours. They were mostly related to concrete linings, concrete admixtures, and blasting design, plus matters regarding North American experience and practice.

When the question period was over, one of the Japanese tunnelling contractors gave me an envelope. Later I opened it. It was my honorarium in Japanese yen.

9

China

China
(1980-1984)

The author was invited to present a paper on Open Pit Mine Stability to the First International Conference held in China following the Chinese Revolution. Through contact with the Beijing University of Iron and Steel Technology, an invitation was later received to present a seminar to eighty Chinese engineers from all over China. As a result of the lecture, the author was invited to meet Vice Chairman Deng Xiaoping, and found conversations with the Vice Chairman to be fascinating. Deng was more interested in producing work for his people than improving productivity. He stated that his most pressing problem was to improve communication within China, where over 1,000 dialects were spoken.

The author was further invited to give two weeks of lectures on Rock Mechanics and Mine Stability at the same university. His wife came with him. Prior to the lectures, several operating mines were inspected where advice was provided on pit slope design, blasting and dewatering to increase safety and productivity. In Beijing, the author was invited to meet with ministers of several departments. They wanted to know how they might improve training and technology transfer. The format of the Canadian National Research Council was described to them.

The Chinese proved to be excellent hosts. They provided a full-time interpreter, the best accommodation available on the trip, and a short-wave radio every night to receive the BBC news or the "Voice of America." The author and his wife were taken to the Chinese opera, an acrobat show, numerous museums, several movies and on a special tour of the Great Wall, Ming Tombs and the Forbidden City.

Chapter 9

China

George Argall, former editor of *World Mining* and *World Coal* in San Francisco, phoned me in early 1980 to state that he was arranging a Specialty Conference in Mining with the China Coal Board and the Chinese Society of Metals. This would be the first international conference to be held in China since the communist revolution.

The conference was to be held at Beidaihe on a beautiful beach on the China Sea, north of Shanghai. This area had been developed for American schools' summer camps prior to the era of Mao Tse-tung. Argall indicated that the Chinese would be allowed to send 200 internal delegates, from whom four papers would be accepted for presentation. The foreign mining community would be allowed 100 attendees and twenty-three technical papers. Argall was to select the author and topic for each non-Chinese paper.

He asked if I would contribute a paper similar to my presentation to the 1980 conference on Rock Mechanics in Surface Mining in Vancouver. After determining that there would be no conflict in my schedule with consulting projects, I agreed. Argall also asked if I could suggest authors in several other specialties that I was knowledgeable in. He was looking for a wide international representation. He advised that the conference would pay all my travel expenses, and that I would have a private room. Most delegates had to share rooms, so this made it easier to accept.

I was teaching Geomechanics in the Mining Department at the University of British Columbia at the time, as well as doing specialist review consulting engineering internationally. As it happened, the Department of Applied Science, which included Engineering, had just made arrangements with several Chinese Ministries to accept a number of senior Chinese engineers as "visiting scholars." These engineers would attend selected classes, have access to the various engineering libraries and laboratories, and make selected field trips to Canadian projects. The purpose was to introduce them to Canadian technology. (I believe this program was facilitated to a large degree by the respect the Chinese had developed for Canada as a result of the outstanding medical service in China provided by Dr. Norman Bethune, a Canadian.) All these visiting students had to have a reasonable working skill in English. Most worked actively in the mining industry.

One of the visiting students to the University of British Columbia was Mr. Wang Ke, a very senior Chinese mineral processing engineer. When he heard that I would be going to China, he was excited and asked if I would be prepared to give a one-day lecture to Chinese engineers in Beijing before the conference started. Since all foreigners had to arrive in China together, I would have to give up one day of touring. I would have enjoyed the tour, but Wang Ke said I would likely be invited back again. As a result, I accepted, and I am very glad that I did.

All of the foreign group gathered in Tokyo and then flew on a Pan American Airlines charter directly to Beijing. There, most were housed in the Friendship Hotel, built by the Chinese, but designed by a Russian architect. It had earlier been called the Peking Hotel, and was actually a hotel complex of several large buildings used exclusively for foreign guests. This system gave the Chinese greater control of the foreigners' comings and goings. Anyone who left the complex had to have a form filled out — in Chinese — with his or her destination. Usually this was given to a taxi driver, with return instructions added. Delegates from some countries who were sponsored by their embassies or consulates made separate arrangements.

The first three days were to be spent in Beijing, with sightseeing trips to the Great Wall, the Ming Tombs, the old Imperial Palace and around the city. Attendance at the Chinese opera and a Chinese acrobat show were planned for the evenings.

The Beijing University of Iron and Steel Technology, where I was to present my seminar, was a cluster of grey, drab four-storey buildings. The drabness was due to the frequent smog created from the extensive burning of coal. Landscaping was sparse and disorganized. I asked two students where to find "Professor Tong Guangxu," head of the Mining Department. They smiled and escorted me to his office.

When arrangements were being made for my seminar, I had requested a skilled translator who had a specialty in geotechnical engineering with emphasis on mining projects. I was assigned Professor Liu, a slim, bright young junior professor who was self-taught in English. Professor Tong, about sixty years old (I guessed), had been trained at the Colorado School of Mines in Denver, Colorado, some thirty-five years previously. He vouched that Liu would be a capable interpreter.

The procedure agreed to was that I would speak a sentence in English, and then Liu would translate after I had finished the sentence. When I wrote anything on the board (plyboard painted black), I would leave space between the lines for him to translate it into Chinese characters.

I had brought several hundred colour slides to illustrate my lecture, most relating to case examples of projects that I had worked on around the world. One half-day was spent on rock properties and the types of failure. This was followed by case examples, each example containing major mining engineering lessons on failures or poor practice being corrected. The site conditions were described, followed by an assessment of the problem, and finally a determination of the solution.

The seminar was held in a conference room in the middle of the Beijing Zoo — right beside the main panda bear compound! I was able to take two breaks from the seminar to watch them. They were more interesting to observe than one appreciates from television. They are coy, playful, and look more like big black-and-white teddy bears than the brown, black and grizzly bears I was used to.

About eighty mining operations staff from all around China attended. Eight were women. Some were engineers, others were geologists, and a few were university professors. During my opening comments, it was emphasized that I wanted to keep the seminar informal. "If you have any questions, please raise your hand and I'll try to provide answers. If you have experiences that you feel the others would be interested in, please tell us about them," I stressed.

After an hour there appeared to be no interest; they simply sat and stared. They did take a few notes. Thank goodness the windows didn't look toward the panda bears. I decided to try a few jokes, which at least in North America, usually got a laugh. I asked, "Do you know the difference between a mining engineer, a geological engineer and a geophysical engineer?" No reaction. I went on, "The mining engineer says 2 plus 2 equals 4.000. The geological engineer says something between 3 and 5, and the geophysical engineer asks, "What the hell do you want it to equal?" Finally a bit of recognition! Five pairs of eyes blinked and two heads turned!

At about ten o'clock I suggested to Liu that we take a break. It took him about five minutes to convince the attendees that the seminar would continue in fifteen to twenty minutes. In the meantime they could go outside for fresh air (in Beijing this meant smog, about three times worse than Los Angeles), have a cigarette or use the toilet. Practically all of them smoked! I left their return up to Liu, and he was partially successful. The last person returned at 10:30.

By noon I was totally frustrated. All morning, the attendees had sat as still as Sian statues. I had described projects in many countries in detail and with good illustrations, but I felt that nothing I was presenting was getting through.

Tong arrived with his car and driver and took Liu and I for lunch to a nearby restaurant. I told them about my frustration and asked for their interpretation of the audience's "no reaction."

Tong said, "Please realize that most of those attending here travel very little except for their university education. They expect to work at their present mine for the rest of their lives. There is very little travel between mines. By describing so many different projects in so many different countries, and presenting such extensive experience, you appear to them as a god. They are afraid not to look at you! They are afraid to ask questions! But I wish to assure they are listening very intently."

I wasn't really convinced. On the way back to the seminar room I struggled to come up with some idea that would encourage them to participate, or at least get them to provide some indication that they were understanding what I was saying. An idea came to me. How would they react if I described a mistake I had made? I thought, "What the hell — give it a try!"

Liu got them to take their seats. I opened with a few comments about how I hoped they had enjoyed their lunch, although I'm not sure they had any, based on the stares I received. I said, "I've shared many projects this morning with you that ended with successful conclusions. Now I am going to tell you about a failure — a mistake that I made." I said this very slowly. Liu translated. I said it again, still slowly, but louder. Many eyes blinked, heads turned and feet shuffled. Finally, I had captured their attention.

I explained:

"In 1968, my company (I had previously explained what a company was) was asked by the manager of Kaiser Resources Coal operation in Canada to determine the pit slope angle for a new surface coal mine they proposed to develop. The investigation required that we drill and obtain samples of the coal to perform strength tests. We obtained a special drill and split core barrels to obtain undisturbed samples of the coal. As soon as the core barrels and samples were recovered, they were coated in wax, wrapped in gauze, re-coated in wax, re-wrapped in gauze, and placed in a wood sample box with the void filled with sawdust. Three such samples, each four feet long, were prepared, and the boxes were tied together. I carried the boxes to the nearest airport (Cranbrook, B.C.), and purchased two tickets to Vancouver, one for myself and one so the boxes could be strapped into the seat beside me. I did not want them sent as baggage for fear of their being bounced around and damaged.

"I flew to Vancouver on a Friday afternoon, then took the cores to the office in my car, which I had parked at the airport. At the office, the boxes were placed in a moist room to minimize the chance of a change in moisture content, as this would result in a change in strength. On Monday morning we retrieved and opened the first box, cut the gauze and wax, and opened the split tube. The sample fell apart! We could not determine the strength! We opened the other two samples — with the very same results. The drilling and sampling had failed!

"I had not been involved with coal before. I had not recognized that coal often contained methane gas, and that the coal sample was slightly less in diameter than the inside of the sampler. As the sample was taken, the lateral pressure at depth was reduced, the methane gas was partially released, and the sample expanded. Over the weekend, enough expansion occurred in the moist room for the sample to disintegrate when the sample tube was opened.

"The solution was to redo the drilling using core barrel liners equal to the length of the test sample, extract the sample as soon as it was taken, and perform the test immediately on site, in a portable field shear or triaxial test apparatus. Of course this second program had to be done at no cost to Kaiser Resources. This mistake cost us both time and money — and embarrassment."

I asked, "How many of you have encountered problems at the mine you work at where a failure occurred?"

One hand near the centre of the room slowly rose! "Professor Liu, would you ask him to describe it?" I asked.

The engineer stated, "Cracks and push up had developed near the bottom of a high rock slope. After much discussion they decided to excavate the lower cracked area. The excavation was nearly completed when large cracks developed above the excavation and slabs of rock toppled down on the truck being loaded. The truck was badly damaged and the driver was injured."

During the morning I had described the various types of rock failure. I asked, "Based on the description this morning, can you tell us the type of failure that occurred?"

He hesitated then slowly answered, "I think it was toppling."

"Excellent," I complimented him. I looked at the class. "Do you agree?" About ten hands went up and many others nodded yes. The ice had been broken! Between then and about three o'clock I reviewed several more case examples, and then asked, "Would any of you describe any problem that you have encountered, and tell us how it was solved successfully?" Three class members came up to the front, drew sketches on the board and described their experiences. The last one was obviously proud of having solved a problem at his mine. He smiled as he finished and looked at me. I said, "Thank you for an excellent example," and I clapped. The rest of the class also applauded!

At about three o'clock I called for a break and said, "When you return, I would like some of you to describe a stability problem you are presently having trouble with, and then we'll discuss it in class."

About that time Tong arrived and asked how it was going. I answered, "The group are really getting involved and seem to be enjoying the reduced formality. I have many more case examples that I brought with me. If they'd be interested, and it could be arranged, I'd be pleased to continue for a second day."

"Oh, that would be difficult. Travel changes would have to be made for most of them," Tong answered.

The Great Wall of China. I was given a private tour to make up for the sightseeing time lost as a result of my lecturing.

Visiting the Ming Tombs. Left to right: Professor Tong Guangxu, head of the Mining Department, myself and Professor Fang Mei, who spent many days sightseeing in Beijing and translating with my wife.

Liu cut in, "They are very interested. I've not seen such interest in a lecture before. I'll help you with the changes."

Tong looked thoughtful, then said, "I'll try. I will be back by four o'clock."

About 4 P.M. he returned and said to me, "If enough of the group agree, the second day can be arranged." He moved to the centre, in front of the class, and asked for their attention. "If you feel this seminar has been of interest and useful to you, Mr. Brawner has agreed to continue tomorrow. We can arrange to change your travel plans and provide accommodation in Beijing tonight. How many of you would like to attend tomorrow?" Practically all hands went up, and many clapped.

The next day was a lot of fun. The interest and interplay were great. Some class members lined up along one side to get their chance to describe a problem and solution. Others described a problem and asked what I would do. Instead of answering directly, I threw the problem open to the group for discussion. I learned later that they felt comfortable with this because most decisions at the intermediate level of their mines were made by committee. In fact, this is why so many decisions in China take so long. When I finished the seminar, I thanked them for their attention and interest. They lined up to shake my hand.

My lectures paid off! Tong advised me that, since I had missed two days of the conference tour, they had arranged a private tour for me to visit the Great Wall, the Ming Tombs, and the Beijing historical museum. Tong and Fang Mei came along with a tour guide and chauffeur-driven Chinese automobile the size of a Cadillac. The weather was good, the scenery interesting and people were friendly. Many stopped to stare at this white tourist with the bright sports shirt and brush cut. (Remember, this was 1980.)

The Summer Palace off Tiananmen Square. Bicycles were everywhere.

I only encountered about a dozen non-Chinese on the tour, and coloured clothing had not yet made its way to China. Most people wore drab grey, khaki or light beige. The executive level often wore black, high-collared suits.

Tourist information had described the Great Wall and Ming Tombs in detail, so seeing them was not a surprise to me. What was a surprise was to see a herd of camels at the tourist entrance to the Great Wall. For 5 yuan I could have had a camel ride!

The best surprise of all awaited my return to the hotel. I had just finished taking a shower when a quiet knock came at the door. Two official-looking Chinese in well-cut dark suits stood stiffly in the hall. One announced in very good English, "Please be ready at six o'clock for us to take you to a reception at the Great Hall of the People. You will be a special guest prior to the official conference dinner."

"I will be ready. Thank you," I answered.

I had brought a good suit, shirt and tie, and so put them on. At six o'clock the two Chinese returned to escort me to a waiting limousine complete with a uniformed driver wearing white gloves — surprisingly formal. We drove through the city, with the driver negotiating his way through thousands of bicyclists. These created major traffic jams during rush hours. In about twenty minutes we arrived at Tiananmen Square, which was largely deserted, turned to the right side of the square and stopped in front of an impressive pillared building — The Great Hall of the People. My two Chinese escorts led me up the steps, along several corridors, and into a small private sitting room with a bar at one end.

George Argall was there speaking with two Chinese through an interpreter. He waved me over and introduced me to the minister of coal mining, as well as the president of the China Coal Society. There were approximately ten Chinese men and four Caucasians standing in several groups. Four translators, wearing white jackets, were also there.

In a few minutes the door opened and Deng Xiaoping, vice chairman of China, walked in with a security guard and personal interpreter. Everyone turned toward him and applauded. He raised one hand for quiet and spoke softly for several minutes. His translator then interpreted his remarks. He extended his welcome to us, expressed his hope for a fruitful conference and exchange of ideas. He commented briefly on this being the first international conference to be held in China in many years. (No mention of the word revolution.) By this time everyone had been served with a drink, some alcoholic and some juice, so the vice chairman then proposed a toast welcoming us to China. Following this, Argall and one of the Chinese ministers introduced the vice chairman to the four of us through an interpreter.

George Argall, publisher of World Mining *(left), proposes a toast to the vice chairman of China, Deng Xiaoping. The author is at his right. In the Great Hall of the People (1980).*

A conference official entered the room and we were given a card with our table number shown for the dinner. He led us into the Great Hall, where there were about 300 in attendance. My card was marked "Table 1," in the front centre row near a large stage. My seat was beside and to the immediate left of Deng Xiaoping; George Argall sat to his immediate right. Mr. Berglund from Sweden and Mr. Hughes from the United States sat at the table with us, as well as three Chinese government ministers. A translator sat between Argall and Deng, but back from the table. He did not eat. He had work to do!

The discussion around the table was free and informal, with most people participating, and dealt mainly with the Chinese mining industry. Several interesting exchanges occurred. Deng indicated, "China must increase production of selected commodities to obtain foreign dollars or trade credits. One of the commodities is oil. Contracts have been negotiated with four countries to explore the off-shore oil potential in the China Sea." He also said, "China wishes to increase the development of electricity from hydro sources, thereby reducing the use of coal, which is responsible for much of the pollution and smog in China. We also wish to release coal for foreign sale. A major proposal is to develop the hydroelectric capacity of the Yangtze River system."

He also noted, "China wishes to develop a nuclear power capability and is developing a working nuclear reactor." He further indicated that, "China is willing to exchange technical visits and information on the basis of equality and material benefit. We are also willing to collaborate with friendly countries and companies in developing mines by way of compensation trade, joint ventures, and the co-production of mining machinery."

The vice chairman obviously understood English, because as soon as he was asked a question in English he answered in Chinese. Argall had, the previous day, been to a nearby open pit iron mine and observed that small Russian-made equipment (trucks and shovels) were predominant. He asked Deng, "Why don't you use big equipment as we do in America to be more efficient?"

Deng answered immediately in Chinese. His interpreter translated, "I have six hundred million Chinese to keep working. [Most women, except the elderly work.] If we become too efficient, there will be unemployment. Then there will be a revolution and I will be out of a job!" He looked around the table with a wry smile on his face.

I asked him, "What do you consider the biggest challenge in China at this time?"

He thought for a minute and then answered, "Communication. This may be difficult to understand since this is your first trip to China. Your alphabet has twenty-six letters. Our basic alphabet has three thousand characters. Our technical people have to know about six thousand characters. It takes our children two to three more years to learn to read and write than your children. Our typewriter is very crude and slow. A good typist can only produce five to six words a minute. We have over one thousand dialects. Many of our people cannot communicate with one another. So I have designated Mandarin as the official dialect throughout China. You can appreciate the problem we have to get enough teachers for many thousands of schools. We have also decided that English is to be taught as the second language, and this presents a major challenge. We can use your numerical computer technology, but have not yet developed computer programs that integrate our Chinese characters. As we open up China to foreigners and trade, we will have to provide travel signs, road signs, and street signs. I have given instructions for these signs to have the Chinese characters translated phonetically, below on the signs, into your alphabet so they can be pronounced in Chinese."

Leaving the special dinner. The Vice Chairman beckoned me to walk with him for the TV cameras. He answered all of our questions without hesitation.

Deng displayed a sense of humour. At one point George Argall asked him, "How many years have you been involved in Chinese politics and government, and how were you so successful?"

He smiled as he answered, "I was not always successful. I learned to be flexible. I was like a rubber ball. I bounced up and down many times."

After the dinner was complete, we left according to table number, with Vice Chairman Deng leading the way — slightly. He told his interpreter to ask me to walk beside him. As we walked through the main door, cameras followed us. A typical politician! The newspapers the next day published our photographs. I later found out we were also on television. The vice chairman did not address the conference.

Typical steam locomotive in use for many trains. Diesel electric locomotives were also used.

Early the next afternoon, we were driven to a private train that would take us to Beidaihe, about a four-hour trip. The train was all second-class day coaches — about equivalent to Canadian Pacific Railway coaches in the 1940s. Two large steam locomotives powered the train. The cars were reasonably clean, until some of the occupants in our car opened windows. Smoke, soot and the Beijing smog infiltrated the car. With this added to the smoking, and many Chinese smoke, the atmosphere was a bit like the fourth day downwind of the Mount St. Helens eruption. It didn't take long for the windows to close!

I sat with Professors Tong and Liu opposite me, and Jan Dunnyworth, secretary from *World Mining*, beside me. Most of our discussions centred around mining, the university and conditions in China. One discussion I remember well. Tong indicated that, "China is self-sufficient in food — a major feat for a country of some one billion people." He indicated, "The climate and soil conditions dictate the predominant type of crops that are grown. For example, wheat and grain in the colder north, corn in the central area, and rice in the more tropical and wetter southern regions."

The trip from Beijing to the resort was mainly through agricultural areas — cornfields, orchards and pastures where horses, cows, sheep and goats were prevalent. During this conversation, we were travelling through an area of cornfields, and Tong commented, "The corn has many uses, more than in your country. When the corn is ripe, the cobs are stripped, and the husk and silk tassels are separated and stored to feed the cows, horses, sheep and goats of the commune. The cobs are laid out in the sun. When dry, the kernels are shelled by hand and stored in pails, sacks or barrels. The centre of the cob is saved and used for fuel in household stoves or fireboxes. The stocks are left in the fields until they are dry, at which time they are cut off, collected and stored for fodder for the animals. That's not all. The short remaining stock is pulled out, beaten on the ground to remove the soil, collected and saved for use as fuel for fires. We find a use for everything." That is a lesson we could learn in North America. Our material waste is unconscionable to Third World countries.

Tong described the Chinese commune system. Each was responsible for cultivating a range of crops, rearing various farm animals and stocking up supplies for winter. Excess supplies were traded with other communes. Everyone shared in the work according to their ability, and received an adequate share for their personal needs. Several communes shared service facilities, health facilities and schools.

The toilets on the train were an experience. There was one for each car, in a room that was about 4 feet wide and 6 feet long. At one end was a water tap and a square metal basin with a drainpipe to a hole in the floor. There were no paper towels. At the other end was a hole about 9 inches in diameter with two foot-patterns marked on the floor. These footprints were spaced about 12 inches apart and in front of the hole. Whether you squatted or stood, the aim was to hit the hole. This wasn't that simple. If you were tall, you lost your balance when you went around a curve. If you were short you had to lean to a precarious position, and if you were on the chubby side you were in danger of missing the hole because you couldn't see it!

It was interesting standing in line waiting your turn. If the person came out with a satisfied look on his face you knew his accuracy was good. If he came out in a hurry with his head down and rushed back to his seat, you knew it was poor. If he took more than five minutes, you knew he had missed completely, or perhaps gone into a trance looking down the hole counting the railway ties as they went by. In North America, there is a sign in the washroom to "not flush while you are in the station." In China however, a line-up developed before every station, because it seemed they knew their chance for accuracy was better. I think every foreigner on the train tried the toilet at least three times! They didn't want to get off the train until they felt they had beat the challenge.

On arrival at the Beidaihe station, we were met by buses which took us to the conference centre. It had been developed as a typical resort, with separate villas and a large number of multiple apartment units. These units ranged from hotel-room style with bath to three-bedroom units. The complex contained the typical amenities — main conference theatre, numerous small meeting rooms, a large bar (which served mainly Chinese beer and liquor and Johnny Walker Red Label scotch), a large dining room, a billiards room, a smoking room and a commercial-size kitchen.

During the British and American period in Shanghai, large gardens had been developed, but in recent years they had not been well maintained by the Chinese. The complex had not been used during the revolution except for one or two villas that were used by Chinese generals. Argall had advised me that the Chinese planned a major program to reopen the facilities especially for conferences. This had largely been completed for our arrival.

Because of the language difference, all foreign papers were presented in English and translated into Chinese sentence by sentence. The Chinese papers were presented in Chinese and translated into English. After the presentations, instead of having questions from the floor the speaker was taken to a private room with a technical translator. Anyone who had questions for the speaker came to that room to meet the speaker. Usually at least ten Chinese came to ask questions or hear the discussions. The presentations were spread over three days.

The conference was a success and resulted in a number of professional conferences being held in China in the next few years. On the morning of the fourth day, a Chinese air force Russian-made jet picked up all foreigners at the nearest airport and flew us to Beijing to meet our charter to Tokyo.

In June 1981, I received an invitation from Dr. Zhang Wen Ji, President, Beijing University of Iron and Steel Technology in China, to provide two weeks of lectures on rock mechanics in mining. Professor Tong Guangxu, head of the Mining Department, who had been my contact with the university the previous year, initiated the invitation.

I approached Dr. Douglas Kenny, President of the University of British Columbia, to determine the potential for travel funding. He advised me that a Canadian government /university program existed to fund the presentation of technical seminars by university staff in "underdeveloped countries." He assured me that I would qualify for travel cost assistance.

Based on this, I sent a letter to Dr. Zhang suggesting, "To provide maximum benefit to Chinese engineers, I recommend that, prior to my lectures, I inspect a number of Chinese mines where stability problems exist." With this experience I could lecture to

北京钢铁学院

BEIJING UNIVERSITY OF IRON AND STEEL TECHNOLOGY

Beijing (Peking), China

June 12, 1981

C.O.Brawner
209B 1571 Bellevue Avenue
West Vancouver, B.C.
Canada V7V 1A6

Dear Prof. Brawner:

As I understand that you are coming to our University and give a short course on Rock Slope Stability, which is not only to let us have an opportunity to learn from your rich experience, but also to promote the friendship and understanding between our two peoples, I represent our University to Extend a formal invitation and express our warm welcome to you and Mrs. Brawner.

About your visiting arrangement from August 10 to September 6 this year, Prof. Tong Guangxu had already consulted with you, you will spend two weeks for lecturing and another two weeks to visit Zhangsha Research Institute of Mining & Metallurgy and 2 or 3 openpits with serious rock slope problems, meanwhile we shall provide you and Mrs. Brawner the travelling and boarding expenses after your arrival in China. I think everything is already arranged. I shall see you and Mr. Brawner in our University and wish both of you will have a pleasant trip.

Yours sincerely,
Zhang Wenqi
Zhang Wen-qi
President

北京钢铁学院

BEIJING UNIVERSITY OF IRON AND STEEL TECHNOLOGY

Beijing (Peking), China

THE UNIVERSITY OF BRITISH COLUMBIA
SEP 29 1981
PRESIDENT'S OFFICE

Dr. Douglas Kenny
President
University of British Columbia
Vancouver, B.C. V6T 1W5

Dear Dr. Kenny:

With your permission and assistance, Prof. C.O. Brawner of your Department of Mining & Mineral Process Engineering came to my University and gave a series of lectures on "Rock Slope in Openpit" this Summer. We had about 70 audiences from different parts of our country in the field of Rock Slope Engineering to attend his lectures. These lectures were very successful and were highly praised for a wide range of his knowledge as well as his rich experience. Prof. C.O. Brawner had also visited three openpits in the south part of our country before giving the lectures, he warmly gave his suggestions and answered all questions, which are also very helpful to those mines. I think all these works done by Prof. C.O. Brawner this time would promote the friendly relations between our two Universities and deepen the understanding between our two peoples. So I write you this letter and express my great gratitude to you, Dr. G.W. Poling and Prof. C.O. Brawner.

With my best regards.

Sincerely yours,

Zhang Wen Qi

Zhang Wen Ji
President & Prof. of
Metallurgy

Chinese conditions instead of just providing international experience which may not apply in China because of different equipment, different skills, different knowledge, and different work habits. For example, mines in China did not have to make a profit! They only had to supply a commodity.

I advised, "I will cover the cost of my return transportation to China, provided you will arrange all travel in China at their cost, and I wish to have the opportunity to visit Shanghai for two days." Then I requested the ultimate. I asked to bring my wife, Pat.

I received a letter confirming the program, including the visit to several mines, specifying the dates of the visit, and advising they would pay all our expenses within China. They had also approved the presence of my wife. Basically, they agreed to my request and arranged everything. I had told my wife about the invitation and that I would try to obtain approval to take her, but she wasn't impressed. Obviously she didn't think it would be approved.

My wife is a meticulous housekeeper, everything clean, tidy and in its proper place. Based on my trip to China the year before, and realizing that it was truly a Third World country, I wondered if she would adjust to the deprived conditions. I showed her the

Photograph at the Changsha Mine Research Institute. Front row, left to right: Professor Liu, my translator; Professor Baoshen, Vice President; myself; the university President and my wife. The others are professors or university liaison staff.

letter and asked, "What do you think? It'll be roughing it. There will be surprises. The food will be different. The sanitation will be poor. However, I'd like you to come. I believe you'll find it a fascinating experience, an adventure that we'll never forget, and one we'll never get a chance to have again. Please consider it."

She thought for just a moment. “It’s sooner than I would have liked, but I can arrange things. If you really want me to go with you, I will.”

I contacted the Canadian Department of Foreign Affairs in Ottawa, advised them of our trip, and asked for information and advice. They provided travel, protocol, security and health information, and I was advised we should contact the Canadian Embassy on arrival. I made an appointment to see the senior commercial officer in Beijing the day of our arrival.

With the letter of invitation I had no difficulty in obtaining a travel visa to China. The Chinese consulate in Vancouver had already been made aware of our visit. For a Third World country, some things can move surprisingly fast.

We flew to Tokyo, then on to Beijing. On our arrival, both Professors Tong and Liu met us. They drove us to the Canadian Embassy where we met the senior commercial officer. I explained the purpose of our visit and showed him the invitation, along with the itinerary Tong had provided. I asked for advice regarding protocol and what “not” to do. The main advice was, “Be friendly, accept all reasonable invitations, do not get involved in politics, dress conservatively, ask permission before taking photographs, do not chew gum, don’t drink too much, and do not pat the Chinese on the shoulder.”

I then asked for a favour — special assistance from the embassy. Near the end of the seminar, I wanted to take the university president, Professors Tong and Liu and their wives, and my wife’s personal translator and guide, Professor Fang Mei, to dinner. I also wanted to invite a representative of the Canadian Embassy. The commercial officer advised me that Chinese wives were not normally asked out, but I said, “My wife will be with me. Please make the invitation,” and he agreed. He said the group would be most honoured to be invited to the International Club, a social club and store specifically for foreigners.

The next day we were taken to the Beijing domestic air terminal to fly to Changsha. At the terminal we were ushered into the “foreigners” room. We were the only ones there except for Professor Liu. A Chinese announcement came over the loudspeaker, which we didn’t understand, but Lui said, “It is time to go.” Outside, the Chinese travellers were exiting from another room and running toward the plane. We walked and so were the last up the steps. Of course, there had been no seat selection, and only three scattered seats were available as we entered the plane. Liu spoke to the stewardess. She took the intercom and, according to Liu, said, “We have special foreign guests. Will some of you move so they can sit together?” About half of the passengers immediately got up to give us their seats. Liu took charge and arranged for my wife and I to sit together near the front of the plane (a 4-propeller British Viscount), and for him to sit across the aisle. As we stepped forward, everyone on the plane turned — and smiled! We felt welcome, and we were soon to re-experience the real warmth of the Chinese people. On our arrival at Changsha, everyone on the plane remained seated until we had left. I said to Liu, “Please ask the stewardess to thank everyone on board.” He said she would.

At Changsha we were met by the vice president of the university, Dr. Lui Boashen, whom I had met at Beidaihe the year before. We were taken to a central hotel and given a two-room suite on the sixth floor, the best accommodation they had. After we had settled in, a knock came at the door and we were given a short-wave radio. That evening we listened to the “Voice of America” and the BBC news from London, feeling almost at home. Remnants of the revolution still existed. All our hotel meals in China were taken in a private dining room with only ourselves present. We weren't to mingle with the locals!

The next day I gave a half-day seminar on Rock Stability in Surface Mining to members of the Mining Department, with Liu interpreting my presentation. Just prior to lunch, my wife arrived and we had a group photograph taken with the university president and several senior staff. I had Liu take two Polaroid photos, and I gave one to the president. He was astonished and delighted with this unique technology. He had graduated from Ohio State University about thirty years prior and spoke English. The ice was broken. We talked about mining, the university research, and the need for their library to have more mining books translated into Chinese. Our discussions lasted more than an hour.

The afternoon was spent in touring the Mining section of the university — the lecture rooms, the mining laboratory, and then a large building where they were trying to simulate the construction of tailings dams. They were studying methods of depositing sand to construct a dam, investigating how to develop steeper slopes using the sand, and experimenting with several different designs of cyclones to separate coarse from fine sand. They were very proud of this facility. I provided a number of recommendations to enhance their investigations. In addition, I offered to send them copies of proceedings from tailings dam conferences that had been held in North America. On my return to Canada, I forwarded these publications and subsequently received profuse letters of thanks.

I inspected their library and found it sadly lacking in information on foreign technology. One book they proudly showed me was *Mining Engineering* by Peele — printed in the 1930s! They advised me that the textbooks they received were purchased by the Ministry of Communication, but no one on that staff was familiar with mining! I made a strong recommendation that their chief librarian contact the major publishers of engineering texts, such as McGraw-Hill, Balkema and Elsevier, for their catalogues, and that he prepare their request for mining texts.

Finally our escorts took us to their prime jewel — the computer room. The room was about 20 feet square, temperature and humidity controlled. We were asked to remove our shoes and put on paper booties. In the centre of the room were several vertical units with what looked like light bulbs and large circular, film-like cassettes. The earliest of computer models. Remember this was 1981. They spent about half an hour demonstrating the operation. We didn't understand much of it but knew that, to them, it was technology to be proud of. We smiled and nodded, and they were so genuine we were actually thrilled with the demonstration. It was a very touching experience.

Very early the next morning, about four o'clock, we were taken to the train. Professor Liu and a protocol man carried our baggage on their shoulders and put us in a first-class compartment with four bunks, two lowers and two uppers. Pat and I were the only occupants.

At about eight o'clock Liu knocked on our door and said breakfast would be ready in ten minutes. We dressed and waited for him to call for us. When we were led to the dining car, there was no one else there. He urged us, "Please hurry."

All of our meals in China were multi-course. This was no exception — sliced oranges to start, followed by a cooked leafy vegetable, hard-boiled egg, Chinese ham,

Chinese biscuits, rice and Chinese tea. When we finished, we stood to return to our compartment, three cars away. As we left the dining room, we passed a line-up of about forty Chinese. They had honoured us with a private breakfast. The same thing happened for lunch.

The train truly was an experience. It was pulled by a steam locomotive, similar to those that had been used in North America decades earlier. There were four classes of "coach." First class consisted of private compartments; second class was day coaches with upholstered seats, third class meant wooden benches, and fourth class was boxcars where people sat on the floor. At each stop Chinese vendors would try to sell fruit, ice cream, soft drinks, cigarettes and trinkets. Some were soft sell, but others became quite belligerent if you didn't buy. After about the third stop we pulled the blind down when we arrived in the stations.

By noon the heat was stifling; it must have been over 100°F. We were served hot tea, which proved surprisingly refreshing. Because of the soot from the engine, we could only open the window a small amount to let the breeze in.

It was fourteen hours we will never forget, allowing us a good look at the variety of the countryside, particularly the agriculture. All of the farmland was a hive of activity, with manual labour the standard. Occasionally small tractors were being used, and they were very versatile, hauling materials and products, pulling plows, discs and harrows, as well as small trailers for workers to travel on. About half of the workers were women. Only the very young, very old or the disabled did not work.

On the road to the remote copper mine. The workers were widening the road. Summer 1981.

We got off the train at a small city called Shangri, where we were taken to a local hotel and given a suite. Liu accompanied us to our room. He asked, "Would you like to take a bath?"

"We'd appreciate that," I answered, and he spoke to the Chinese attendant.

In a few minutes a coolie arrived with two pails of hot water carried on the typical Chinese bamboo shoulder sling. Of course he didn't speak English, so we pointed to him to put them on the floor. My wife checked the bathtub. It was dusty, appeared as if it hadn't been used in weeks. There were two taps but only cold water, which was brownish in colour. We let the tap run for several minutes, and it gradually became clearer. Pat took a towel and washed the inside of the tub. There was no drain stopper, so we tried rolling up some toilet paper — at least they called it that — but it was thick and rough. We eventually ended up using Kleenex. Pat ran some cold water, after which I poured in the hot water. The bath was actually very refreshing — even though I had my bath in "second-hand" water.

I had also asked Liu if he could get me a bottle of Scotch whiskey, saying I would pay for it. He said, "I think so, but it won't cost anything." About a half hour later he returned with an entire case of Johnny Walker Red Label! I said, "Thank you very much, but I only want one bottle." I opened the case, took out one and said, "That's enough, thank you." He was surprised, but took the case and left.

All the windows were wide open when we arrived, and had no screens to keep the flies and mosquitoes out. Our bed was a wood bunk with a mosquito net hanging over it, and a fibre mat for a mattress. There was no air conditioning, and by 5 A.M. we were slept out. We looked out the window to the street below. Day was breaking, and we could see many people sleeping on the sidewalk and on balconies. Some were already up and doing *t'ai chi* exercises. This was a common occurrence wherever we went.

We heard a knock on our door at about 6 A.M. Approximately twenty minutes later we arrived for breakfast at a small restaurant, where Liu translated our order. The food was good: Chinese bacon, which was thicker than ours, two hard-boiled eggs, a banana, toasted bun, and hot Chinese tea. We asked for several apples to take with us.

Outside we were introduced to three other Chinese who were to accompany us, and the driver, who would take us up to a copper mine about 100 miles away. The vehicle was an eight-passenger Toyota van which looked like it had just come from the sales showroom. It was meticulously clean, inside and out, literally sparkling. I said to Liu, "Please ask the driver if we can take his picture with us and his van."

The driver beamed. He stepped over to the vehicle, removed several specks of dust, then turned around. We stood on each side of him with the van behind, and Liu took two pictures. I tried to get the driver to smile, but, although he would smile at us, he would not do so for the photo. This we found to be the Chinese way.

The Polaroid camera had been a real winner, delighting those we photographed. But this time it scared us! The driver had put the photo in his shirt pocket, and about every fifteen minutes he would take the photo out and admire it, while still trying to negotiate a crooked road! Finally one of the Chinese told him to "watch the road."

The drive to the mine took about three hours, initially on a paved road, then gravel, and finally on a one-lane mountain road with passing or meeting turnouts about every half mile. It was through a very picturesque area, with very high hills and valleys. In the valleys and about halfway up the hills rice paddies were common. The steeper areas were terraced, following the same topographic elevation.

Road maintenance was unusual. We didn't see any road graders to smooth the road surface, but at one location we came upon two Chinese workers, one with a broom and one with a shovel. They were filling the potholes and wash-boarded surface with these primitive tools. It was windy outside and the van was very quiet, so they still didn't see us as we approached. Suddenly the worker with the broom turned, used his broom rather aggressively, and several small stones hit the front of the van.

The driver immediately stopped the van, opened the door and all the Chinese, including Liu, piled out. They each took turns scolding the workers for flinging rocks at the van. The driver spent several minutes looking for marks on the front. I don't know if he found any, but he yelled at the workers again anyway, getting the last word. The van was his responsibility.

The poor workers. They must have felt like the Chinese Revolution had started again. They shuffled their feet, looked at the ground; one removed his hat. They just stood there silently, stunned and overwhelmed by the events. Both Pat and I felt extremely sorry for them. However, we realized we must not in any way intercede. As we drove away, I looked back. The two were still standing on one side of the road, arms limp at their sides. They stared after the vehicle until we went around the next curve.

As we neared the mine and adjacent town, we again encountered roadwork; this time construction to remove a rock outcrop on a sharp curve. We had to stop for a few minutes. The workers were drilling holes to place explosives and blast the rock. Their tools — steel drill rods with a star rock-bit mounted on the end. One worker held the drill rod in both hands while the second hit the rod with a sledgehammer. You can imagine the teamwork and accuracy required. One miss with the sledge and the drill rod holder would have a smashed hand!

A short distance further on, large waste rock from the mine was being broken into pieces about 2 inches and smaller for use as road-surfacing material. A truck had dumped a load of the rock, generally 6 to 24 inches in size, and about ten Chinese women with hand hammers were breaking the rock! As Deng Xiaoping had said the year before, "I have to keep my people employed." Job efficiency was not a priority. The one positive thing I noticed was that they wore gloves and safety glasses.

The van entered the town and drove to a three-storied building, which we concluded was a guesthouse for management and visiting officials. We were taken to our room, which had twin bunks with thin mattresses. Not only was there a short-wave radio, but also a TV! I turned it on and found three channels. Two were Chinese and the third was giving an English lesson. I learned later that a lesson was given every day from 5:30 to 6:00 P.M. on that station.

That evening we were guests of the mine director and several senior engineers in a private room adjacent to his office. Again it was a seven-course dinner with chicken as the main course. We sat around a circular table with the courses placed in the middle on a revolving tray. The Chinese sitting beside us served us, and then went on to display a habit that we had not encountered before. They ate the chicken with their fingers, and when they finished a piece with a bone, they tossed the bone over their shoulder onto the floor! We didn't have the nerve to do that, and instead put our bones on our plates. After a dessert of litchi nuts and tea, they offered a toast with Chinese

Maotai. Pat does not drink, but our hosts were courteous and allowed her to toast with tea.

That evening we were escorted to a theatre to watch a Chinese movie. The mine was in a remote area where very few foreigners visit, so that wherever we were taken, everyone stared at us. At the theatre, Liu sat between us and translated. The movie was a love story along the lines of our soap opera programs. During the midmovie intermission, Liu excused himself and bought ice-cream cones for us.

The next morning after breakfast, we were taken to the administration building where Pat was introduced to two Chinese ladies who would escort her for the day. It was all sightseeing and drinking Chinese tea.

Entrance to an iron mine near Wuhan. Mining operations were not well coordinated and most equipment was Russian made. Mines were required to meet production goals rather than profit goals.

I was taken to the engineering office and provided with an overview of the mine geology and mining program. The primary ore was copper, which was being mined at a rate of about 12,000 tons per day. The open pit was about 600 feet deep. The rock slopes were developed with catch benches every 30 feet. It was explained that the mine was having many bench failures, with rock ravelling down the open pit walls. As a result, rock had damaged one loading shovel and several trucks. In addition, several operators had been injured, one by a rock that went through the windshield.

Following this meeting, I was taken on a tour of the open pit. One thing struck me. The mining equipment, Russian made, was much smaller than in North America. Several trucks, shovels and Caterpillar tractors were inactive, and I was advised that they were awaiting repairs. Preventative maintenance was not standard in China.

After inspecting the rock faces, I concluded that the failures were due to overblasting — using too much explosive per hole to break the rock. They had proposed flattening the slopes. This would improve stability, but would increase the amount of waste rock to be excavated. I was aware that profit did not mean anything to them, but I also understood they had a production quota to meet. Flattening the slopes would mean that much more rock would have to be drilled, blasted, loaded and hauled. To meet their quota of ore, they would require more mining equipment, which they did not have.

Wuhan open pit mine. The north wall followed a smooth shear zone. Rock ravel was dangerous. Scaling every bench was recommended with a cable net catch fence at the toe of the slope.

I explained that by modifying the blasting to reduce seismic energy near the final wall, they could steepen the slopes, reduce the waste rock excavation, improve safety, increase ore production and exceed their quotas. I asked Liu to translate, "Would this be beneficial for your operation?" He understood my proposal and asked the question with excitement evident in his voice.

The director said, "That would be very good. If we exceed our production goal we will receive a bonus." I didn't have the nerve to ask what that might be. "Can you tell us how to do that?" he asked.

Returning to the mine office, I sketched on a paint board in a meeting room how they could reduce the blasting damage. The sketch showed a blast plan with hole size, hole location, spacing, depth, sequence of delays, *et cetera.* Everyone in the room copied the sketch. Before I left the mine, I wrote out my recommendations, leaving two spaces between each line for Liu to translate my writing into Chinese.

Construction of a tailings dam in central China. There were over 30 wheelbarrows being used. The entire group moved together as an army platoon. The soil was dumped, spread and compacted by the workers' feet.

Our next visit was to Wuhan, a large city of about 4 million people located in central China on the Yangtze River, where we were to visit a nearby iron ore mine. Wuhan is a major port with much heavy industry, and consequently air and water pollution were a problem. Traffic jams — of bicycles — were observed in the major streets. A large university is located there, and arrangements had been made for me to present a morning seminar to the students and staff of the Mining Department.

We were given a two-room suite on the top floor of a hotel with a short-wave radio and television set. We could receive two channels. It had no air conditioning; instead there was a portable electric fan to create a cooling current of air.

Liu took sick and so was unavailable to translate. The Chinese called a travel agency to send a translator. A young lady arrived shortly, but, only a few minutes into my presentation, she advised in exasperation, "I am very sorry; I do not understand your technical words. I only translate tourist information."

The seminar was cancelled, and, to the credit of the professors, on short notice, they instead organized a city tour and invited along a structural engineering professor who had studied in the United States. He had helped design the major Yangtze River Bridge at Wuhan. At the bridge, he gave Pat and me a detailed description of the design and construction of the foundations. One had to see the size of the river, the enormity of potential floods, and the size of the bridge to appreciate the challenge. I complimented

The downstream slope of the dam being vegetated to resist erosion. All work was done by hand. Seepage from the downstream slope was serious. Pumping wells were recommended.

him on its successful completion. He was especially proud because it was a Chinese project with very limited international involvement, only a review of the design.

Even at that time there was talk about damming the Yangtze to develop hydropower, and I asked him about it. He answered, "Most of our electricity is now developed from coal-fired thermal plants using low-grade coal. You have seen our smog and pollution problems. They get so bad in Beijing that on occasion the airport has to be closed due to inadequate visibility." At the same time, many people in the cities in China burn coal for heating and cooking. We in North America simply have no comprehension of the serious coal-related pollution in China. They have to either develop hydro energy or nuclear power, or both, in order to reduce pollution.

The next morning we were driven to an iron mine near Wuhan, where I was asked to inspect 600- to 800-foot-high open pit slopes where many rockslides had developed along one wall, and a tailings storage dam where major seepage was occurring from the lower half of the dam. My review of the rock stability in the open pit revealed that the ore contact with the waste rock was a through-going shear zone (a zone with disturbed rock) that dipped about 42 degrees into the pit. They were using standard safety bench design, and every new bench was failing along the shear zone. I pointed out the impact of the adverse structural geology, and advised them to forget the benches and instead mine to the shear zone using a revised variable-depth blasting pattern, which I would develop for them. They must then scale loose rock from each bench face.

Processing of ore involves crushing the rock to fine to medium sand size. The non-ore bearing sand is a waste material which is pumped as a slurry into a pond behind a constructed tailings dam. We spent about an hour inspecting the tailings dam at the site. It was about 300 feet high, and clearly a disaster. The dam was built from local soil not adequately compacted and overly steep. Pond water was sitting directly against the upstream slope, and seepage was occurring from many zones below mid-height. Some of the lower seeps were actually carrying dirty water. I was surprised that the dam hadn't already failed.

Following a discussion to determine their capability to drain subsurface water, I developed the design details for several wells on the downstream slope. In addition, I designed a downstream filter system using sand, sandy gravel and fine rock as a loading berm and slope cover. The filter would allow the seepage, but contain the soil within the dam. The Chinese were extremely attentive during the discussions.

The field inspection was fascinating. About 100 Chinese labourers were adding another 10 feet to the top of the dam with soil from a nearby borrow pit. To do this they were using wheelbarrows, filling them with hand shovels in the borrow pit, then walking the wheelbarrows in single file to the top of the dam. One man, apparently a foreman, instructed them where to dump each load. They then returned single file to the borrow pit. It resembled a procession of ants. Another group of labourers spread the soil in a lift about 8 inches thick, then stomped on it to obtain some compaction and increase the density and strength.

As we were inspecting the seepage areas, the entire line-up stopped for about a minute to stare at us. Suddenly a loud voice yelled something in Chinese. They all lifted their wheelbarrows and continued on. As we were getting ready to leave the dam site, I took several pictures. The labourers again stopped to watch. I waved to them as a goodbye. About half waved back, then resumed their monotonous tasks.

I again left a summary of my designs and recommendations with the mine staff, to be translated by Professor Liu. We visited one other mine near Wuhan, and then flew to Beijing where we returned to the Friendship Hotel complex.

On Monday morning of the third week, I started my two weeks of seminars on Rock Slope Design, Stability and Stabilization for Open Pit Mining. They were held at the Beijing University of Iron and Steel Technology in one of the classrooms, with about seventy engineers and geologists attending, and Liu interpreting. I had provided him with my typed lecture notes so he could study them in advance. I have many colour slides that I use for lecturing, as well as a similar number of overheads, and these were extremely helpful. Anything that I wrote on the paint board, Liu wrote below in Chinese.

One lesson I have learned regarding translations. Always ask for a translator who is trained in your field, so that he or she is familiar with the technical terminology. Express a preference for alternate rather than instantaneous translation (*i.e.*, where you break after each sentence you speak to await an immediate translation, as opposed to where your speech is translated simultaneously via a translator in a booth to an audience wearing earphones). Always check out the translator's capability — ask him or her to translate and then ask for a restatement of what you just said in English.

On Wednesday of the second week of lectures, Professor Tong took me for lunch and advised that three government mining ministers wished to meet with me at our hotel that evening. I asked, "Do you know what they want?" He said that he did not, so I asked, "Can you come to translate?" He answered, "Yes, I will be there."

At about 7:30 that evening, the group arrived. They included the ministers of gold mining, coal mining and of iron and steel, as well as Tong. For about twenty minutes we discussed pleasantries, following which I asked, "You have come for some mining discussions, I presume?"

The minister for iron and steel started the discussion, "You visited four of our mines and the Changsha and Wuhan Mining Research Centres. At each location you reviewed the existing programs for only a few days, yet you provided the administration with many very good ideas to improve those programs. We would like you to teach us and our people how you can do that so quickly!" He spoke in Chinese and Tong translated.

I was quite taken aback. This discussion was not what I had expected. Since my experience is geotechnical engineering and all the recommendations I had made were geotechnical, I said, "I can only speak about your tailings dams programs and open pit slope stability."

One thing I had learned about the Chinese was that they made decisions by consensus, or by committee. This process often dragged out a long time, in some cases arriving at a decision only by attrition. I continued, "I suggest that you consider setting up two technical committees — one for tailings dams and one for open pit mine stability." I was thinking about the success of the Canadian National Research Council committees on Muskeg, Soil Mechanics and Permafrost. I described this function and operation, emphasizing that their committee membership should be about ten and include at least five mine operating personnel. They should inspect and review related mining operations at least every two years, and hold an annual conference at various sites around China, with the proceedings published. They were very interested, and a lively discussion continued for about three hours.

Before leaving the university, I asked Liu to translate the notes that I had given him, and see that each attendee receive a copy. In addition, I later sent about 100 selected colour slides of international mine geotechnical case examples with a brief description of each.

During our stay in Beijing, we were taken out for four special evenings. The first was to see the Chinese opera, which was interesting because it was so different. The second was a Chinese "Magic Show," which was quite similar to its North American counterpart, but one trick stood out above all others. A fisherman stood up on stage and cast his line with a lure on the end into the crowd. Within a few seconds he reeled in a flopping fish — apparently alive — about one foot long. The crowd yelled and whistled their approval. He did the same thing two more times. Everyone was fascinated.

The third evening we were taken to a Chinese acrobat show. This was outstanding. We had seen a group put on a similar program in Vancouver, but the Beijing show was longer and more varied. The fourth evening we attended a classical musical program. We had not realized the keen interest the Chinese have in western classical music.

During my seminars, my wife was taken by Professor Fang Mei to many places of interest — a rug factory, jewellery factory, carving factory, art gallery, museums, hospitals, schools, kindergarten, *et cetera.* She saw much more of Beijing than I did. During the middle weekend, Liu and Fang Mei took us both to the Great Wall, Ming Tombs, the Old Imperial Palace and the Forbidden City.

On the second Monday in Beijing Pat came down with a very bad cold. She had chest pains, a sore throat and a fever. The university helped arrange an appointment with a doctor, and Fang Mei escorted her. The doctor gave her three different herbal pills. When Pat discussed payment, he wanted U.S. dollars, so she gave him ten dollars. He was pleased, and it was the best ten dollars she has ever spent medically. The cold and all symptoms were gone in twenty-four hours!

We had to be careful when taking pictures. In one instance, in a city square, we saw a Chinese woman with a baby in a wicker baby carriage, one unlike any we had seen before. Using sign language, I asked to take their picture. She nodded approval and smiled. I took two with the Polaroid camera and gave her one. A small crowd had gathered. She was absolutely delighted and started to show the photo to bystanders. A number wanted me to take their picture, and several became quite aggressive. We had a special guide and translator for this tour. They said, "You must leave now, quickly!" Then took us by the arm and led us into a nearby building and through an official-looking door to escape the crowd. Lesson learned.

During the second week, Tong asked if there was anything else we would like to do in the evening. I was aware that China was a world leader in table tennis and asked if we could see their top table tennis team in action. He said, "I will inquire." The next day he advised me that the Chinese were preparing for the world championships and would not allow outsiders! As if I were a table tennis spy! This, however, was the only request they did not grant. They had been marvellous hosts.

On the last Friday, the dinner I had asked the Canadian Embassy to arrange took place. All those invited, including the wives, attended, as well as the assistant commercial officer of the Canadian Embassy. We met at the International Club, a special club in central Beijing for foreigners, particularly businessmen and dignitaries. The club had a main dining room, lounge, guest rooms and meeting rooms. It was adjacent to the Friendship Store, a department store established for foreigners.

We sat at the traditional round table with a centre turntable. After all the introductions, I spoke and thanked the staff for their hospitality and assistance, particularly Professors Tong, Liu and Fang Mei. I also thanked the wives for attending. They actually seemed at ease, and to be enjoying themselves. The assistant commercial officer also spoke and extended a warm welcome from Canada. He placed a flag-holder in the centre of the table with both the Canadian and Chinese flags. Finally Tong spoke and thanked me for my seminar at the university.

As they were our guests at dinner, it was our place to do all the serving. However, we only assisted with the first course, sort of a ceremonial gesture. They laughed as we apologized for our lack of skill at using chopsticks. We learned that you haven't reached any degree of skill with chopsticks until you can pick up peanuts!

After dessert and tea, I presented each of the professors with a copy of the "Conference Proceedings of the Second International Conference on Stability in Open Pit Mining," for which I had been chairman and editor. At the conclusion I arranged to pay for the dinner. The total bill for fourteen people was only $56! The food was excellent. What a deal!

On the final Saturday morning we flew to Shanghai, with Fang Mei accompanying us. She was thrilled because it gave her a chance to see her brother and sister, whom she otherwise would not have seen. We were met at the airport by a special tourist guide, then taken to the Jingjang Hotel. This hotel had been in its prime during the era of major British trading influence. It had an indoor swimming pool and games room, at that time likely the only ones in China.

We were overwhelmed by the city. It was massive — with over 10 million people and 6 million bicycles! The first day we were taken to a beautiful Chinese ornamental garden in the centre of the city, and for a tour of the Shanghai waterfront and docks. The scale of activity was enormous.

On the second day we were taken through the International Trade Center Building, a massive multi-storey structure that had been designed and built by the Russians when these two countries were on good political terms. In this building, most non-agricultural products that China was producing for international consumption were on display, including tools, rugs, artisan objects, electrical supplies, sports equipment, clothing, shoes, *et cetera*. Each product had its own section on a floor.

One display that caught our attention was the piano section. About twelve different styles were on display. At a beautiful grand piano, a Chinese pianist was playing classical music. He was excellent, so much so that we stayed for about twenty minutes listening to him. That afternoon we left Shanghai for Tokyo and home.

We were very impressed with Shanghai. While it was definitely a Chinese city, the past Russian and British influences were evident. All in all, we have great memories of our experiences in China, and of the Chinese people. I made one further lecture tour in China in 1984, but I made that trip alone.

10
Russia

Russia
(1974)

Following the presentation of a paper on Stability in Mining to the Society of Mining Engineers in Denver, the author was surprised by a request to take the place of a member of an eight-man U.S. technical exchange delegation to Russia. He was to replace a rock mechanics engineer who died suddenly of a heart attack.

In Moscow, many humorous situations developed concerning sightseeing, accommodation, air travel, tour guides, shopping and official dinners. These incidents included bribery for service, a propaganda lesson on the way to tour the Russian space and mining museums, trading blue jeans, gum and pantyhose, and eating raw eggs to prepare for toasting with vodka at a dinner with ministry officials.

The author's paper was translated and printed in Russian, including titles for all colour slides, an effort unique among the U.S. delegation. Following presentation of the paper, questions were not addressed from the floor but in a private room one on one with a translator. None of the questions received related to the lecture, but rather, were directed towards geotechnical problems each Russian was encountering on their projects. They were after free consulting advice.

During non-conference hours, the author, along with other delegates, attended the Moscow Circus, the Bolshoi Theatre and a Russian nightclub.

Chapter 10

Russia

My first contact with Russian engineers was not in Russia, but in Canada. Technical exchanges had begun in the early 1950s between Russia and Canada, and one common area of engineering concern was arctic engineering. The National Research Council in Ottawa played a major role in developing these exchanges. Canada had a keen interest in the far north regarding oil and gas exploration and development, mineral exploration, access to the northern Inuit villages and, of course, military protection, the latter in partnership with the United States.

During construction of the MacKenzie and Dempster highways up to Inuvik, many streams had to be crossed. Experience with culverts under those roads showed that they froze up completely during the winter, with the spring thaw creating major runoff and flooding before the ice within the culverts melted. This required a new assessment of drainage control.

Armco Steel Company had developed a wide-span culvert — up to about 30 feet across — which had a large width to height ratio. The great width provided enough space for a complete winter ice build-up without plugging the entire culvert, allowing flow through the culvert during the initial thaw period in the spring. The secret to successful construction of this culvert was to provide adequate lateral stability. In addition, the base of the culvert had to be comprised of dense non-frost heave susceptible material, or a gravel pad, on bedrock. Non-frost-susceptible sandy gravel or well-graded crushed gravel was compacted to high density immediately outside the flanks of the culvert. This material was placed in thin lifts, compacted to a specified minimum density, and raised uniformly on each side. If all this was done, the program proved successful.

The Russians became aware of these special culverts when they attended a permafrost conference in Ottawa sponsored by the National Research Council in 1972. They requested the opportunity to inspect them. I had an interest in northern geotechnology and offered to act as an escort. A geotechnical engineer with Parks Canada was also selected. He spoke Russian so could act as a translator.

We met three Russian engineers in Edmonton and then flew on a commercial flight to Inuvik, near the mouth of the MacKenzie River on the Arctic Ocean. Inuvik is a unique town developed specifically to function under arctic conditions where permafrost, with an active annual upper freeze-thaw layer, occurs. The majority of structures in Inuvik are founded on piles with air space below the main floor. Separate heating of structures is often replaced with a central heating facility from which heating is piped via humidors to those structures. Sewage and water is also piped in the humidors. Arctic engineering is a highly specialized profession.

The Russians had cameras with them, and we advised them that they could take photographs from the plane. However, since it is against the law in Russia for foreigners to take photographs from an aircraft, they refused to do so. I asked the airline stewardess if the Russians, one at a time, could go up to the cockpit. She asked the pilot, who agreed, but the Russians also rejected this suggestion. Finally, the first officer came back and personally invited the Russians up front, one at a time. This time they accepted and even took photos from the cockpit. North American freedom was a major surprise to them.

We overnighted in Inuvik. The next morning we drove out to the airport and boarded an Otter aircraft to fly to two sites with large-span culverts on the MacKenzie and Dempster highways. About every 30 to 50 miles, where a long, level tangent exists, the highway is widened so that it can be used as an emergency airstrip. We were able to land near two culverts.

After landing near the first culvert, the pilot taxied to a small turnout area where he turned the plane toward the ditch and stopped. We all climbed out, and the Russians walked to the front of the plane and began to laugh, talking among themselves and joking. We didn't understand what it was all about. We asked them to come with us to inspect the culvert. They clambered around the culverts and took some photos, but did not seem overly interested. As we returned to the plane, they started to laugh again. The Parks engineer asked (in Russian) what they were laughing at. They answered, "The pilot isn't very smart to park the plane so he can't turn around." We decided not to say anything.

We got back on board the plane and the pilot started the engine. The Russians watched intently. Slowly, the plane started to back up! The Russians looked at each other in dismay. They were witnessing something they had never seen before — an airplane in reverse. They didn't know that the Otter can reverse the propeller pitch sufficiently to back up. This simple manœuvre caught their attention more than the culverts.

That evening we were all invited to dinner at the home of the government agent in town. He showed them through his house. They paid little attention until he showed them the automatic dishwasher and clothes washer and drier, conveniences they were not used to seeing. They weren't satisfied until they opened the dishwasher during the wash cycle, whereupon it of course shut off.

By the time we eventually parted company with the Russians, we had established a mutual understanding. I felt, however, that they were more interested in the scenery than engineering.

I have provided engineering reviews of several reports on Russian projects for North American consultants. I have not, however, been involved directly in engineering inside Russia. Nevertheless, an interesting opportunity arose in October of 1974 which resulted in my only visit to that country. I had just completed giving a presentation on "Stability in Open Pit Coal Mining" to the American Society of Mining Engineers' annual conference in Denver, when a gentleman came up to the side of the podium and asked if he could speak with me. He said, "It's important and urgent."

I picked up my slide tray from the projectionist and accompanied him to the coffee shop in the convention hall. He introduced himself as James Barclay from International Conferences Inc., and then said, "My company has been hired by the U.S. government to arrange an exchange mission of mining engineers between the United States and Russia under an initiative introduced by President Richard Nixon. Eight engineers, each with a different mining specialty or product, agreed to the technology exchange. However, I was advised yesterday that the specialist in rock mechanics and mine stability who was selected to go has died of a heart attack. The mission is to be in Moscow in one month. I heard your presentation and your topic fits our requirements. The program and travel involves seven days, with all your expenses paid. While the timing is tight, if you get your passport to me quickly, there is time for me to make all your arrangements, including obtaining a Russian visa. If you can come, it'll be greatly appreciated."

I said. "I'm interested, but I'm a Canadian."

He said, "Under the circumstances, I'm sure that's not a problem."

I had my appointment book with me and looked at my schedule. I had two conflicting dates, but felt that I could change these. I said, "It looks okay, but I want to contact the Canadian Department of Foreign Affairs in Ottawa for advice and information. I'll try to contact you by noon tomorrow with my answer." In those days, the Canada / United States relationship with the Soviet Union was still restrained. I wanted to obtain any necessary authorization.

He said, "Okay. We'd greatly appreciate your involvement, so I hope you'll agree." He gave me his business card and phone number while in Denver.

I had travelled from Vancouver to Denver only to give my paper. As a result, I had to leave almost immediately for the airport and the return flight. On my flight home, I considered what would be involved in the trip and what questions to ask the Canadian Department of Foreign Affairs in Ottawa. I could phone from my office the next day about 8 A.M., catching staff before noon Ottawa time, and advising them of the request for me to be part of the U.S. technical mission. (Eight Russian mining engineers were to come to the U.S. under the exchange agreement.) I wrote down a set of questions, a list that included the following:

1. Does the Department have any objection to my going?
2. If not, please advise me about the likely mission protocol.
3. Any special formalities with Russian immigration or luggage, *etc*?
4. Any travel or sightseeing restrictions, for example, photography?
5. How should I handle arrangements for tickets, taxis, ordering food and tipping, translating, *etc*?
6. Could they provide the address, name and phone number of the Canadian Embassy and officials in Moscow whom I may need to contact?
7. Are there any hand signals or body signals that would embarrass or antagonize the Russians?
8. There will likely be several formal meetings. What is the protocol regarding dress, gifts, toasts?
9. Any special advice regarding shopping?
10. Where is the best place to change money?

In the event that I would go, I decided to have my lecture, colour slides, overheads and business cards translated into Russian. I also decided I would ask Barclay if I could set up a corporate display for Golder Brawner, with photographs and descriptions, at the conference hall in Moscow. I had no idea what the potential was for developing consulting projects in Russia, but I felt I had to try to take advantage of the trip in developing whatever professional exposure I could.

The next day I phoned Ottawa. After spending half an hour on the telephone, I finally spoke to the appropriate official. I received a positive reaction to the prospect of my trip, and a promise to airmail a list of dos and don'ts, along with general information on protocol. The official advised me to be sure to contact the Canadian Embassy in Moscow soon after my arrival, informing them of my location, program and travel plans. He gave me contact details.

When I asked about gifts, he said, "The Americans will likely take watches, pen and pencil sets, small calculators, *et cetera.* Why don't you take books on hockey or picture books of Canada? That will be different."

I asked about toasts and toasting. I had heard stories of the Russians proposing many toasts — just to get their guests drunk. He said, "I'm aware of the large number of toasts, but I don't know if that's the reason. At any rate, you *must* participate in the toasts, or they will consider it disrespectful and be offended. I recommend you eat two raw eggs just before you go to the function! You should be able to get them from the dining room. They'll coat your stomach and reduce the amount of alcohol getting into your blood." I made special note of this suggestion. He also said, "If you don't like vodka, ask for wine."

As soon as I got off the phone to Ottawa, I phoned Barclay and advised him that I could go. He said he would teletype the itinerary travel plans and all instructions, and agreed to the corporate display for the conference. I told him that I wanted the full flavour of Russia and therefore wanted to fly on Aeroflot from London to Moscow, return. He was surprised, but agreed again. Today, knowing about the poor record of the Russian airlines and airplanes, I wouldn't do that. He asked for me to send my passport by "special delivery" airmail to him. There was no courier service in those days. I also indicated that I intended to take some books as gifts, and that I would print 100 copies of my lecture and a special corporate brochure, in Russian. He said, "I'll send you labels for Russian customs clearance."

That afternoon I contacted the Russian Department at the University of British Columbia and arranged for my paper, business card and corporate brochure to be translated into Russian by an engineer. I had learned in the past that it was essential to have a technical specialist in engineering, or the translation would be very poor. I also arranged with Ray Chow, my executive assistant at Golder Brawner, to prepare a poster display with Russian descriptions that I could bring along in a parcel as luggage. He did a great job. The display stood 5 feet high and 5 feet wide, but fitted into a container 2.6 feet by 2.6 feet by 4 inches.

The information package from the Department of Foreign Affairs was bulky. The major instructions — they were stronger than recommendations — were quite specific, for example:

1. Take an assortment of small items to be given as a token of friendship to lesser officials, or for tipping, *i.e.,* ballpoint pens, chewing gum or silver dollars.
2. Take plenty of business cards. [I found out later there is major emphasis placed on your business title in Russia. My card did not show my title, and, as a result, I was the only one who did not receive a gift from the Russians. I guess they figured I was less important than the others. As it turned out, the gift was a miniature dragline in

a clear glass cube. Big deal!]

3. Keep the address and phone number of the Embassy and hotel with you at all times, in Russian. Note that there will be no readily accessible phone books in your hotel room.
4. Do *not* fraternize with the Russian women. You will be staying at an Intourist hotel with special areas allocated to non-Communist foreigners. Russian ladies are trained to casually meet foreigners and find out who they are. If they conclude you may have information useful to Russia, they are trained to seduce in pursuit of that information. These women are very attractive and are bilingual.
5. Make all arrangements for taxis, tours, special events tickets, exchanging money, *etc.,* at the Intourist offices in your hotel.
6. Any arrangements for tours, guides or translators must be made one day in advance. Have the address of your destination translated and printed in Russian, along with the time you wish to be picked up. Note that normally the room rate also includes a car, driver and translator. These also must be ordered in advance.
7. If you wish to go shopping, contact the Embassy and they will arrange to take you. Shopping in Russia is complicated and slow.
8. Do not take photographs of the airport, train or subway stations, industrial plants, bridges or other security risk areas. Do not take photos out of airplanes over Russia. They will confiscate your camera and film!
9. Only go out at night with a group.
10. Do not exchange currency, or exchange or sell your clothing with the locals. At this time blue jeans are a commodity much in demand by the Russians, and they make a practice of accosting you in the elevators with a request for such clothing.
11. Do not take any religious literature or a bible with you.
12. Do not take pornographic literature, *i.e., Playboy.*

In due course, all arrangements were made, including the translated information. My passport with a Russian visa, airline tickets, itinerary and customs labels arrived. My first flight was to Chicago to meet up with the group; then it was on to London.

I was the only one who flew Aeroflot to Moscow. The others took Pan Am. I was seated before the pilots came on board, but when they did, I noticed that they were dressed as air force officers and carried pistols in a hip holster. No wonder they have few hijackings of airplanes in Russia! The plane was not a design usually seen at home — four jet engines mounted at the rear. The interior finish and cleanliness of the plane was much below North American standards, but service on board was adequate, and the flight attendants spoke reasonable English.

My seatmate was an Englishman who worked for a medical supply company in England. He said he was born in Russia, spoke Russian and had immigrated to England in his teens. I noticed he had a lawyer's-size briefcase, and so I asked, "What's with the large briefcase?"

He answered, "What do you think I've got in it?"

"I don't have any idea," I said.

The Russian aircraft I flew on from London to Moscow. There were four jets at the tail. The pilots were air force pilots and wore pistols.

He reached down, pulled the briefcase from under the seat and opened it. It was completely full of pantyhose and fancy nylon stockings! He said, "Service in Moscow is really slow, especially in restaurants, and you have to bribe agents for tickets to the theatre and sporting events to get good seats. Most of these people are women. In order to get fast service and good seats, I slap a pair of pantyhose on the table or counter and they look after me immediately. In restaurants they almost fight amongst themselves to serve me."

In hindsight, I should have tried to buy some of the pantyhose from him. In Moscow, I soon saw that Russian-made stockings did nothing to flatter the women's legs, and, just as he said, the service in Moscow was terrible.

On our arrival at Moscow airport, we walked from the airplane to a bus which took us to the immigration entrance. All my documents were in order, so immigration clearance was reasonably rapid, but I had to wait extra time for my baggage. The packages with the corporate posters and books arrived at a different location. Finally I took my bags and packages to the customs counter, where customs officials opened my briefcase and suitcase and went through everything. One official wanted to open my camera, even though the film counter was at "6." Fortunately, translators were available and I was able to convince them that they were photos taken in Canada of my family. They also questioned my bottle of vitamins. After about ten minutes I received clearance.

My travel package included a taxi voucher for the trip from the airport to the Rossia Hotel (pronounced Rue-see-a). Being mid-November there was snow on the ground. In the residential area, the buildings were all multi-storey, apartment/condominium-like structures; I did not see a single family dwelling along the way. Occasionally there

The Rossia Hotel in Moscow where we stayed. Non-communist travellers were assigned to one wing. The hotel had an Intourist office and tourist store.

were church spires and market squares. As we neared the centre of the city, I observed imposing office-type buildings. The main boulevards were very wide and could accommodate eight lanes of vehicles — but there was little traffic. At numerous street corners I saw several armed individuals dressed in what looked like army uniforms, obviously there for security reasons. The hotel was only a couple of blocks from the Kremlin, and covered a complete city block. It stood about six storeys high. Upon leaving the taxi, I gave the driver a package of Wrigley's spearmint gum and a Canadian flag lapel pin. He returned the trace of a smile. The Russians were obviously not a happy people. Very seldom did they smile.

I checked in at the front desk and obtained a key for a room on the fourth floor. A middle-aged bellhop took me up to my room. Stepping off the elevator, we were met by a heavy-set matronly uniformed woman who took the key and opened the door. The bellman took my bag in and set it down, and I gave him some gum. The lady motioned for me to follow her. Near the elevator, there was a common room with her desk, a key rack behind, a TV nearby, and about three sofas. There was also a small four-burner hotplate with a coffeepot and kettle. She put my key on the board, and I understood that she would be "Keeper of the Key."

I went back to the room, unpacked my clothes and then looked around. The room was equivalent to a medium-quality hotel in Canada or the U.S. The bathroom was standard but the furnishings were old fashioned. There was a standard-size bed with oversize pillows, two chairs, a writing desk with a telephone, but no phone book. There was also a very ordinary chest-of-drawers. A large scatter rug and two smaller ones lay on the floor, and a chandelier light with four bulbs hung from the centre of the ceiling. A wide double window looked over the street. I could see an Orthodox church nearby, and a corner of the Kremlin in the distance.

I opened the box with the books and papers I had brought. The Vancouver photo books were missing! The Russians had confiscated them. I thought, "Why would they do this?" Each book contained about 150 colour photos of scenery and people in action around Vancouver — parks, gardens, beaches, buildings, museums, sports, ethnic areas and the local mountains. Then it occurred to me. There was an air photo of the Vancouver airport, and several photos of Vancouver harbour. Could this have been of interest to them? (I later reported this loss to Barclay's assistant, but she said it wasn't worth the trouble to report it; no action would be taken.) I had also brought ten copies of the Gordie Howe story. The Russians are great hockey fans, but these books were still in my possession.

The central square near the hotel and Kremlin. The time was February. It was cold but there was no snow.

In Chicago we had been told to meet in the foreigners' dining room at 6 P.M. At about 5:30 I went down to look around. I determined that the hotel had four wings, each for different nationalities. Our wing was for so-called free world foreigners. Other wings were for Russians, Chinese and communist country travellers, and one for other groups. There was the check-in area, a large dining room, an already busy bar with numerous Russian women present, an Intourist travel area, an Intourist store and a gift shop.

The Intourist store was quite large and unique. One could buy a great variety of items, including TV sets, stereos, jewellery, appliances, musical instruments, *et cet-*

era, but only with foreign currency! Russian rubles were not accepted. Anyone with foreign currency could buy there, including the Russians. I learned that most large items in Russia, particularly appliances and televisions, were very scarce, and Russians normally had to place an order and wait for as many as twelve months to receive their purchase. As a result of this scarcity, a thriving black market in currency had developed. Rather than attempt to exchange currency in the open, the Russians would wait near the elevator until a likely looking foreign traveller entered. One or two Russians would rapidly follow, and after the doors closed and the elevator started, they would push a floor number. On arriving at that floor, they would push the "door close" button and begin negotiations. They normally offered rubles at a 10 percent premium relative to the Intourist bank, actually a good deal, and many travellers apparently exchanged money. By obtaining foreign currency, the Russians could buy at the Intourist store and avoid the long wait. The authorities seemed to turn a blind eye to this trading.

Our group met in the dining room with Barclay and his travel and seminar coordinator, a woman about forty years old who spoke Russian. Besides myself, there were representatives from Caterpillar (makers of mining equipment), Detroit Diesel (large engines), General Electric (electric motors), Marion (draglines), Harza (engineers and contractors), as well as representatives from two mineral-processing equipment manufacturers. The travel coordinator had alerted the dining room staff, so that we were given menus immediately. There were ten pages of main courses, with all items listed in Russian, English, French and German. It took nearly twenty minutes for everyone to decide. A waitress duly took our order. After about fifteen minutes she returned and advised us that six of us would have to reorder, as they were out of those items. Barclay stepped in and asked, "What do you have?" The waitress listed only five from a list of over forty items on the menu. Welcome to Russian efficiency!

The dinner was barely edible. Everything was only partially cooked and almost cold. Barclay said, "It will do no good to complain. This is about as good as it gets." Everyone drank a little extra wine, which came from near the Black Sea and was actually quite good. The chap sitting beside me was quite paranoid about the possibility of "bugs" (listening devices) in his room. He informed me that, shortly after checking in, he went through everything in his room — the dresser, the bed and mattress, the table drawers and the bathroom. He also checked the chandelier, the light switch and plug-in receptacles. Finally he even rolled up the carpet. In the centre of the uncovered floor, he noticed a knob in the floor. He turned this and heard the chandelier in the room below fall to the floor! He said he immediately recognized what he had done, rolled the carpet back, left the room and went for a walk. I didn't know exactly what he was afraid of, unless he talked in his sleep, but I didn't have the nerve to ask.

After dinner Barclay described the following day's program. The meetings were to be held in a government building, which had a conference hall. Each morning and afternoon session would have four speakers — three Russian and one American (I was treated as an American) — with the American speaker leading off each session. Questions could be directed to each speaker after his presentation. We were to have breakfast at 7:00 and meet at 8:00 for taxis to take us to the seminar site.

The next morning was quite cold (–5° C), but everyone had a winter coat. I was the only one without a hat. I hadn't brought one because I didn't think we would be outside much, and it would only be something extra to keep track of. We arrived at the hall together and were met by a Russian who spoke English with a London accent. He looked at me without my hat, shook my hand and said, "You're tough."

Alternate translation of the presentations was provided throughout the sessions. This is much more difficult than instantaneous translation. The presentation that I had planned was automatically cut in half. Before the first presentation, I put up the Golder Brawner display in the adjacent entrance room. All descriptions were in Russian, and it attracted considerable attention, but few questions.

Nothing terribly exciting occurred during the technical exchanges. I gave my presentation the second day, and was the only "American" to have Russian descriptions on my slides, tables or overheads. In addition, each Russian received a copy of my paper in Russian. Many expressed their thanks, and several Americans also commented favourably. I felt that they were embarrassed at not having their information translated.

The presentations by the Russians were very theoretical, with a lot of complicated formulas. On the other hand, all of our presentations were either case examples, a summary of recent developments, or practical topics such as safety and preventative maintenance. We generally concluded that little was learned from the Russian presentations.

The question period was unusual. I was taken to a room with only a small square table and four plain chairs; no other furniture. An attractive young woman sat at the table, and, as I entered, she motioned for me to sit down. She explained that questioners would come in one at a time, sit down and ask questions. She would translate.

I asked, "Do you have any engineering experience?"

She answered, "Yes, I am trained as a mining engineer." She would therefore understand our technical terms.

I had eight Russians come in with questions. *None* were about clarification of anything I had said in my presentation. They were all working on mining engineering projects throughout Russia, and each presented me with a summary of a problem that they were uncertain how to deal with. Most dealt with controlling groundwater, or drainage of rock and soil slopes. They had no experience with the use of horizontal drains to improve stability. I discussed their problems as I understood them, and in some instances forwarded technical information to them on my return. I never received an answer back or a thank you!

I had heard that the Russians made a habit of "picking your brain" while giving nothing in return, and I concluded that, in this instance, they were simply after free consulting. Shortly after my return, I received an invitation to present lectures at Moscow University. In exchange, as I had done with my return visit to China, I requested the opportunity to visit several Russian mines prior to the lectures. They refused this request, so I did not accept the invitation. Had they accepted my request I would have gone.

I felt uncomfortable with the question and answer format. Although the translator said no, and I couldn't bring myself to look under the table or chairs, I am convinced to this day that our discussions were recorded. Regardless, the translator was very professional, and appeared to do a good job. On completion of the session, I thanked her for her assistance and complimented her appearance. She obviously misunderstood my intent — which was only to say thank you — and positively bristled. "I am not that kind of person," she harrumphed, turned on her heels and left.

I came to understand that Russians do not give compliments or say thank you easily. They are very reserved — until they become angry. Then all hell can break loose! We watched several fights break out between the taxi drivers in front of the hotel, and one at a restaurant that we visited. We also noted the large number of street people, many of them drunk. Some appeared to sleep outside, despite the cold temperatures.

During dinner the second evening, a Wednesday, we discussed what we would like to do Thursday evening. Four decided they would like to attend the Bolshoi Ballet, and four others the Moscow Circus. I was more interested in the circus, so went with the others to the Intourist desk to arrange for tickets. The representative from Detroit Diesel, Ken Allen, asked the Intourist attendant for four tickets to the Ballet. She opened the desk drawer, looked for a few seconds and said, "*Nyet.*" Allen asked about Friday, Saturday, Monday and Tuesday. She looked in the drawer each time and said *nyet* for each day.

The Caterpillar representative, George Sawyer, then came up to the desk and placed a case with an attractive pen and pencil set in front of her. He asked, "Do you have four seats to Thursday's Bolshoi Ballet performance?" She opened the case, touched the pen and pencil, opened another drawer and put them in. She looked into the ticket drawer again, looked up and said, "Yes we have some seats in the lower balcony!"

Sawyer turned to Allen and said, "That's the name of the game over here once they realize you're not Russian." I wondered what she did with all the "loot" she took in. The salesman I met on the plane to Moscow was right: no bribe — no service.

After dinner, five of us went into the bar. There was a Russian woman singing and a three-man group playing western music. They wore extremely gaudy outfits, almost Liberace in style, but they were actually quite accomplished musicians. We sat together and ordered drinks. After we were served, an attractive Russian woman came over, pulled up a chair and introduced herself as Olga. She wore a low-cut blouse, midlength skirt, and her makeup was a little overdone. One of our group struck up a conversation with her. After about ten minutes the two of them left for another table. I don't know what developed later, but he didn't show up for breakfast the next morning, and never spoke to any of us about it.

After dinner I went up to the fourth floor. I was chewing gum as I approached the Keeper of the Key. She took my key off the board and said, "Gum!"

I said, "Key, please."

She said, "Gum," again.

I got the message. She wanted some gum in return for my key. I nodded and waved for her to follow me. She opened the door to my room; I went over to my briefcase and took out a package of gum. She stated some words in Russian followed by "gum" again. She apparently wanted more. I shrugged and said, "*Nyet.*" She frowned, gave me the key, then turned slowly and left. I had successfully met my first Russian confrontation. I'm unsure what I would have done if she had pressured me further.

The next evening we took a taxi to the Russian Circus. Held in a large arena-style building, it was generally comparable to the Barnum & Bailey Circus in North America. There were animal acts, trapeze acts, jugglers with balls, dishes, swords and flaming torches, acrobats and gymnastic events. Two large brown bears put on an excellent act of climbing ladders while balancing balls, play wrestling and jumping over objects.

The most outstanding event was a display representing the launch of the Russian space vehicle Sputnik. An actual rocket simulation was performed, with the rocket igniting in a belch of flame and smoke. It made two turns inside the arena and then disintegrated in a shower of sparks while dropping a parachute with a dummy to the centre of the arena. It was spectacular. The smoke remained for about ten minutes. The arena was about half full of people, and the Russians watching, although they were in general a very subdued group, reacting with little emotion throughout most of the acts, applauded the rocket simulation heartily. A few even whistled. We all enjoyed the show.

No seminar program was scheduled for Saturday or Sunday, so we enquired about museums, displays, art galleries, *et cetera*. We were told that excellent displays of the Russian space program and mining industry were open to the public on the outskirts of Moscow. We checked at the Intourist office, and were advised that Saturday morning would be a good time to go. Three of us, Art Lang, Jim Slade and I arranged for a car and an interpreter for 8:30 A.M. Saturday.

After breakfast, we checked at the Intourist transportation desk and met our guide and translator. The driver was a man about fifty years old, and the interpreter/guide was a woman about thirty-five, who spoke reasonable English. Her name was Katrina.

The weather had warmed up and threatened rain as we climbed into the car. During the drive, we asked a number of typical questions:

1. Where did she learn her English? She said, 'In the Moscow library using a Russian–English language dictionary and reading the *New York Times*. In addition, the library provided English language lessons.'
2. Was she married? 'Yes, I am married to a Russian army officer.'
3. Did they have any children? 'Yes, a girl who is eight years old.' They planned on no further children.
4. Did they live in a house? 'No, we live in an apartment rented from the government.'
5. How large was it? 'Two rooms.'
6. What was her salary? She didn't answer.

At that point she commenced a review of details about Russia and Moscow, including population, industries, culture and education, but nothing about politics, cost of living, travel, *et cetera*. She sounded like a phonograph record, one reciting a propaganda summary.

About fifteen minutes into the drive, it began to rain. The driver stopped the car, opened the glove compartment and took out one windshield wiper. He climbed out and attached the wiper on the driver's side. Then he got back in the car and turned the wiper on.

Art Lang asked Katrina, "Why did the driver keep the windshield wiper in the glove compartment?" Again she didn't answer.

Shortly after, we arrived at the Ministry of Space display. It had stopped raining. Not having seen the outstanding display at the Kennedy Space Center at Cape Canaveral, Florida, I was impressed. A full-size duplicate of the Yuri Gagarin space capsule was set up in the central area of the display. Details of the space flight and recovery, Gagarin's training, and the early Russian space program were described on display boards — in Russian. Katrina translated the descriptions. Approximately six full-scale rockets were also mounted throughout the area. One missile launcher was mounted on a very large mobile truck, an impressive vehicle.

Photographs were allowed! Jim Slade asked Katrina if she would allow her picture to be taken with us, with the space capsule behind. She showed surprise, and Jim asked, "Don't foreigners ask to take your picture?"

She answered, "No."

He said, "We want to be friends. You're an attractive woman and we want to show our families in the United States that Russians also want to be friends." He stood beside her in front of the space capsule, put his hand on her arm and said, "Look at the camera and smile." She did!

From that point on she asked questions about America, which we answered. We in turn asked personal questions about her, her family, her friends, and also about living in Russia. She answered all of them and frequently smiled. Her entire personality had changed. I think it was because we had complimented her.

At the Museum of Mining on the outskirts of Moscow. Our female tour guide translated the printed descriptions for us.

She said she would like to travel to America; however, "It cost too much and it would be difficult to get travel permission." At that time in Russia, the locals had to obtain police permission to travel even so far as the next city.

We next went to the Ministry of Mining display. It included the layout and description of many mines in Russia. It also had small-scale models of numerous mining machines, including drills, loaders, draglines and underground excavators. One dragline model was motorized and could be manipulated from a control box. Children were lined up to work the model. Even a few adults got into the act.

On the ride back to Moscow it started to rain again. The driver stopped and repeated the exercise of putting on the windshield wiper. Again Art Lang asked Katrina, "Why does he take the windshield wiper off?"

She asked the driver in Russian, then translated his answer, "The factories only make windshield wipers for two weeks each year, so if my wipers wear out they are difficult to replace. As a result it is necessary to steal them from other cars, and they would also steal mine. Therefore I always take them off and leave them locked inside when I am out of the car." Made sense. Back at the hotel I noticed that none of the taxis had hubcaps either!

When we reached the hotel we gave Katrina a US $15 tip, and she gave us the biggest smile of the day. Art Lang asked if there was anything she would like from the United States. She said, "I would like some movie magazines." He asked for her address and said he would send her some. I hope he did. Her assistance resulted in an enjoyable day — once she forgot the propaganda line.

The GUM department store in Moscow. You had to wait in line for everything. At this time there was a shortage of many items.

We got back about 3:00, early enough to visit GUM, Moscow's "state department store." At that time it was reported to be one of the largest stores in the world, multi-storied and extending beyond one city block. The store contained areas for practically all products found today in North American department stores, but for many commodities, such as televisions and kitchen appliances, for which the Russian demand was greater than the supply, only demonstration models were on display. Customers had to order the product and wait their turn, often one to two years.

The purchasing process was noteworthy. There were very few physical displays. If you wanted to order, say, a shirt, you stood in a line-up which could be thirty to forty people or longer. There you gave the details of your requirement — style, size, colour, *et cetera*. The clerk checked the supply. If they could match your order, you were given a supply slip. You then took this to the supply line, again with some thirty to forty people. Upon reaching this clerk, the product was brought to you. Following your inspection, if it was what you wanted, you agreed and were given a payment slip. With this slip you went to a cashier to make payment and obtain your shirt. Again the line could consist of thirty to forty people. It was conceivable that the entire process would take several hours.

Most Russians carried shopping bags at all times. There was such a shortage of so many commodities that, when rumours of a new supply of an item came in, they would immediately try to purchase the item, even if they didn't need it at the time. It was described as uncontrolled rationing.

On Sunday we took a walk around the central part of the city, including a stroll into Red Square, near the Kremlin. Many people were there, as were several squads of armed army personnel. The Kremlin is a massive structure comprised primarily of an office complex. At the time we were there, it was painted a drab yellow, and the window sills white. Situated at one end of the square was a large church, and at another end was the Lenin Mausoleum, housing Lenin's tomb. There was a long line-up heading into the tomb, which we observed to be moving slowly.

As we approached the structure, a tourist guard stepped toward us and motioned for us to follow him. We concluded there was a special line for visitors. He took us in a side door, to where several large photographs of Lenin hung on one wall. Though it was very dark inside, we could see the crypt. However body features were still difficult to make out. A guide was speaking in Russian, but unfortunately we could not understand what she was saying.

On Monday night Barclay had arranged for us to go together in a van to visit a Russian nightclub. Our group all sat at one table. The evening started off with a buffet din-

Lenin's tomb. We were given a private tour. However, all tourist information was provided in Russian.

ner and wine. Next to us was a group of American tourists, about thirty altogether. Beyond them were numerous separate tables of Russians with more women than men. There was a large dance floor, and soon an eight-piece band and two singers came on stage and began to perform. Practically every second piece was a lively polka; the band also played the occasional waltz.

Several of the Russian women came to our table and asked us to dance. They were husky, and you felt like you were in a wrestler hold, but they could dance. At one point, when the band played the theme song from Dr. Zhivago, an American woman became quite excited and ran up to the stage saying, "That's Dr. Zhivago! That's Dr. Zhivago!" The bandleader turned toward her, said, "*Nyet*," and then stopped the music. I don't know why. We stayed at the club until about 11 P.M.

The evening before our departure, the Russians planned a special dinner for us. It was to be in a private conference room in the Mining Ministry Building. Prior to leaving the hotel I went to the dining room and asked for two eggs to take with me. The senior waiter, who spoke some English, just stared at me. I repeated, "I wish to order two uncooked eggs to take with me."

He finally said, "What you want eggs for?"

I felt like saying, "It's none of your bloody business," but I resisted the urge and said instead, "I need them for a stomach problem."

He seemed unconvinced, but went into the kitchen. He returned with two eggs — not even in a bag. I returned to my room, broke the eggs into a glass, held my nose and gulped them down. Then I sat still for a few minutes. They stayed down!

The dinner tables were set up in a U-shape, with Russian dignitaries and a translator at the head of the table. They included the minister of mining and five other senior officials. Four of us sat along each side, with lesser Russian officials between us. Speaking through a translator, the minister opened the dinner with words of welcome, followed by a brief summary of Russian achievements in mining. He thanked us for our technical presentations and indicated that he believed the exchanges would be beneficial. He hoped they would continue. He then introduced the other Russians in attendance. This was followed by each of us introducing ourselves and emphasizing our area of specialty. At the same time each of us presented a gift to the minister. I gave a set of Canadian maple leaf cufflinks and tiepin, plus a book on Gordie Howe. The American gifts included a painting, Indian carving, Alaskan soapstone carving, Texas agate necktie, among others. No two gifts were the same.

The minister then proposed a toast to start the dinner. Small glasses of straight vodka had previously been placed beside our plates. Everyone stood up and "downed the hatch." Most of our group coughed, or at least took a deep breath, and the Russians smiled at our reaction. Four more times during the dinner the minister proposed toasts, and we were all obliged to join in. At the completion of the main course, the minister suggested that the head of our group should return the toast.

We hadn't selected anyone for this "honour," but Art Lang took control, stood and gave a few complimentary remarks. Next followed dessert and coffee, along with Russian brandy. Another series of toasts followed. Several of the Russians exceeded their liquor capacity and could no longer stand up, but none of our group overindulged. We learned to down only part of the glass each time. Finally the minister got up, thanked everyone for coming and left. We all followed. I felt some of the effect of the vodka, but concluded the raw eggs had done their job.

Our last seminar day was Wednesday. That morning we had breakfast together: tomato or orange juice, cold hard-boiled eggs, thick bacon or fatty sausages, rice, and thick hard bread, which could be toasted. Not a very stimulating breakfast. We were told to be packed and ready to go by 10 A.M., as planes were to depart at about 2 P.M.

We arrived at the airport about two hours prior to departure. A special check-in line was arranged for the others at the Pan America counter. I was taken to a special line at the Aeroflot counters. At the time, Aeroflot was the only Russian airline, serving both domestic and international travellers. They had approximately ten counters. I was directed to the departure area and then taken by bus to our plane.

On board, I sat beside a young American mountain climber. He and five others had been climbing mountains in the Ural range. At the same time, young Russian climbers had gone to Alaska to climb Mt. McKinley. He talked about their experiences and explained that they had been looked after well. He said the climbing was challenging and that he would like to return again.

We were served a meal on the plane. The climbers, who were sitting close together, were talking about souvenirs when one noticed the cutlery had the Aeroflot symbol on the handle, and so they each decided to retain a spoon. After we had landed in London and were taxiing to the gate, a flight attendant's voice came over the intercom: "We are missing six spoons. No one will be allowed off the plane until the spoons are returned! I will be coming down the aisle. Please give them to me." Six embarrassed young men obliged.

Russia — unfriendly, drab, conservative, corrupt, poorly maintained. I have had several other consulting opportunities there since, but find I have no urge to return, and so have declined. In spite of this, I wish there had been time to explore more of Moscow, its culture and its history.

11

Latin America

Latin America (1969-2003)

The author's initial experience in South America was in Chile. He describes conditions for consulting mine geotechnical engineering during the Frei, Allende and Pinochet presidencies. Assignments developed from a series of seminars sponsored by the Canadian government and given in Peru, Chile, Argentina and Brazil through the mining association in each country.

Mine stability studies were first performed for the massive open pit at Chuquicamata in northern Chile, In spite of negligible rain in the Atacama Desert, the water table was encountered at 500 feet. This led to pit wall stability problems. At the nearby Exótica open pit mine the ore body extended under a high waste pile. The author was called upon to determine how close the pit could come to the waste pile.

Further problems are described at the El Salvador underground mine and El Teniente tailings dam. The author also evaluated a dump pocket rockfall which nearly killed four workers at the Omai mine in Guyana. Several years later a tailings outflow from the Omai tailings dam occurred. The author describes the ensuing investigation and resolution of the problem. The long-term solution involved construction of a new tailings dam.

Chapter II

Latin America

The Canadian International Development Agency (CIDA), in the late 1960s and '70s, had a program to develop Canadian trade outside North America. A company that wished to expand their business overseas could apply for financial assistance in providing international professional services or selling Canadian products. Pending a successful application to confirm the merit of the application, CIDA would pay all expenses involved with travelling and presenting a business development strategy in the offshore country. If the effort was successful and resulted in contracts, the company was required to reimburse the government. If the effort was not, repayment by the company was not required.

Golder Brawner Associates put together a proposal in 1969 to present a one-day seminar on Geotechnical Stability in Surface Mining to the mining associations and companies in Peru, Chile, Argentina and Brazil, at no charge to those countries. The intent was to present the technical capability and international experience of Golder Brawner Associates in a low-pressure multiple exposure forum. The presentation made use of several hundred colour slides, and included extensive educational information with related case examples.

The Canadian Embassy in each country was contacted to determine the potential interest of local mining companies. Significant interest was expressed, and each embassy offered to host the seminar. They advised us that they would contact local mining companies, prepare and send invitations, host the program, as well as offer drinks and sandwiches after the seminar.

The proposal, with letters from each embassy, was forwarded to Ottawa. The reaction was swift! Within one week, approval for the program was received. However, the approval was contingent on a specific schedule for the seminars. On the date of acceptance, I burned the South American telephone lines, calling each of the embassies and suggesting dates within a two-week period, two months in the future. Within three days a seminar schedule was developed: Monday in Lima, Thursday in Santiago, the following Monday in Buenos Aires, and Thursday in Belo Horizonte, Brazil. The Canadian Embassy suggested this latter location, since it is the centre of iron ore mining in Brazil. Two days were planned in each city in the event companies might wish to discuss potential project interest on the second day.

I speak neither Spanish nor Portuguese, but each of the embassies agreed to provide simultaneous translation and all the necessary seminar props required — overhead projectors, slide projectors, flip charts and microphones. We planned to provide a set of summary notes — English on one page and Spanish or Portuguese on the opposite page. Foreign engineers are very appreciative of a copy of reports or presentations provided in their language as well as English. It shows a measure of respect and consideration. This is recommended to anyone working on a project in a foreign country. The cost of translation is modest, but it must be provided by a specialist in the related field.

The seminar presented the basic theory of geotechnical stability and design, and included many actual practical case examples that defined the problem, presented the investigation details, the redesign and the stabilization program. The goal was to emphasize experience — a soft rather than hard sell approach. We had also developed a small corporate brochure in Spanish for anyone interested.

In Lima I spoke to over 200 people. Many stayed after the presentation to ask questions, but no Peruvian work developed immediately. Instead, after the seminar, a Bolivian engineer who was in Lima on mining business asked me if we could develop the design of a tailings dam for a tin mine in his country at Mina Matilda. I said yes, and met with him the next day to arrange the details of the program.

On returning home later, I was advised by CIDA that, should we obtain work in South America, some companies might not pay our full invoice. To ensure payment was obtained, we took out insurance with the Canadian Export Bank. The charge was 2 percent of our invoice. This is a sensible financial safeguard when working in Third World countries. If our bill was not paid within six months, the Canadian government would pay 90 percent, and then seek payment from the foreign company or country.

In Santiago, about 180 mining people attended. Representatives of the government asked me to meet with them the following day. They wanted me to evaluate rock slope and tailings dam stability at the Exótica, Chuquicamata, El Salvador and El Teniente mines, all at that time privately owned by U.S. companies. The government wished a review of safety and mining practices.

Rolfe Bechnke, Director of Mining in the Chilean government, arranged the program. He invited me to his home for dinner that evening. He and his wife were a charming couple. She collected teaspoons, and before I left I had to promise to send her spoons with B.C. and Victoria insignias — I did!

In Buenos Aires, the embassy arranged my seminar and a special dinner with Argentine mining executives. The dinner was one of the most wonderful eating experiences of my life — six courses taking four hours. The waiters were dressed formally — tuxedos, black bowties, ruffled shirts and white gloves — and the Argentinean steak was superb.

The results from the seminar, however, were negative — for six years! Then, out of the blue, came a call to do the geotechnical site consulting for the Sierra Pintada uranium-mining project, an experience that has taught me patience in doing business in South America.

Our plane from Buenos Aires arrived late into Rio de Janeiro, and to catch the connecting plane at the domestic terminal I took a "private" cab. What a mistake. Upon arrival at the terminal, the driver demanded the fare in U.S. dollars at double the normal rate, and wouldn't give me my luggage until I paid. My first reaction was to tell him to "go to hell," but my experience in international travel said, "Pay it and smile; live to see another project!" I paid and haven't made that mistake again. Always arrange to take a well-marked regular cab, and negotiate the fare before you get into the taxi. In countries with poor infrastructure, or a history of kidnapping, arrange in advance for a chauffeured limousine.

A representative from Cia Vale do Rio Doce met me at the Belo Horizonte Airport. The Canadian Embassy had arranged that the Brazilian company would sponsor the seminar. Prior to my seminar, I was wined, dined and entertained in true Brazilian hospitality. Unfortunately, not speaking Portuguese I was unable to understand all the festivities.

Following the seminar, I was taken to inspect the company's major open pit mine. Though it was massive, the geology was straightforward. The footwall dipped very uniformly. Provided they mined to this angle, there would be no stability problems. One of the largest iron ore projects in the world, and the site conditions at that time were favourable to stability. I quickly realized there was no work for us there.

Later I determined that Brazilian companies were not potential clients — Brazil required foreign consulting companies working there to be at least 51 percent owned by Brazilians. It wasn't a starter, since we were not interested in foreign ownership. I had not done my homework in trying to develop work in Brazil. I am surprised the Canadian commercial officer in Brazil did not point this out to me.

Following my return to Vancouver, I received an official invitation from the Chilean Ministry of Mines to inspect the four mines noted previously to evaluate stability and safety. I flew on an overnight flight from Vancouver to Santiago via Los Angeles and Lima. Herman Schmidt, a ministry representative, met and escorted me to the domestic terminal and a flight to Chuquicamata in the Atacama Desert. We arrived there about 1 P.M. In less than twenty-four hours I had flown from sea level to 14,000 feet, and found I had to stop to catch my breath after only a single flight of stairs. Fortunately, the afternoon was spent in the office going over mine plans, stability and mining issues.

That evening we had dinner at a local restaurant, where everything was fine until I finished a second glass of pisco-sour, the national drink. At 14,000 feet the shortage of oxygen provided a very cheap drunk, and an early morning hangover. It taught me to try to take the altitude change in two steps, spread over two or three days, and to go light on the liquor.

The desert was unusual, not a leaf or blade of anything where irrigation wasn't applied. I concluded that it was like the moon. Water had to be piped 40 miles from the upper Andes, where snow and glaciermelt provided water.

The Chuquicamata open pit mine had encountered the water table at about the 450-foot depth. After that, water in the blast holes made blasting very difficult. I was asked for suggestions to lower the water table, and I offered two options. Drill near-horizontal drain holes, 200 feet long, into the toe of the slope, in which perforated pipe could be placed to lower the pit slope water table, or install vertical pumping wells in blasted, re-drilled holes about 60 feet deep around the pit bottom. Design details for each were left with them so they could decide.

The Exótica open pit was being developed at a site about 4 miles away. Inadequate exploration drilling had been performed to define the extent of the ore body, and later it was determined that the ore body extended under one of the Chuquicamata waste dumps, which was about 500 feet high. Removal of the waste dump was obviously uneconomical. As a result, the question directed to me was, "How close to the toe of the waste pile can we develop the pit crest?" The pit crest was about 450 feet from the toe of the waste pile at that time.

I spent a full day mapping the open pit slope and estimating the strength of the rock by using a knife to gouge or scrape the rock. The rock immediately below the overburden was weathered and weak, and I concluded that failure would be through this near -horizontal weathered rock zone. Fortunately, there was no groundwater to deal with at that depth.

Chuquicamata open pit copper mine — northern Chile. Groundwater was encountered in the base of the mine. A review of the pit slope stability was performed.

Exótica open pit mine — northern Chile. Ongoing exploration indicated the ore body extended under the waste pile in the background. An investigation was performed to determine how close the open pit could come to the waste pile.

El Salvador underground mine — northern Chile. Two areas of the mountain where high-grade ore was mined. The pillar between the ore bodies became highly stressed to make mining of the lower-grade zone much more difficult.

The original main tailings dam for El Teniente Mine in the Andes. This dam developed considerable leakage as a result of the El Cobre earthquake. Pumping wells downslope were recommended.

The new tailings dam under construction in the foothills of the Andes. Tailings were to be piped from the mill site, nearly 50 miles.

Stability analysis indicated that, if the waste pile was within 250 feet of the pit, the safety factor approached 1.0, the point where failure would be expected. The recommendation was made to "develop the crest no nearer than 300 feet to the toe of the waste pile, and monitor horizontal movement at 75-foot increments between the pit crest and waste pile toe. The number of blast holes per delay near the final pit wall should be reduced to minimize seismic-induced stresses."

On the third day we flew to El Salvador, a very large underground mine. The ore bodies there included two separate high-grade zones with a lower-grade zone in between, all under a high mountain peak. The high-grade zones were being mined first. With this sequence, the lower-grade zone became a large pillar below the mountaintop. Hence, this pillar carried most of the load of the upper mountain and, as a result, rock stresses became very high. The mine designers neglected to take this into consideration. Underground drifts 8 to 10 feet wide in this zone began to squeeze due to the high rock pressure. Special steel sets had been brought in from Germany to try and stop the closure of the drifts, to no avail.

The mine was advised that mining would have to be developed from well below the base of the pillar, with the ore extracted above by blasting and caving. Rock stresses induced by underground mining are often misunderstood, which can result in serious mining problems. Today, computer modelling of stresses has reduced this problem.

At El Teniente, south of Santiago, mine management was very concerned about the stability of the Barahona tailings dam. It had earlier withstood the El Cobre earthquake (Richter 8.2), but had developed serious seepage from the downstream face and toe area. The earthquake had obviously disrupted many of the underdrains that were developed during original construction. The seepage water during my visit was clean, which indicated that piping, or removal of fines, was not occurring. There had, however, been some piping which occurred immediately after the earthquake because cones of sand existed at many of the seeps.

I recommended: "Install two rows of pumping wells to lower the phreatic surface in the lower slope and toe foundation area, one along the toe to bedrock and one 75 feet up the slope through the base of the dam." Upstream piezometers later indicated the dewatering was successful, and stability greatly improved.

Several years later Codelco decided to construct a new tailings dam about 50 miles away on the lowlands below the Andes.

On my return to Santiago, I spent a day sightseeing. The city is very European, with most of the population originating from Spain and central and Eastern Europe. Santiago and the country were prospering under the democratic leadership of President Frei. There was, however, a socialist groundswell developing. About a year later, a Communist, Salvador Allende, was elected president. Within months, the Communists confiscated private companies, large landholdings, and the large U.S.-owned copper mines, without payment. Within several years the previously vibrant economy was transformed into an inefficient and bankrupt one, unable to provide adequate food. The only industry to keep the country afloat was mining. A country that had been the economic jewel of South America, exporting copper, fertilizer, machinery, textiles and agriculture, now had to import food from Argentina and Brazil! Allende was an affront to democracy.

Approximately two years after Allende's election, Golder Brawner was retained to provide an updated geotechnical review of the mines. This visit was depressing. The foreign mines were now being run by a new government agency called Codelco. Whereas during the previous visit I was able to propose new ideas and technology to improve safety and productivity, this time my recommendations concentrated on trying to hold the mines together. The biggest problem was the loss of the senior U.S. mining engineers, who had all returned home. The Chileans were not adequately trained to be senior managers or mine operators. Production was emphasized, but mine planning, profit and particularly safety became less important.

Prior to my trip, I received a request from the mining ministry to present a seminar on open pit mine stability during my visit. The seminar was to be held one afternoon in the downtown Sheraton Carrera Hotel.

About eighty people attended, including four senior army officers who carried pistols in holsters and sat directly in front of me in the first row of seats. This was a daunting and unique audience. The room was divided in half by a row of free-standing folding doors, and I gave the presentation from a stage that was raised about three feet above the main floor, providing a good view of the audience.

About halfway through my lecture, a gentleman near the centre of the room stood up to leave. As he walked behind the last row, he fainted and fell against the divider doors. They went crashing loudly to the floor, sounding like machine guns. Bang! Bang! Bang! Bang! The army officers jumped up, pulled their guns and turned toward the back of the room, crouching and preparing to shoot! I hit the stage floor so as to present a smaller target. After only a short time, the commotion was under control, and I continued my presentation, a little less confident.

After three years of destructive Communist rule, General Pinochet overthrew Allende in 1973. Rolfe Bechnke, who had first brought me to Chile, was appointed a senior manager to assist in running Codelco, the state mining agency. Three weeks after the revolution, he asked me to come to Santiago to advise on rehabilitating the mines, which were still retained under state control.

On my arrival, I checked into the Sheraton Carrera Hotel, located on the south side of the main city square, kitty-corner from the Presidential Palace where Allende was killed. My room was on the sixth floor, facing the square. Upon entering my room, I noticed bullet holes in the ceiling and upper walls that had not yet been filled in. The bellman apologized for the lack of repair, saying, "We haven't had enough guests to pay for the repairs yet." I was glad I wasn't there three weeks previously when Chile's future was decided by fighting and bombing in the central square just outside my window.

Two days were spent in Santiago on an update of the status of the mines, and to review instructions and terms of reference. These were actually very straightforward: "Recommend a stability program at each mine to improve safety and productivity to that of pre-Allende conditions."

I had brought a carousel of colour slides with me so that I could provide a training lecture at each mine, and Bechnke further advised me, "Chuck, the existing management has seriously neglected safety and stability. I'm assigning a trusted mining engineer to go with you; Sr. Sergio Oyanadel, whom I wish you to provide with advice on upgrading stability and geotechnical programs."

Oyanadel was a pleasure to work with — intelligent, clear thinking, a good listener and communicator, and above all, a well-trained engineer. He also spoke good English. We spent three days at each of the major mines.

One problem was obvious. Mining was the only industry under Allende that continued to bring in foreign currency, and so, to keep the mines productive, each year the miners had received a substantial wage increase, unusual for a socialist regime. After the coup, and Allende's fall, the miners went on strike for more money, but Pinochet said no. However, the miners refused to return to work. A platoon of soldiers was sent to Chuquicamata to halt the strike. They did — by lining the union leaders up against a concrete wall behind the administration building, and shooting them.

No doubt Pinochet was ruthless initially in taking over control, but many Chilean engineers told me he had to be. One said, "We must vaccinate ourselves against Communism and never let it happen again." I am surprised today that Socialism and Communism are again creeping back into Chilean politics. Many people appear to have forgotten the unfortunate Allende experiment.

Prior to leaving Santiago, I had dinner at the Bechnke's again. During dinner, Rolfe asked me, "Is there any chance that my son, who is taking second-year engineering at the University of Santiago, could get a job in Canada between semesters?"

Here was a chance to repay Bechnke's professional and business courtesy. I answered, "Rolfe, I will personally give him a job as a technician in our Vancouver office, and pay him a typical technician's wage. When would he be available?"

"November to the end of February," he said.

Rolfe's son Claudio worked in our Vancouver laboratory for his first three months. He spoke good English, fitted in well, and seemed to enjoy the change in environment. Near the start of his last month, Golder Brawner obtained a power development project in Newfoundland which required field testing and engineering. I approached him, "Claudio, would you be interested in an assignment in eastern Canada doing field testing under one of our senior engineering staff?" He hesitated and I added, "It'll be a chance to see another part of Canada. Think it over tonight and let me know tomorrow."

He obviously phoned his father that night and discussed it with him. The next morning he came to my office and said, "Mr. Brawner, I would like to go to Newfoundland. My father said it would be good experience."

Pleased, I said, "Great, we'll make all arrangements for your trip."

While there, Claudio was introduced to the Newfoundland dog. He was so taken with this monster of an animal that he bought a black, two-month-old pup and brought it back to Vancouver. He arranged for the necessary shots and exit papers and then took the dog back with him to Chile.

The next year, Rolfe Bechnke called to offer another assignment. He said, "Bring your wife and stay on for a holiday." I accepted, and Pat came with me. Following a day of rest in Santiago, we were invited to the Bechnke's home for dinner. As our taxi approached the front gate an enormous black dog ran to greet us, put his front paws and monster head on top of the gate and barked his welcome. Claudio came to take control of the dog and welcomed us. He was so proud, exclaiming, "He is the only Newfoundland dog in Chile!"

The sightseeing was memorable. With the Bechnkes, we toured the city and surrounding area, including the lower Andes. As well, we were introduced to Victor Peterman, a young engineering friend of the Bechnkes, who took us for a weekend to Vina del Mar, a local ocean resort area. It was a very relaxing weekend at Chile's top ocean retreat. The "well to do" in Chile have servants, and the Petermans brought their cook/maid to their seaside house, who spoiled us by providing us with breakfast in bed. We returned the favour six years later by entertaining the Petermans in Vancouver and at our summer home in the Okanagan, in Oliver.

That assignment was one of many to follow in Chile over the next few years. That experience taught me that corporate promotion is more than phone calls, brochures, sports tickets or free lunches. It taught me the meaning and benefit of professional friendship.

In 1990, Cambior, a French-Canadian mining company in Montreal and Val d'Or, Quebec, was developing an open pit gold mine in Guyana, South America, about 120 miles south of Georgetown, near the Essequibo River. Dr. Louis Gignac was the president of the company. He phoned me and said, "Mr. Brawner we are developing a gold mine project in Guyana, and we've had a serious rockfall accident in a dump pocket which nearly killed several workers. We've stopped work until we can get an outside specialist opinion. You have been recommended to me. Would you be available to help us immediately by going to the site, evaluating the problem and advising us what to do to stabilize the excavation slopes?"

I looked in my Daytimer and determined that I could reschedule a trip and leave the next afternoon. I advised him, "I can leave tomorrow afternoon. What routing do you recommend?"

He answered, "Via New York on British West Indies Airlines. We can make reservations for you."

It then occurred to me that Guyana was a communist country and entry might be difficult. I asked, "Do I need a visa?"

He exclaimed, "Damn, I forgot about that!" He was quiet for a minute then said, "Plan on catching the early afternoon BWIA flight from New York to Georgetown the day after tomorrow. I will see that arrangements are made for you to get on that plane — by whatever means are necessary. We'll advise them at the check-in counter to be ready for you."

I caught a flight to Toronto late the next afternoon, overnighted at the Airport Hilton, and flew to John F. Kennedy Airport in New York mid-morning. Arriving two hours before the flight to Guyana, I went directly to the BWIA check-in where I was advised that they had been alerted to my arrival, but that no travel approval had been received as yet. They said, "Come back in an hour."

Nothing had arrived thirty minutes before flight time. The check-in agent gave me a pass and said, "Go directly to the gate and wait there."

Fifteen minutes before flight time, a BWIA staff member called me to the check-in counter with a telex message that authorized me to board the flight. I was advised I would be first to deplane and would be met by a company representative. Gignac had kept his word!

A pleasant flight, the first-class fare and service were excellent. We made one stop in Port of Spain, Trinidad, where I got off to stretch my legs. What a jolt to experience the stifling heat and high humidity.

On arrival at Georgetown, the senior flight attendant arranged for me to exit the plane first. At the bottom of the steps I was met by Norman McLean, Human Resources Manager for Cambior. He introduced himself and said we had to proceed through a special entrance. It was nearly dark, and we felt our way up two flights of stairs to a private room, dimly lit and sparsely furnished. McLean asked me for my passport and plane ticket and said, "Mr. Brawner, please wait here and be quiet." I felt apprehensive but obeyed his instructions. Twenty minutes later he returned and said, "Everything is arranged, but I'll keep your documents and return them when you leave Guyana."

He drove me to a small hotel in downtown Georgetown, helped me check-in and saw me to my room. He said, "Georgetown is dangerous at night. Please do not go out. Be down at the desk at 7 A.M. and you'll be taken to the municipal airport to catch the Cambior plane. You'll have breakfast at the mine."

McLean picked me up on time. It was already hot and humid, so I lowered the car window and put my left arm out. He advised, "I suggest you keep your arm inside. You could lose your wrist watch within five blocks!" I pulled my arm in and rolled the window up halfway.

It was about five miles to the small municipal airport. The road was full of potholes and many houses were shack-like, often on stilts with storage areas underneath. It was apparent that the country was very poor, very Third World. The aircraft was a ten-seat twin turboprop with fixed landing gear. The runway was narrow, and the cloud cover was very low, only about 100 feet above the airport. I thought we wouldn't be flying out that morning, but I was wrong!

Another eight passengers showed up. The baggage was loaded, we climbed in, and in a few minutes we were airborne. We flew for about fifty minutes — in the clouds. Suddenly the pilot reduced throttle and we started down, still in the clouds. Abruptly we broke through the clouds with treetops on both sides, just below our wing tips. Seconds later, we landed on a smooth gravel airstrip. I had flown numerous times on small charter aircraft in northern Canada, but never in such low visibility. I asked the pilot, "How can you fly with such low cloud limits?"

He looked at me, smiled and waved me over to the cockpit. He showed me a small box on the top of the instrument panel and an electronic panel on the dash. "This is GPS, called Global Positioning System. Tied into satellites above, I know my location and elevation within several feet!" I was amazed. This was my first contact with this breakthrough location technology. I was later to use it as an observation system for slope stability monitoring in open pit mines, with an accuracy of plus or minus 10 millimeters. Imagine, such accuracy from satellites 150 to 300 miles above the earth.

The camp was built beside the large Essequibo River. I was met at the airstrip by the Chief Engineer and driven to the dining room. After breakfast, I checked into a room, changed into work clothes, and was taken to the site of the crusher building excavation. The crusher is a facility where mine trucks dump gold ore blasted to as large as 24 inches for crushing to about 2 inches in size. The crushed rock falls from the bottom of the crusher to a conveyor system that carries the rock to the mill where it is again crushed, this time to finer silt and sand sizes. This secondary crusher structure requires a pit developed to about 40 feet deep, and it was in this pit that four workers were excavating the soft rock when the cave-in occurred.

The upper 30 feet of the pit were excavated in "laterite," a material which was originally granite but which, over millions of years, has weathered to a soil-like material that is clayey in behaviour. In laterite, all of the rock structure, the joints, bedding and shear zones, have become extinguished. To the geologist, it is classified as a soft rock, and to the geotechnical engineer, it is classified as a cohesive, clayey compact to dense soil. It can be dug by hand with a shovel, with some difficulty. Failure is through the intact material.

Below 30 feet, the soil-like texture remained, but the geologic structure still existed. This material is described as a saprolite, and contains the original discontinuities — joints, bedding and shears. These discontinuities are weaker than the soft rock, and represent potential failure surfaces.

During excavation below 30 feet, two joints formed a wedge in the side of the excavation. The excavation had totally undercut this wedge, and a block containing about 10 cubic yards of saprolite fell to the floor, covering over one-half of the excavation bottom. Fortunately all four workers were on the other side of the excavation at the time and were not hurt. Had they been under the wedge they could have been killed.

Unfortunately, the original geotechnical investigation had not differentiated between the laterite and saprolite, which behave very differently. This is a major engineering lesson to be learned in tropical environments.

The joints and bedding were mapped, and it was determined that a parallel wedge would extend deeper if the excavation were deepened the remaining 4 feet. I proposed to the design engineers that they instead raise the crusher facility 4 feet. Otherwise two sides of the pit would have to be stabilized with specialized earth anchors, and this would delay the project ten days to two weeks, the time to obtain the anchors and clear customs. The design change was agreed to. By not having to excavate and stabilize the extra depth, a further week was saved. When a project such as this is being constructed with borrowed money, any saving in time is a financial bonus!

In 1998 I was requested to inspect and evaluate the open pit slopes. I recommended modifications to the blasting program and to steepen the slopes from 45 to 58 degrees. Scaling of every bench was required.

Overall, I was most impressed with the Cambior operation. Norman McLean had been a general in charge of the Guyanese armed forces, and it was a brilliant move by Gignac to hire him. He knew everyone of importance in Guyana and was able to facilitate many matters expeditiously. I had no trouble leaving the country.

Scaling of loose rock in the Wenot open pit at Omai. The bucket of the backhoe was replaced with the ripper tooth from a Caterpillar tractor. By using controlled blasting and scaling the bench face the pit slope was steepened from 45° to 58°. Millions of dollars were saved by reducing waste rock removal.

In the summer of 1996 my oldest son, Randy, invited me to a weekend of golf, and to attend a special Peter Jacobsen golf tournament near Portland, Oregon. We were playing the 17th hole when a marshal approached us and asked, "Is either of you Chuck Brawner?"

I answered, "I'm Brawner."

He said, "There's an urgent phone call for you from Montreal. You can take it in the clubhouse."

With more than thirty years of experience behind me at that point, I was able to guess the problem. Another geotechnical failure! I thought, "Another twenty minutes isn't likely to help me catch an airplane," so I answered, "Please go back and tell them I'll come to the phone within thirty minutes. Please obtain their phone number." After all, there was only one more hole to play and I was heading to a good score.

Thirty minutes later I called from the clubhouse. It was Cambior once again, and Gignac said, "Chuck, we've had a serious failure of our tailings dam in Guyana. It's extremely urgent. How soon can you come?"

It was 2 P.M. I said, "I'll head to the Portland airport and take the first plane to Vancouver. Hopefully I can get to Georgetown by tomorrow evening. As soon as I have arranged my flights, I'll contact your office."

I called my Vancouver travel agent from Portland and arranged for reservations and tickets. Arriving in Vancouver in time to pick up the required belongings, my passport, extra U.S. currency and ticket, I then caught an overnight flight to Toronto, and a morning flight to Georgetown via Barbados. This time I did not need a visa. I reached the Georgetown airport in time to catch a special charter flight, arriving at the mine just before dark.

I emphasize the use of U.S. currency. In Third World countries with limited tourism, the Canadian dollar is difficult to exchange. Always carry U.S. dollars and Travellers cheques, plus one-dollar bills for tipping. The American dollar was the only international currency accepted everywhere at that time.

The tailings dam was constructed with coarse blasted waste rock as a downstream section, with one layer of sand between the rock and an upstream saprolite clay section. The sand was supposed to act as an interfilter, preventing the clay particles from migrating through the sand as a result of seepage. With proper gradation, the sandy filter should not pipe into the waste rock.

I had asked the pilot before takeoff from Guyana to fly over the dam prior to landing so that I could see and photograph evidence of the failure from the air. It was dramatic. About six sinkholes and a crack about 1,000 feet long had developed all along the crest. However, the main portion of the dam was intact, and had not been breached. At the far end of the dam, a major seepage flow was exiting from near the lower toe area. This flow initially entered a gully where it was carried to the Omai River, actually a creek in size. The outflow continued in the Omai River for only about 1 mile, where it entered the large Essequibo River, flowing from there about 100 miles to the Guyana seacoast.

The outflow through the tailings pond carried diluted cyanide-contaminated water. The amount of cyanide was great enough to kill fish in the Omai River, but it became highly diluted when the flow entered the Essequibo River. At that point it was no longer dangerous. The problem was first recognized at night when a truck driver moving waste rock observed an unusual flow of water from the downstream slope. Senior mine staff located at the site were alerted, and as soon as possible made an inspection. It was immediately recognized that a seepage "break" had developed through the dam.

Rejean Gourde, Vice President and General Manager, was resident at the site, and he instructed that wire mesh be placed across the mouth of the Omai River to halt fish movement, and that the mill be shut down. By 9 A.M., calls had gone out to the Cambior head office and Golder Associates. A chartered jet was arranged from Montreal to take Gignac, Rick Firlotte and Dave Campbell, partners in Golder Associates, to Georgetown. They had arrived at the mine the day before I did.

I was met at the Omai airstrip by Gourde, who drove me directly to the dam site. It was frightening. The entire upstream portion was disturbed. Viewed from the ground, as compared to the air, the sinkholes and longitudinal cracks were much more pronounced. I could see water flowing into several of the sinkholes. Below the far end of the dam, at its low point, a major flow was exiting through the coarse waste rock into the gully.

Omai tailings dam failure in Guyana. The central sand filter material piped into the downstream coarse mine rock. The upstream saprolite shell collapsed into the resulting void. Rapid construction of a downstream dam limited pond water outflow. A new dam was designed and constructed.

Firlotte and Campbell had already recommended that an emergency earthfill dam be constructed as soon as possible in the gully, several hundred feet below the dam outflow area. This was done by dumping waste rock downstream, followed by finer crushed rock and low-permeability soil upstream across the gully. Within two days of the break, the flow to the Omai River was contained.

In the morning, a detailed site visit was performed, followed by a review of the design. In the afternoon, Gignac and Gourde led a discussion on how to proceed. Action that was taken included:

1. Discontinue mining and milling
2. Advise the Guyana Mining Ministry of the incident
3. Prepare an initial press release
4. Request Golder Associates to initiate a forensic investigation
5. Establish water-quality monitoring stations
6. Assemble design, construction, quality control reports, climatic data *etc.* for detailed review. This would also be required for subsequent government assessment reports, liability and potential litigation
7. Determine what to do with the pond and dam and future tailings disposal

It was also decided that the pond should be lowered about 10 feet. This would take the level below the downstream emergency dam storage height. Since the mining was shut down, we recommended that the pond water be diverted temporarily into the adjacent 800-foot-deep open pit mine. Mine staff were reluctant to do this since the water would have to be pumped out before mining could restart.

I recommended the water be siphoned. Kevin Atherton, the Chief Engineer, asked, "Are you sure it'll work? What size pipe diameter would we need?"

I answered, "You can use your tailings distribution pipe." This was an 18-inch-diameter pipe. "I recommend you dig a shallow trench across the dam at the closest point to the open pit. Lay the pipe with the upper end below the pond water elevation you want. Extend the pipe through the trench, with the outflow end at a lower elevation than the inlet, emptying into a drainage ditch that continues to the open pit. The entrance pipe elevation must not be more than about 18 feet below the top pipe elevation for efficient operation of the siphon. Install closure valves near each end, and an inflow and air port at the highest elevation. After the installation is complete, close the inlet and outlet valves. Then open the top inlet port and fill the pipe full of water. After it's full, close that port. Next, commence opening the lower closure valve to start the flow, and, seconds later, commence opening the inlet port. This must be timed carefully to develop the necessary vacuum to start the flow. Flow will continue without any pumping."

The siphon was subsequently installed, and worked as planned. It is recommended that a siphon system be stored at every tailings dam, available for use in any urgent situation where an emergency spillway does not exist. This is inexpensive insurance.

Louis Gignac was a very conscientious professional. When Cambior completed the original feasibility studies for the new mine, and approached the Guyana government for approval and permits, there were no environmental or mining regulations in place in the country. Samuel Hinds, the prime minister and a chemical engineering graduate from the University of New Brunswick, Canada, asked, "How can we be assured that Cambior will be a responsible corporate citizen?"

Gignac offered, "We have several mines in Quebec, Canada, where comprehensive provincial regulations exist. I propose that Guyana adopt these regulations, and Cambior will adopt them at Omai."

Hinds agreed, and this concept has merit for other Third World countries. In some instances, moderate adjustments may be required to adopt them for different climatic,

Construction of a siphon to discharge water from the pond to the Fennel open pit to lower the pond level. The siphon was successful. Omai Gold Mine.

economic and political conditions.

During our meetings, Gignac stated that the government must be kept fully informed. He invited government inspectors to the mine to view conditions, and made any information that existed available to them. To assist in evaluation of the health aspects of the tailings breach, Hinds in turn contacted the World Health Organization, who sent a medical specialist in chemical poisoning to the site.

The news media were having a field day; international newspapers and TV stations were extremely vocal. Critics were claiming that the cyanide had poisoned the Essequibo River, and that this had killed fish in the river extending into the ocean. It was also claimed that domestic animals and birds that drank the water were dying, and that locals who used the river water were in danger.

In spite of assurances that the dilution on entering the large river would create no health hazard, a massive local outcry occurred. Omai had a business office in Georgetown, and locals threw dead fish at the building, claiming cyanide had killed the fish. It turned out that the fish thrown had been caught by ocean fishermen, and were not even river fish. Environmentalists who act in this manner destroy their credibility. It is unfortunate that emotion rather than scientific fact and experience fuels many environmentalist actions. The annual cost of such irresponsible behaviour is billions of dollars and years of delays.

Gignac decided to invite news media to the site. The access road to the mine had been closed, so access was only by air. He arranged a forty-passenger charter flight from Georgetown to the mine. At the mine, a site inspection was conducted, as well as a presentation describing the accident and the procedures being taken. The key presentation was made by the World Health Organization doctor, who emphasized that the natural decay period of cyanide is about thirty days, and that the major dilution on entering the Essequibo River resulted in no downstream health or fish hazard. A woman reporter from *The Miami Herald* made a loud display of disagreement. The WHO doctor responded by drinking a glass of water from the river.

During our site meetings, it was recommended to Gignac that Dr. Norbert Morgenstern, an internationally recognized geotechnical engineer, also be requested to join the investigation team. He arrived two days later.

One of the main questions Gignac wanted addressed was, "Can the tailings dam be successfully repaired?" An entire afternoon was spent discussing options and procedures. In the end, Gignac made what I consider to be a difficult but correct decision. It would likely have been less expensive to repair the existing dam and allow early recommencement of mining. However, he concluded, "If we repair the dam, there will always be the question by the media and outsiders whether the dam might fail again. We'll develop a new tailings pond, and turn the existing pond into a storage site for waste rock." He authorized Rick Firlotte of Golder Associates to immediately commence a study to determine a new tailings dam and pond location, and to develop the design.

The forensic investigation involved Omai, Golder Associates, the Guyana Mining Ministry, and an external panel headed by Dr. Steve Vick from Colorado. The Guyana government also ordered a government inquiry headed by a Georgetown judge. The inquiry convened after the forensic investigation was completed. It was concluded that internal erosion developed within the filter sand because filter graduation design criteria were violated, and the filter had been inadequately compacted. The initial failure was believed to have initiated by piping at or near a diversion conduit under the centre of the dam which had not been properly filled with concrete, and whose filter backfill was undercompacted. Erosion within the sand produced upward collapsing cavities within the saprolite. The water level rose to saturate the upper dry filter sand. This destroyed the surface tension cohesion of the sand, which collapsed, removing support for the saprolite blanket locally, causing sinkholes and grabens to develop. Water broke through into the rockfill zone and into the gully.

Three major lessons were learned:

1. Underdrain conduits are undesirable. If used, they must be designed and constructed with extreme care.
2. Internal sand filters must be densified sufficiently so they do not consolidate when they become saturated, although physical compaction is usually difficult to obtain. I recommend flooding each lift with water and letting the downward seepage flow develop the necessary density, similar to how receding wave seepage develops density sufficient for small planes to land on beaches.
3. Filter gradation criteria must be adhered to.

I was retained by Gerald Viellette, Vice President, Cambior from their Denver office, to provide advice on the San Pedro gold project about 25 five miles from San Luis Potosí, in north-central Mexico. Silver had first been mined at the site during the time of the Spanish, some 400 years ago. Both surface and underground mining had occurred. During the last century, geological investigations included sampling along old exposed surface and tunnel walls. Fine-grained gold was encountered, in addition to the silver. During the mid-1900s, the Anaconda Company redeveloped the site, primarily as a gold mine, with secondary silver. However, underground mining eventually proved to be uneconomical.

In early 1990, Cambior became part owner of the property and investigated its potential as an open pit mine. Extensive drilling proved positive, but several conditions were encountered which focused on potential geotechnical problems. One open pit "glory hole" about 300 feet deep existed with near-vertical faces within the overall ore zone, and mining adjacent to this face would require extreme care. As well, underground exploration revealed several large mined-out caverns, and mining through these by open pit means presented stability and safety concerns. The old mining plans of these areas were incomplete. Finally, the ore body extended a short distance under the town, including under a beautiful 400-year-old Catholic church. Cambior offered to move the church, but this was unacceptable to the Mexican authorities.

Typically, open pit mining is performed with very large mining equipment. In the area of the "glory hole" steep slopes, it was recommended that smaller equipment be used to maximize excavation flexibility. It was proposed that the outer zone, near the steep face, be blast-casted into the pit bottom, where access through a tunnel was available. Additional mining to develop a 50-degree face could be done with the smaller mining equipment.

To mine in the area of the underground cavities, the procedure pioneered by Drum Cavers of AGRA Consultants and myself at the Hedley Mascot gold mine in British Columbia was proposed. On this project, core samples were tested for compressive strength and modulus of elasticity. Knowing the dimensions of the openings, the required roof thickness was determined to safely mine above the opening. Several drop raises were then developed. (The procedure is described in this book's chapter on Canada.) After filling the cavity, the pit floor could be safely mined downward in lifts through the cavern area. Based on the success at Hedley Mascot, this program for San Pedro was reasonable.

Old mine plans revealed some mining and tunnelling had been performed under the old church, and one sinkhole had occurred near its front entrance. This had been filled with soil and covered with a concrete slab. The exterior walls of the church had cracked due to differential settlement. Heavy concrete — rock pilasters had been constructed against the outer walls, spaced about every 20 feet, to provide extra support for the walls and roof.

Special arrangements had to be made with the bishop to inspect the interior of the church, which was used periodically for Sunday sermons and weddings. Though it was quite plain inside, the church had a beautiful altar with an image of Christ at the front. The major concerns for the integrity of the church were the vibration due to blasting which could cause damage to the structure, sinkhole collapse under the church, and damage from flyrock.

A structural engineer from Mexico City with extensive experience in the repair of old historic buildings, Dr. P. Girault, was retained to evaluate the structural condition of the church and to assist in determining procedures to maintain stability.

To evaluate the potential damage from blasting, several blasting tests were proposed with small control blasts at various distances from the church. These would be monitored with geophones located at various distances from the church, and at several locations on the church itself. Crosshole seismic profiling between vertical boreholes on several cross sections was performed to try to locate any underground openings created by the mining. This investigation revealed anomalies interpreted to be cavities about 30 feet below the church floor. A further investigation using vertical ground-penetrating radar was also proposed for when access could be obtained to work inside the church.

Stabilization measures being considered were:

1. Fill the underground voids with pumped concrete and grout. Prior to this work, angle drilling was proposed under the church, to proof the seismic and radar results.
2. Construct reinforced concrete cross beams directly under the floor, between pilasters, in hand-excavated trenches.

A further evaluation was to determine how close the open pit mine crest could come to the church. Final feasibility and permitting had not been completed by mid-2000, at which time the property was sold.

In January 1996, Dr. Ray Benson, Senior Vice President of Klohn Crippen Consultants Ltd. of Vancouver, phoned and asked, "Chuck, would you be willing to help with advice for rock excavation and slope stability on a hydroelectric project in Colombia?"

I asked in return, "Can you tell me something about the project?"

"It's called the Sagamosa hydroelectric project, and it's near Medellín," he replied. "A dam is to be constructed as a concrete-faced, rockfill structure. An underground powerhouse and a separate surge chamber are also proposed, as well as twin diversion tunnels. The dam will be about 130 meters high; the underground powerhouse rooms about 26 meters wide, up to 160 meters long and about 50 meters high — large openings. The rock is generally a competent sandstone with about 15 percent siltstone, the latter including some thin claystone layers. A question for you is, can the rock in the powerhouse and surge chamber be ripped with a Caterpillar? If so, damage to the rock walls will be reduced as compared to excavating using blasting. A second concern is rock slope stability adjacent to intake and tailrace cuts."

One side of my head felt the project sounded challenging, and the other said "Whoa, that's drug cartel country." I questioned Ray, "What about our personal safety?"

He replied, "I've been there twice before and was always escorted. I felt safe."

I answered, "In that case I'm interested. When did you plan to go?"

Ray said, "Near the end of February."

Ray and I met in Miami and flew together to Bogotá on American Airlines. In Bogotá we were met by Francisco Nandares, a design engineer, who took us to the Tequendama Hotel to overnight. The Tequendama is a charming hotel with an old Latin American atmosphere. We had an enjoyable dinner there together, and the next morning we flew to Medellín.

After spending most of the day at the engineering offices of Sedic Ltd., and meeting with Dr. Guillermo Henao, we checked in at the Hotel Intercontinental, which is built on the side of the mountain overlooking the city. Henao said, "You will be comfortable there. However, members of the drug cartels have homes higher up on the mountain. They are intense rivals and frequently engage in gun battles late at night. If you hear gunshots, that is what it is. They will not be a danger to you."

We were escorted wherever we went, during all meals and on our return to the hotel. I was a bit apprehensive, yet at the same time a little excited, realizing danger existed so close.

The second day we were driven to the dam location. As we neared the site, which was within a moderately deep canyon, we were advised to stay with Nandares and not go across the bridge to the other side of the canyon. He warned, "The mountain on the other side of the river is guerrilla territory."

We parked the car at a field office and were taken along a trail above the river to an exploration tunnel. This tunnel was driven into the mountain about 420 meters, near the proposed underground powerhouse location. We spent several hours looking at the rock's geologic structure, particularly the shear zones, joints and claystone layers. We also were shown the locations where samples were taken for testing.

After completing the inspection of the tunnel, we were taken by riverboat to the other side of the river to look at several isolated drill sites, as well as a shorter tunnel. Again we inspected the rock and structural geology. I took a number of photos in the tunnels and of the canyon and river. I never saw any of the guerrillas! At least I don't think I did.

We drove into the nearest town for lunch, and afterwards were taken to a field storage and testing site. There we inspected rock core of various sizes, and the rock-testing facilities.

That evening Ray and I were guests of Dr. Ernesto Bacci, General Manager of Sedic, and Dr. Henao at an excellent specialty restaurant in the city. The surroundings were tastefully developed, the atmosphere was quiet and reserved, the food was excellent and the service gracious. We could have been in New York, San Francisco or Vancouver. Except for the atmosphere outside! Upon arrival at the restaurant, Dr. Bacci turned his car over to a parking guard who came complete with automatic rifle. We entered the front entrance with an armed guard on either side of the door. I didn't know whether to feel safe or intimidated.

Most of the talk was engineering related, particularly comparing experiences on interesting projects. We had no difficulty finding topics. Very little talk centred on life and conditions in Colombia. I concluded that if one went about his professional business and didn't interfere in Cartel affairs, life went on. There was, however, the perpetual guard or security personnel somewhere in the background.

We spent the next day going over engineering details — geology, rock strength, design considerations, and finally the potential for ripping rock in the powerhouse and surge chamber. Based on my initial evaluation of rock and geologic conditions, I considered that there were sufficient massive sandstone zones with few structural discontinuities that some of the rock could not be ripped, and would require blasting. I suggested that seismic testing be performed to relate the velocity of sound with rippability, as has been documented in design charts by Caterpillar. I further suggested that light blasting could possibly render the rock rippable. If such trials could be entertained during the project and accounted for adequately in the bidding documents, I recommended that such trials be performed. Since some blasting would be required, I developed an applicable blasting specification including controlled wall blasting, and provided it to Sedic. They were pleased with the detail.

The last evening and next day, Ray and I prepared a draft report which we presented verbally to the Sedic engineers. Ray later forwarded a final typed version.

We flew direct from Medellín to Miami on ACRES Airline. The check-in at the Medellín airport was slow, through about six security check stations. Many travellers took advantage of a service that involved covering baggage with a sealed plastic wrapping, a good practice since it makes it difficult for baggage handlers to secretly place drugs in a suitcase and have a baggage handler in Miami recover it. Immigration in Miami was straightforward and efficient. I was surprised, since we were arriving from the drug capital of the world.

I am glad I accepted the offer of work on the Sagamosa project, and would return for further work on this project if asked, but will decline other opportunities in Colombia.

Sadek El-Alfy, Vice President of Crystallex International Corporation, contacted me in April of 1998 regarding open pit slope stability at their Albino mining concession, near La Clarita, Venezuela. I had worked for El-Alfy on pit slope and ore dump pocket stability problems at the Iron Ore Company in Labrador City, Newfoundland, about five years previously, work which had turned out successfully. He had since moved to Seattle to head up engineering for Crystallex.

El-Alfy advised me that Crystallex had purchased a small open pit gold mine where they were mining and milling ore from surficial weathered, clay-rich saprolite to a depth up to 50 meters. The mine had been operating since 1994. Several slides had developed in the pit slopes which were associated with steeply inclined throughgoing joints. Also, serious spalling of the bench faces was occurring. One failure had infringed on the access ramp. Exploration had located a high-grade ore zone in underlying hard andesite, and the access design for an underground program was being developed.

I met El-Alfy at Dallas airport for a flight to Caracas. The Caracas airport is about twenty-five miles away from the city, over a mountain. We were met there by a Crystallex driver who drove us to the Hilton Hotel, where we checked in and had a pleasant dinner together. El-Alfy reviewed the Crystallex project and its plans for the next two years.

The next morning we took a charter from the local city airport, headed to an isolated emergency airstrip about twenty miles from the mine where a driver from the mine was to pick us up. The weather was clear but very windy. After two hours of flying, the pilot advised us that he was using so much fuel he would not be able to return without refuelling. This forced him to make a detour to a larger airport to refuel, a stop which delayed us nearly two hours. We eventually arrived at the agreed to airstrip, but the driver had left. There was no town nearby, so El-Alfy decided we should fly to a town about thirty miles on the other (east) side of the mine.

When we arrived there was no one to meet us at the airstrip for this town, and no terminal. El-Alfy noticed a boy playing nearby, called him over and gave him some local money to find a taxi and driver in the town and instruct him to come to the airstrip. After about fifteen minutes, a battered taxi arrived. It took several minutes to negotiate the cost of the trip to the mine. However, the driver wanted cash in advance, and this necessitated a trip to the local bank.

The drive to the mine was partly through a driving rain, so heavy that the wipers could not clean the windshield fast enough. The driver stuck his head out the side window to drive at 50 miles per hour!

The sun broke through before we reached the small town near the mine. Water was everywhere, and the roads, constructed mostly of clay, became very slippery. The company had taken over one of the local motels; we stayed there one night, then transferred to a second motel for the rest of our stay.

The camp "mother" had been the wife of a senior government official who had recently died. She took on the responsibility for the Crystallex camp in order to keep busy. I sat beside her for our first dinner. She was very well educated and gave me a lesson on Venezuelan politics and history which revealed the typical ups and downs of most South American countries. Emotion often overrides reason and responsibility.

Just before dark, a helicopter appeared and sprayed the town for mosquitoes. The camp mother said that most residents of the town had various stages of malaria. As soon as I got back to my room, I sprayed myself with mosquito repellent and put on a long-sleeved shirt.

The next morning was spent with the geologist and a review of reports. Over millions of years, the volcanic andesite and flow materials at the site had weathered to a clay-rich saprolite. Remnants of shear zones and jointing remained in the material, and these discontinuities had led to the pit slope failures. Planar, wedge and toppling movements were observed.

Joints in saprolite are very slippery. During and after rain, the standard mine haul trucks would spin out on the grades. When El-Alfy became aware of this, he brought in 4-wheel-drive articulated Volvo trucks to maintain haul cycles. This greatly reduced downtime and mining costs.

The saprolite was very similar to that at Omai, Guyana. I used that experience to develop recommendations to redesign the pit slopes. At the locations of the two slides, the back slopes had become oversteepened. It was recommended that the crests be set back by excavating with backhoes which could reach down below the crest. During excavation, a spotter was used to watch for cracks at the crest, which would indicate impending instability. If cracks developed he alerted the operator.

In areas where structural geology did not control the stability, a bathtub slope design was recommended which recognized the increased shear stress with depth:

Upper 20 meters - 50 degrees

20 – 40 meters - 45 degrees

Below 40 meters - 40 degrees

Since no rock ravel occurs with this material, it was also recommended that no benches be developed. This reduced failures and allowed more rapid runoff. An increase in the pit dewatering capacity was required to deal with high rainfall inflow.

Mining down through upper laterite with saprolite below. Failure is occurring in the saprolite along relic joints and minor shear zones. When wet these materials require all-wheel-drive vehicles for trafficability. Crystallex Gold Mine, Venezuela.

Our return to Caracas and Dallas was uneventful, but before El-Alfy and I parted company at Dallas, he said he had sat beside the same man as he had on the flight down, and this man described a frightening ordeal. On arriving at Caracas with only carry-on luggage, he was approached by a man who asked if he needed a ride to the city. Assuming he was a licensed taxi operator, they agreed on the price. Near the top of a mountain, the driver pulled off the highway where a turnout occurred and stopped the car. Suddenly both doors were opened by two men with guns who ordered this man into the centre of the back seat and demanded money, any Travellers cheques, his watch and rings. They threatened to kill him if he didn't obey. Of course he did obey, and the car then returned to the highway. They asked what hotel he was staying at, and after he told them, not a further word was said until the car stopped two blocks from the hotel. They ordered him out and instructed him not to look back or to alert anyone. He did so, walking to the end of the block before looking back. They were gone! With the tension now broken, he began to shake, and had to lean against a storefront for some time before he could regain his composure. Eventually he realized that he still had his life, and that the thieves were good enough to leave him with his passport, airline ticket and briefcase.

This is an all too common occurrence in Third World countries, and the lesson is simple. Always arrange for a driver and car through a registered tour company, the hotel, or your embassy. Never agree to travel with soliciting drivers.

In August 2001 I received another phone call from Sadek El-Alfy of Crystallex, telling me that a rockfall at their gold mine in Uruguay had seriously damaged a D8 Caterpillar dozer. Luckily, the operator had climbed off momentarily to have a cigarette, and so was uninjured, but the accident caused serious alarm. He added, "We'd like you to visit the mine as soon as possible to evaluate stability. We've closed that area of the mine until we resolve the issue."

I looked at my diary and advised, "I can adjust my schedule to go early next week." We discussed travel arrangements, and the provision of reports related to pit slope design and rock stability.

The following Monday I flew to Chicago and connected to United Airlines for a flight to Buenos Aires and on to Montevideo, an overnight flight of twelve hours. The flight was comfortable, however, on a new Boeing 777 jet with sleeper seats. They are a great innovation.

In Montevideo I was met by Diego Sarroca, who assisted me rapidly through immigration and customs. He was to take me to the mine, and our route took us through the city, along the waterfront, to reach the highway north to the mine. We passed several miles of excellent beaches before coming to a large port area that is famous for sheltering the German battleship *Graf Spee* from British warships during the Second World War. After a short period, the government ordered the battleship to leave. Instead of doing battle with the British, however, the German captain chose to leave port at night for deep water and scuttle the ship. A small museum on the east side of the city has been established to commemorate the event.

The drive north to the San Gregorio Mine, near the Brazil border, took five hours, driving at an average of 70 miles per hour. Uruguay is often referred to as a "big farm," and the countryside was indeed gently rolling with extensive farm and pastureland. Near the city, the agriculture was mainly vegetables and fruit orchards; beyond this it was mainly ranching — cattle, horses and sheep.

Accommodation was available at the mine in several furnished ATCO-type trailers, one of which had a kitchen and dining area. A local woman provided cooking whenever visitors stayed at the site.

The next morning El-Alfy met me and I was introduced to Bira de Olivera, the Mine Manager, and Faustino Lopez, the Geologist. After discussing the rockfall problem, the mine operations and geology were reviewed. Three open pits were being mined. The central pit was the largest and deepest — about 500 feet. A large wedge between two inclined shear planes had moved, and cracks had opened near the pit crest. The total height of the wedge was about 350 feet. Rock had been dislodged along one edge of the large wedge and rolled and bounced to the pit bottom. Numerous other rocks looked very precarious.

There was no bench access to the unstable areas, and Olivera indicated that the mine had no capability to scale or remove the unstable rock. One of my concerns was that vibration from adjacent blasting, or water pressure build-up within the slope due to heavy rainfall, could destabilize more rock. There was also the possibility that the large wedge could fail, covering the ore zone and closing that end of the pit. Thousands of ounces of gold could be lost.

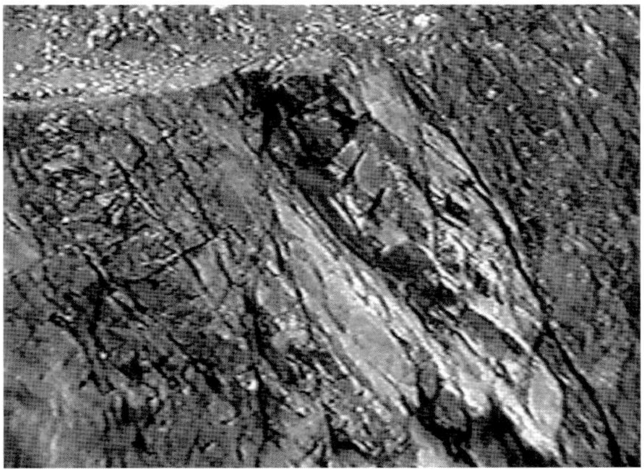

Blasting unstable rock near the top of an open pit slope with a 100 mm gun supplied by the Uruguay army to improve stability. San Gregorio Mine, Uruguay.

a) Unstable rock face along the flank of a major rock wedge.

b) Exploding shell scaling loose rock.

The unstable rock and potentially unstable wedge had to be dealt with. I turned to Olivera and asked him if Uruguay had an army! He looked at me with some alarm, as if to indicate, "What does the army have to do with rock stability?"

I have had experience with shooting down snow avalanches in the Canadian Rocky Mountains and White Pass Railway, as well as rock talus slopes along the Canadian National Railway, using howitzers and exploding shells. I explained this option to him, saying, "With binoculars we can locate the most critical rocks, then shoot at them to break them up and cause them to roll to the pit bottom." He indicated that he would contact the army to determine whether they had the artillery capability.

The result of that contact was favourable. Army officers came to the mine, inspected the site and indicated they had the howitzer capability. They added that they would welcome the opportunity for "target practice." Olivera requested the army submit a proposal describing how they would perform the exercise, the equipment they would use, and a cost estimate. This was subsequently done.

I provided instructions to locate and designate appropriate targets, marking them on a blown-up set of photographs of the area. Unfortunately, I was unable to coordinate my schedule with that of the army, and I was unable to visit the site until shortly after the artillery exercise. The procedure was to first shoot paint bullets with a "sighting rifle." The bullet exploded and left a red paint mark. The howitzer was then sighted at the paint mark if it was the actual location, or at a designated location relative to the paint mark — *i.e.,* 12 inches at 50 degrees from the paint marks. The howitzer was a M40 recoilless rifle, which shot a 4.2-inch-diameter shell. The shell penetrated about 4 to 6 inches into the rock before exploding. A total of twenty-two shells were fired; the first three into the toe area of the wedge to test stability. The wedge did not move. The remainder of the shells were directed at designated rocks along the edge of the wedge that were considered unstable. The exercise was considered a success, with an estimated 1,000 cubic meters of rock scaled in this manner. Monitoring of the upper crack indicated the large wedge did not move, in spite of the vibration caused by the multiple explosions. This was a good test of stability, and mining was allowed to proceed after rock clean-up in the pit bottom.

To my knowledge, this is the first use of this procedure to scale slopes at an open pit mine, and it provided an interesting addition to my experience in South America.

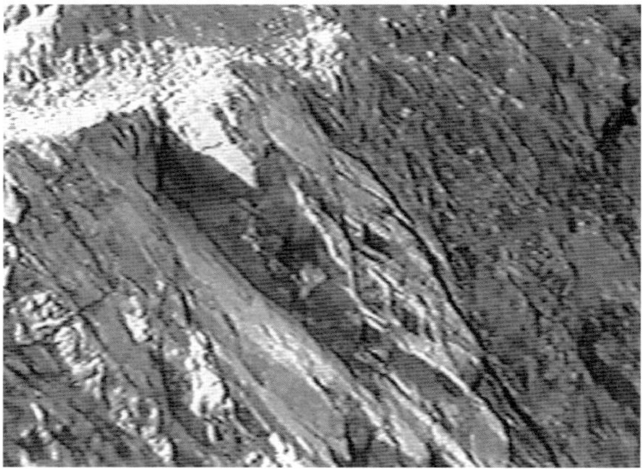

c) Stabilized area of rock adjacent to the rock wedge.

d) 400 mm recoilless army rifle which shot penetrating shells that explode on impact.

12

Africa and Saudi Arabia

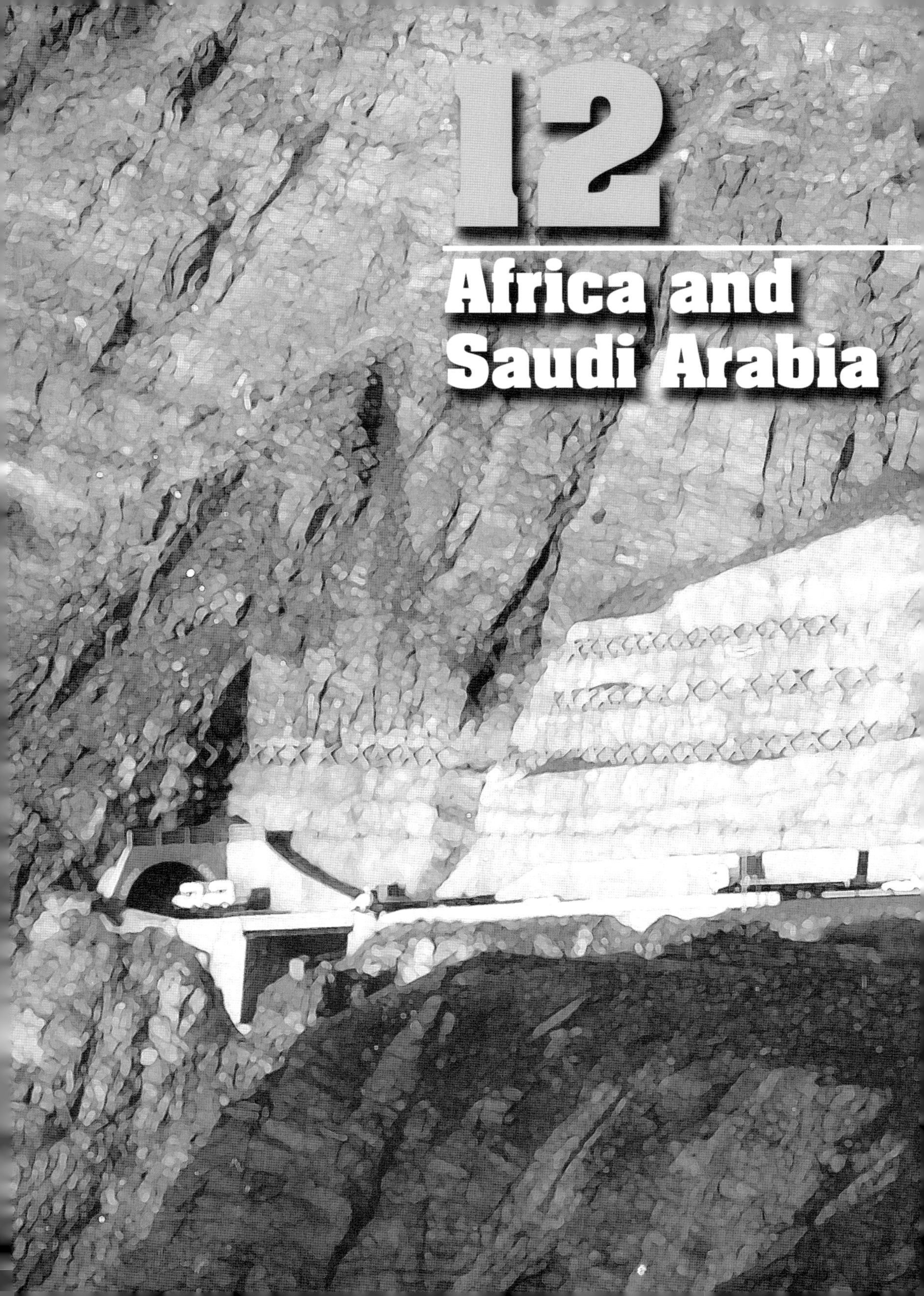

Africa and Saudi Arabia (1969-2003)

Through rock mechanics discussions with Dr. Evert Hoek of Imperial College London University, the author was introduced to the African mining industry. A major project developed at the Chambishi Mine in Zambia. It involved stabilizing a serious slope failure, designing a drainage tunnel and assisting with the design of underground mining access from the existing open pit. A major lesson learned was that a number of exploration drill holes should always be drilled well below the base of an open pit to determine if adverse structural geology or excess water pressures exist which may impact mine stability as the pit is deepened.

An unusual assignment developed for Placer Dome Mining, in South Africa, where five massive underground tunnel plugs were required to control mine flooding from an adjacent mine. The plugs were designed to resist up to 5,800-foot head of water, a world precedent. The tunnels were constructed using the prepacked technique, where the boulders are placed and the voids are filled with sand-cement grout. This procedure, which is a South African standard, results in less shrinkage than for massive concrete.

The author was retained by JECOR of the U.S.A., a technical assistance organization, to inspect and evaluate rock slope stabilization on the Sha'ar Descent highway in western Saudi Arabia, where a large number of rock bolts were failing due to faulty construction. While in the country, the author presented a one-day seminar on Rock Stabilization for Highways to Saudi Arabian Highways Engineers.

Chapter 12
Africa and Saudi Arabia

During one of my visits to Imperial College, London University, I stayed with Dr. Evert Hoek and his wife, Theo. Hoek, assisted by a number of graduate students, had taken on a major research project on slope stability in open pit mines in central Africa. Some sixteen mining companies sponsored the research, which was to extend for several years. Three large companies with operations in Africa — Anglo American, Roan Consolidated and Rio Tinto — had major involvement. Much of the practical fieldwork was performed in mines in Zambia.

Commencing in 1969 at the college, Hoek and I held a number of discussions regarding rock slope stability. He advised me that several large open pit mines in Zambia were experiencing serious stability problems, and asked if Golder Brawner would be available to provide consulting on the projects. At the time, several of Hoek's graduate students were completing their Ph.D. degrees. Their research was performed at these mines and directed toward developing field geologic and geotechnical modelling, mechanism of failure investigations, and mathematical evaluations of stability.

I said, "Evert, I'd be pleased to make our consulting services available."

Hoek arranged for me to accompany him on a trip to Zambia to meet officials of Anglo American and Roan Consolidated. Inspections were subsequently made of the Chambishi, Mufulira, Nchanga and Luanshya mines, and stability was discussed at each of them. Arrangements were made for several consulting assignments. To meet the additional staff requirements, on my return through London, I offered graduate students David Pentz, John Sharp, Brian Stimpson and Nick Barton jobs. The first three accepted. When they completed their studies, they moved to Vancouver, all arriving within six months. Golder Brawner paid all their moving and travel expenses, and they soon became key players on our rock mechanics and mine stability team.

The rock mechanics consulting assignments at Golder Brawner had expanded internationally during a very short period. Since we also had staff with earth- and tailing-dam specialization, as well as spoil pile stability and groundwater expertise, we could offer a broad range of geotechnical services in mining. To meet the expanding demand, the world was scoured for qualified mine geotechnical engineers, and many excellent engineers and geologists were hired. Bruce Kennedy and Richard Talbot came from Annaconda, Ken Rippere from Kennecott; Wilson Blake and Rudy Kvapil from Zambia; Adrian Brown, Kevin Rosengren and Ken Mathews from Australia; Ross Hammett from England and Manuel Socoras from Venezuela. In the late 1970s, Evert Hoek joined Golder Brawner in Vancouver as a partner.

Local senior staff included: Robert Wilson, Norman McCammon, David Campbell, Glen Gilchrist, Norman Norrish, Herbert Hawson, Bill Hodge, Bjorn Carlsen, Peter Stacey, Duncan Wyllie, Richard Butler, Trevor Fitzell and Jim Marlon Lambert. It was a powerful team.

On my second trip to Zambia, Golder Brawner had been requested to evaluate the potential to change from open pit mining to an underground operation at the

Chambishi Mine, near the town of Ndola. En route to this assignment, I stopped over for a day in London to meet with Evert Hoek. The next day I took a flight from London to Lusaka, the capital of Zambia, on Zambia Airways, which flew DC-8 jets with British crew. The plane left London at midnight, and stopped at Nicosia, Cyprus and Nairobi, Kenya, on the way.

After taking off from Cyprus, I fell asleep. As the plane approached Egyptian airspace, the captain's voice came over the intercom, thus waking me up. It was nearly daylight. He said, "I regret to advise you that we've been ordered by the Egyptian Air Force to land immediately at Cairo. They have given me no reason. If you look out the left windows, you will see two Egyptian Air Force fighter jets that are to escort us. I intend to follow their instructions. I'll keep you informed."

On such long trips I flew first class, and there were only two other passengers in this section. They were black, sitting directly behind me, and they became very excited. However, they spoke a language I couldn't understand, apparently some African dialect, so I didn't learn what they were worried about.

After we landed the captain again spoke, "I will get off the plane to determine what this is all about. The first officer is in charge while I'm gone. I recommend that you do not speak to anyone who comes on board unless I or the first officer are there."

Nothing happened for at least a half hour. I had concluded that, being a Canadian businessman, I had no reason to believe I was a target. I also concluded that what was occurring was not a hijacking. Accordingly, I lay down again, and went back to sleep. In the older jets the middle armrest between seats could be removed easily. I had done this, and so had two wide seats to stretch out on quite comfortably. When lying down, I was always sure to wrap the seatbelt around me so that it could be seen.

The next thing I heard was the pilot announcing, "Please fasten your seatbelts and prepare for landing at Nairobi." I had slept right through whatever happened at Cairo Airport, and the subsequent takeoff! I sat up, turned to the passengers behind me and asked, "What happened in Cairo?" Instead of answering my question, one blurted, "You slept through everything. How could you do that? Weren't you afraid?"

Somewhat dismayed, I answered, "There wasn't any violence; I'm not a criminal; I don't have much money with me, and I wasn't worth kidnapping."

After we landed at Nairobi I asked the captain what had happened at Cairo, and he answered disgustedly, "Zambian Airways officials had not renewed our over-flight permit for this month!" Thanks a lot!

On arrival at Ndola after a one-and-a-half-hour flight from Lusaka, I was met at the small airport by a protocol officer and driven to the "Guesthouse." There I was met by the "house-lady," a matronly English woman of about sixty. She welcomed me, instructed a houseboy to take my luggage, then showed me around the house and to my room.

It was very British and very comfortable. There was a large living room, a smoking room with billiard table, a dining room and a well-stocked bar. I didn't see the kitchen. She said that there were six bedrooms. Outside was a swimming pool, a tennis court and well-kept gardens. About two blocks away was the golf clubhouse and an eighteen-hole course where only whites were allowed to play.

It was about four o'clock, so I didn't plan to go to the mine that afternoon. The house-lady said dinner was served at 6 P.M., breakfast at 7 A.M. and lunch at 12:30. They would give me a wake-up call at 6:15.

David Pentz had arrived several days before I did. He was to assemble the relevant geology, rock mechanics data, mine design and plans that had been developed by the Chambishi staff for the proposed underground development. At about 5:30, Pentz arrived with the Mine Manager and Chief Engineer. We had a drink in the bar, followed by dinner, which was served very formally by a black waiter.

During dinner the Manager talked about the Zambian mining industry, the history of the mine, and the changes that were being made following the departure of the British and the ascension of Kenneth Kuanda, the first black president. He emphasized that people who had not lived in Africa did not understand the African tribal culture, and he gave a number of examples. Most Central and South African countries have two separate tribes. South Africa has many more. Invariably the tribes have political, cultural and language differences. Working in a mine, where teamwork is essential, it is wise to have members of the same tribe working in the same group. Otherwise you can expect disharmony and sometimes violence. He said that most of the blacks have large families, and if one member is much more successful than the others, he must assume more family responsibility. The less thriving may even try to live with the successful family member. Because of this, there is less incentive to succeed, or to seek a promotion.

Expatriates, usually from Britain, supervised most of the mines. During the early 1970s, it was common for the top 15 to 20 mine supervisors to be white and working on a fixed-time contract, which was usually renewable. These senior staff brought their families, and they were provided with good furnished housing, staff including a cook, maid, driver and gardener, and an automobile. The children attended a private school, and there was a private golf club. With a number of mines nearby, the number of expatriates in the mining region exceeded several hundred. Life was easy. Many of the blacks were jealous.

When Kuanda came to power, he wanted more blacks in senior positions and as mining engineers, so he established a program to select about forty bright students and send them to a variety of international universities to obtain mining engineering degrees. This included Canada, the United States, England, Germany and Russia. The intent was to assign some of them to be foremen in the underground mines. In this capacity, it was proposed that they would supervise black miners, most of who were more than twice the age of the young engineers. The older miners refused to work for them. Part of the reason was cultural, and part was that the young engineers had no practical experience, and jeopardized safety. While I was there, the young engineers were working in non-supervisory jobs.

The Chambishi pit was about 600 feet deep. Along the west side of the pit bottom the ore body suddenly changed from near horizontal, to near vertical. The plan was to develop openings from the open pit into the pit wall, and use declines at 8 to 12 degrees for the mining areas to be accessed. This negated the use for expensive shafts, and would allow for greater productivity. The initial engineering question was to determine the thickness of the outer pillar required to maintain pit wall support.

Rock core was obtained from selected boreholes for strength and modulus testing. Stress modelling was performed and pillar dimensions were calculated to provide theoretical safety factors of at least 3.0. This was a very cumbersome process, with all calculations done by hand calculator. (Computer programs for this type of problem had not yet been developed.) When minimum pillar dimensions had been determined, the mine engineering staff developed the underground mine design.

Several months later I received an urgent call from the Mine Manager. He said, "We've had a major slope failure. We'd appreciate it if you can come as soon as possible to evaluate stability. The mine is closed, as the failure has disturbed the haul road into the pit."

I arrived three days later. About half of the east side of the pit had slumped up to 25 feet, and a zone of push up had developed in the toe area of the slope. A portion of the access ramp had also slumped to destroy pit access. (This leads to the conclusion that a second access road is a desirable operating design in open pit operations.)

Chambishi open pit copper mine in Zambia. A major slope failure occurred on a thin seam of weak argillite. This destroyed the main access into the pit. Drilling indicated artesian conditions below the argillite. A drainage tunnel was recommended to reduce water pressures in the slope.

Geologic cross sections were only available to 10 to 20 feet below the ore body, so several deep geotechnical boreholes were drilled to obtain detailed geology and rock samples. About 25 feet below the base of the ore zone, which dipped at about 38 degrees and levelled out near the pit bottom, a 6-inch-thick argillaceous layer was encountered. The samples obtained were disturbed with slickensided layers — obviously the failure zone. Piezometers were installed in the rock above and below the failure zone. There was a differential positive head of 15 to 25 feet of water just below the argillite, allowing us to conclude that water pressure was likely the main cause of the failure. (A further lesson is to always drill some exploration holes well below the ore zone to locate potential deeper stability or groundwater problems.)

The pit was to extend northward. Exploration drilling in that area encountered similar geology — the argillite and excess pore water

Preliminary investigation and early development to change from open pit mining to underground mining. The ore body dips steeply under the far pit wall. Chambishi Mine.

Kalalushi open pit mine in west Zambia. The water table was very shallow. This caused serious blasting problems. A dewatering well is under construction in the foreground.

pressure. To stabilize the slide area and develop a stable slope for northerly expansion, a drainage adit was proposed, running perpendicular to the slope at a 1 percent grade, and extending about 20 feet beyond the argillaceous layer. At that point, adits behind that layer would extend in both directions. The entrance of the adit was to be outside the active slide zone, and further drain holes, 40 feet long, were to be fanned out from the adit every 100 feet, increasing the effective drainage.

While on this same trip, Pentz and I were asked to visit a newly developing open pit copper mine at Kalalushi, about 200 miles west. We were flown there in a small twin-engine charter airplane, picked up at the local airstrip, and driven to the mine. As we neared the open pit, we were stopped at a checkpoint by a black guard, who told us all to get out of our vehicle. A second guard began to spray the wheels and lower portions of the 4-wheel-drive unit. I asked the driver, "What is he doing?"

He answered, "He's spraying for tsetse flies." The guard had only a small hand sprayer, so it took him about twenty minutes.

The open pit had thus far only been developed three benches deep, a total of about 90 feet. The pit bottom was very wet. All blast holes filled quickly with water; some even flowed artesian. The problem was obvious. In order to continue mining, the water table would have to be lowered substantially below the pit bottom.

A phase-one investigation program would require four boreholes to a depth somewhat below the ore body, about 600 feet. Permeability tests would then be performed between packers in the boreholes, which would typically be expanded to test 10-foot sections of the rock. Details of the program were developed, and the field tests were performed by a South African consulting firm. The test data was evaluated to determine the number, location and design of the dewatering wells that would be required. It was proposed to drill the wells full depth, but to only set the pumps at depths sufficient to dewater to just below the mining depth. As the pit deepened, the pumps were to be lowered incrementally.

Several years prior to this trip, I had met Dr. Jerry Jennings, a geotechnical engineer and head of the Civil Engineering Department of the University of Wittswatersrand, Johannesburg, South Africa, at a rock stability seminar in Edmonton, Alberta. He had said there, "If you ever get to South Africa, please contact me at the university. We'd appreciate you giving a lecture." While on the Zambian trip, I decided to accept his offer, and so flew from Lusaka to Johannesburg. I gave two lectures, one on landslides in soils to civil engineers, and one on rock slope stability to mining engineers. As a result of these lectures, both Pentz and I were invited to present papers to a major South African mining conference in late 1972, an invitation we accepted.

My travel agent was instructed to arrange flights by the shortest and quickest route, which was via Toronto, New York and Rio de Janeiro. From New York, we were to fly on Varig Airlines, refuel at Rio, and then cross the Atlantic — a total flight time of twenty-two hours.

Upon arrival in New York, we checked in at the Varig counter. The check-in agent said, "We're sorry, but your plane is two hours late. However, we'll hold the plane from

Rio to Johannesburg for you!"

"Our ticket shows the same flight, " I said. "What's the problem?"

The agent replied, "We've had mechanical problems and will be using a flight from Buenos Aires. But we'll hold the plane."

Our plane was actually three hours late, and it was full. The flight was to be eleven hours. About one hour into the flight, the captain informed us in broken English, "I regret to advise we have a technical difficulty and have to return to New York. We are too heavy to land and will have to go to a lower elevation to drop some fuel. There is no danger!"

After landing back in New York, the ground crew took one hour to repair and refuel the plane. Pentz asked the chief flight attendant to reconfirm our flight time across the Atlantic, and, shortly before take off, the flight attendant came back on board and said to Pentz, "I'm sorry; they are not going to hold the plane. You'll have to stay in Rio de Janeiro two days before the next flight."

Pentz almost yelled, "Get my bags off the plane. I'm getting off!"

The flight attendant glared and stated firmly, "That's impossible. It would take too long."

Whereupon Pentz stood up, took him by the lapels and shook him, this time yelling, "You had better have your president meet this plane. We can't make that connection!" He was livid.

I stepped between them and said to Pentz, "It'll be okay. I'll phone Jennings from Rio and have him change the time of our papers." By this time it was after midnight. Pentz had obviously frightened the flight attendant, but nothing further transpired.

When we landed at Rio de Janeiro about noon the next day, a ground attendant came on board and announced, "Please remain seated. We have five people to disembark first." He asked Mr. Pentz, Dr. Roper, Mr. Jenkins, a young lady whose name I can't recall, and myself to come forward. The five of us were going on to South Africa. We went to the front of the plane and down the steps, to where a black Cadillac was parked. A chauffeur opened the doors, and we were driven to a special entrance in the terminal. Inside we were met by a well-dressed gentleman who introduced himself as Varig's manager in Rio de Janeiro.

"I apologize for your inconvenience, he said. "We have arranged for you to stay at the Copacabana Beach Hotel for two nights. We have also arranged for you to tour our beautiful city, to attend special shows tonight and tomorrow, and to pay for all your meals and expenses. We have arranged for you to fly first-class to Johannesburg in two days. We will also pay for you to phone South Africa."

Dr. Roper was going on a safari, and Jenkins, a senior executive with Dow Chemical Company, was on his way for business meetings. The young lady was from South Africa. She only attended the tours with us. Thus we had a free two-day holiday in Rio de Janeiro, and it was actually very enjoyable. Varig were great hosts. We were able to alert Dr. Jennings and arrange for our presentations to be on the third day of the conference.

Once in South Africa, however, the excitement wasn't over. We wanted to visit the South African Chamber of Mines in Pretoria, and hence rented a car at the airport, in Pentz's name. That evening Pentz wanted to visit personal friends and took the car. At about three o'clock the next morning, I got a call from the police station. An officer said, "We are holding Mr. Pentz as a result of a car accident. Do you know him?"

I answered, "Yes. He's one of my engineering partners. We're attending an engineering conference from Canada. What's the problem?"

"We have found a rental car that has crashed, and it was signed out to Mr. Pentz. We picked him up earlier at your hotel," he explained.

Pentz was released, into my custody!

It seems that when Pentz returned to the hotel after seeing his friends, he was asked to leave the key to the car at the hotel desk. Apparently, one of the hotel's young employees took the key and went for an unfortunate joy ride, slamming into a traffic signal post. We received another car the next day.

In 2000, Placer Dome Mining from Vancouver purchased 50 percent of the South Deep gold mine in South Africa, located about 40 miles south of Johannesburg. An adjacent mine, owned by Harmony Mining, was running out of ore. It was a deep mine with a major water inflow problem; they were pumping about 3 million gallons of water per day to the surface.

At levels 5,000 and 5,800 feet below the surface, five old exploration adits connected the Harmony mine and the South Deep mine. In closing their mine, Harmony wanted to shut off their pumps. The result would be that the South Deep mine would flood through the connecting adits.

Two options were available: South Deep could continue the pumping at a cost of millions of rand per year, or the adits could be plugged. Placer decided on the latter course. I was contacted by Warwick Morley-Jepson, Projects Manager at South Deep, and asked if I would be available to review the design. He said that Placer Dome had advised him that I had supervised the successful investigation, design and construction of a major permanent plug at the Marcopper Mine in the Philippines, controlling outflow through a tunnel that connected a tailings-filled open pit through a mountain to the Makulapnit River. On that project, at my request, two other internationally recognized consultants had assisted me: Professor Stuart Littlejohn of England, and Dr. Donald Bruce of the United States. Both were specialists in underground water control and grouting.

I asked Morley-Jepson, "Can you provide me with more details?" He spent about five minutes describing the issues. The adits were about 12 feet by 12 feet. The plugs would have to resist up to 5,800-foot head of water — a precedent for water pressure on underground plugs in mining. The mine life was expected to be about ninety years. Hence, that would be the maximum design life. Some of the mine water was very acidic, so the plug material had to resist that condition. A pillar of rock would be left between the two mines, and this pillar would also have to resist high water pressure.

Morley-Jepson further advised that he would courier several reports to me introducing the problem, if I was interested. I replied, "I'm very interested. The problem represents a world-class challenge. I'd recommend a panel or 'Review Board' approach." I described how I had worked with Littlejohn and Bruce at Marcopper, and would also recommend that again. I continued, "I'll prepare a proposal and fax it to you for review."

One week later I received approval. Littlejohn and Bruce agreed to work with me on the project. A first step was to visit the mine. We agreed on the date for a site visit, allowing four days there. Prior to the visit we reviewed the reports forwarded to us.

It's a tough trip from Vancouver to South Africa, overnight from Vancouver to Europe, and a second overnight from there to Johannesburg. Fortunately, I was able to rent a day-room at a hotel in the London airport and get six hours' sleep between flights.

Preliminary design of the plugs had been done by Steffen Robertson and Kirsten (SRK), and a construction proposal had been developed by Cementation Skandia Mining Contractors. The concrete plug at Marcopper Mine in the Philippines had been designed and constructed using North American procedures. In South Africa, the Chamber of Mines provided mining research and developed engineering standards, which included plug design. In North America, water control plugs were generally constructed of mass concrete. In South Africa, however, prepacked construction was used, involving the placement of hard, cleaned boulders (plums) as the plug, and grouting with sand cement to fill the voids. Concrete shrinks when it cures, however, leaving a void at the top of the plug that must be filled with grout after the plug has cured. With the South African procedure, the hard plums fill about 50 to 60 percent of the space, and do not shrink. Hence, the void at the top of plug is much less. This procedure is no longer used in North America because of the high cost of manual labour to place the plums. South African labour costs are much less, and engineering in the Third World in general can often afford to use much more labour-intensive procedures.

Olly Olivier, Project Manager of South Deep, met us at the airport in Johannesburg. He advised, "It's not safe to travel by yourselves. We've arranged for you to stay at our guesthouse."

Several meetings were held with SRK and Cementation Skandia staff to review their original proposals. Following that, we went underground to inspect the plug sites — five separate locations. This inspection was difficult. We had to walk a total of about 3 miles to see all the sites, and the entire area was unventilated. The air was rank, the temperature near 100°F, and the humidity high. I couldn't use my glasses and camera because the lenses fogged up.

The final two days we reviewed our initial evaluation and made numerous recommendations. They included the following:

1. The plug sites must be located in competent rock which would require scaling, mapping, measuring, rock testing and wall grouting. Blasting during the original adit construction had damaged the rock in the tunnel walls.
2. Detailed materials specifications, construction procedures and a method statement were required.
3. Detailed Q_a/Q_c (quality assurance and quality control) testing would be required.
4. The hydraulic gradient along the plug must not exceed 50.

5. Given that the plugs would be "State-of-the-Art," it was recommended that a test plug be constructed.
6. The length of the plugs must be increased from the original design.
7. Detailed construction drawings and specifications were required.
8. Backup equipment was required for every phase of construction.

Rison Consultants provided site geological mapping and inspection of the pillar rock between the two mines, and hydrology studies. SRK performed stress modelling of the pillar area. Cementation developed construction details for the test plug, including temperature monitoring, and both consolidation and compaction grouting. SRK supervised sample recovery and testing of the prototype plug. The test plug turned out to be a very valuable exercise. Crews became trained in all the procedures, including the Q_a/Q_c testing.

On state-of-the-art projects, such a test program should always be considered. Based on the test program, the final design and construction can proceed with more assurance. The five plugs were completed in July, 2003. This project has been a great experience, again illustrating the benefit of a review panel on difficult and complex projects. The final step is the requirement for the review panel to "sign off" on the project, confirming that it meets the international "standard of practice." Following completion of as-built drawings and reports the sign-off report is expected in September 2003.

In October of 1994, I was contacted by JECOR in Washington, D.C., to determine if I was available to provide a one-day seminar on "Rock Slope Stability in Riyadh, Saudi Arabia," and, while there, inspect rock slope stability on several highway projects. JECOR is an American agency that provides technical non-military advice and assistance to underdeveloped countries. They would pay all expenses and generally look after me while I was in the country. I agreed to go.

I flew overnight from Vancouver to London, spent a day at the Ariel Hotel near the airport to rest up, and the next day flew on Saudi Arabian Airlines to Riyadh. As part of my contract with JECOR I flew business class. This made it easier for me to travel with only carry-on luggage.

I had been alerted by JECOR as to items to not take into Saudi Arabia, which included the Bible or other religious information, pornographic literature, and alcohol. I followed this advice. I was nearly the first off the airplane and, with no baggage to pick up, I was the first through immigration and customs. My luggage was thoroughly searched in a very efficient manner. The customs officer, dressed in North American clothes, smiled and in perfect English said, "You may proceed; welcome to the Kingdom of Saudi Arabia."

Upon leaving the customs hall, I stopped and looked toward those waiting for arriving passengers. I was to be met by a driver. Slowly scanning the mass of people, I spotted a dark-haired man holding a JECOR sign high over his head. I walked toward the railing confining these people, pointed to him and waved. He nodded, smiled and indicated I should go to the left. Near the end of the railing we met.

I introduced myself, "I'm Chuck Brawner, here to give a seminar."

He answered, "I am Abdullah, your driver. I have one more person to meet off this plane. Please wait for me against that wall." He pointed behind him to a group of chairs.

The second person was a professor from the University of Texas who was a specialist in traffic engineering. He took over half an hour to go through immigration and customs.

The Riyadh airport is unusual. There are three terminals: one for international travel, one for domestic travel, and one solely for Saudi royalty. The number of Saudi royalty is very large, some 2,000, plus their families. These people are wealthy and travel a lot; the most senior own private jets, a few of which are Boeing 747s! Occasionally, these are seen in Las Vegas, where the Saudis go to gamble and play.

The drive into the city was uneventful. We were advised by Abdullah that JECOR was located within a single compound which contained the JECOR offices, rooms for business meetings, a communications office, a restaurant with a bar, and an apartment complex. The compound was off-limits to Saudis without invitation. All major countries with extensive dealings in Saudi Arabia had separate compounds, which were considered an extension of their home country. The British compound, for example, had a British pub, curious since alcohol was technically prohibited in Saudi Arabia.

Life for non-Muslim foreigners in Saudi Arabia is difficult. While Saudi dress is not required, certain formalities are expected. Only a minimum of skin is to be exposed, meaning long sleeves and socks when wearing sandals, for example. Signs of male–female fraternization in public are frowned upon, even holding hands with your wife.

I checked in with security, showing them my passport, visa, and contract with JECOR. They showed me to my room, actually a two-bedroom suite. My partner in the seminar was Larry Pierson, State Geologist, from Salem, Oregon, and he was given the second bedroom. The attendant showed me around the premises and indicated the meal hours. Pierson arrived later that day. We had worked on many rock stability projects together in Oregon, where Pierson had asked me to provide a second opinion. He is an excellent geological engineer, and we got along well together.

In the contract with JECOR, it was recommended that we inspect some of the major highways in Saudi Arabia so we could include recommendations for local conditions in the rock slope stability seminar. This recommendation was accepted by the Saudi Ministry of Communications, and by Dr. Howard McCann, the senior JECOR representative in Riyadh.

The next morning we met with McCann, and were also introduced to Jamal Elkaissi, a Lebanese–American civil engineer on the JECOR staff who was to be our guide and translator for the field inspection. The program for the field trip and seminar was discussed. We were advised that we were to meet with the assistant deputy minister of communications to discuss our interpretation of site conditions. The ministry had hired a rock engineer, Dr. Kari Saari from Finland, as a consultant to them, and he accompanied us on our inspection.

The next morning we drove to the domestic airline terminal, boarded a Saudi Arabian Airlines Lockheed L-1011 jumbo aircraft and flew to Abhã, a city 6,000 feet up in the Asir Mountains, which extend along the west coast of the Red Sea. The plane was full, including many women in full Arab gowns and veils.

On arrival, we had a quick lunch and then drove to a construction camp to meet a Saudi engineer. He was the government inspector on a project to repair major highway damage, which had occurred during a record rainfall some months before. This included relocation of a section of highway.

The highway descended about 6,000 feet from Abha toward the Red Sea. The Saudi climate is very dry, and thus the mountains were lightly vegetated and covered with only a thin mantle of soil. The rock was deeply weathered and subject to severe erosion during occasional high-intensity thunderstorms. The design speed for the highway was 70 miles per hour, and, as a result, very high rock cuts up to 350 feet were developed at some locations. On the project inspected, benches were developed down the slope, intended to catch rocks which would ravel. (This procedure is no longer used in North America because the benches fill with rock and tend to act as a ski jump directing rockfall out onto the highway.) Considerable use was made of shotcrete to stabilize rock faces that were cracked due to blasting and naturally jointed. Where large blocky rock structure existed, rock bolts had been used. Relatively narrow catch ditches existed along the inside shoulder. The rock faces were quite ragged, indicating to Pierson and me that the contractor was using blasting procedures that developed excessive seismic energy.

We took many photographs and made extensive notes. We did not discuss our interpretation of site conditions, the design or construction with the Saudi inspector. There were many recommendations we could have made, but there was no point in confusing or embarrassing him. We would discuss this with the assistant deputy minister on our return to Riyadh.

Two unrelated observations were noteworthy. At several locations we encountered large monkey colonies perched on the rock cuts, watching us and chattering away. We didn't have any food so they kept their distance. As well, a 48-inch steel pipeline was observed along the route at several locations. We were advised that it carried water from a large desalinization plant on the Red Sea up to Abha and the surrounding area. Fresh water is a major problem, and a very expensive commodity in Saudi Arabia.

That evening Elkaissi drove us to a new, large and luxurious hotel near Al Baha. It was almost empty. We were advised that when the summer heat descended on the country, the hotel was full of people seeking the cooler temperatures at 6,000 feet.

The next morning we drove to the office of the contractor who was making repairs on the section of the highway running down the east side of the mountain, called the Sha'ar Descent. The upper portion of the highway was spectacular, dropping 2,100 feet in 8.5 miles. It had been designed by Italconsult from Rome, constructed by the Chinese contractor Ret-Ser Eng.Ag, and completed in 1982. The project included 32 bridges, 11 tunnels, 351 culverts and 12,000 feet of retaining walls, plus extensive shotcreting of slopes and thousands of rock bolts.

However, after twelve years of use, a number of serious problems had developed. A surprisingly large number of rock-bolt installations had failed. They were manufactured by Dywidag, and epoxy grouted to develop the inner anchorage. They were tensioned to about 70 percent ultimate strength, but were *not* grouted between the anchor and face plate! The anchorage was failing and, with the sudden release of tension,

The Sha'ar Descent highway. In this photo many rock bolts have failed and a rock slide destroyed a bridge.

many bolts suddenly shot out of the drill hole, across the highway, and down into the canyon. It is a miracle no one had been killed. The anchor plates were a large reinforced concrete cross. These fell onto the inside of the highway.

Not grouting the rock bolts full length after tensioning was a serious error. Full-length grouting reduces corrosion and maintains tension in the bolt for the long term. Also, tensioning the bolts to such a high stress was excessive for the circumstances. This is a common error when stabilizing rock slopes.

Sha'ar Descent highway east from Al Baha, Saudi Arabia. The highway was designed by the Italian consulting firm Italconsult and constructed by a Chinese contractor. The highway was completed in 1982 through very rugged terrain.

Near the centre of the Upper Descent, a bridge had been taken out by a rockslide from above. The rock slope had been stabilized with tensioned rock bolts, but again these bolts were not grouted full length. The bolts were installed in a pattern spacing of about 12 feet by 12 feet, and they were all 16 feet long. By using all the same length bolts, a zone of high tension developed, which then cracked the rock at the inner tip of the anchors. Over time this became a shear plane. Good practice calls for random variable-length pattern bolts to ensure that such a continuous crack does not occur in the rock.

Dames & Moore, geotechnical consultants, had recommended that a bypass tunnel be developed at the site. The assessment by Pierson and myself indicated that the original alignment could be redeveloped, provided the upper rock slope was properly stabilized. This would be much less expensive than a tunnel.

Near the lower third of the Upper Descent, a major switchback existed. At several local points along the switchback, rock slope instability involving some 50,000 to 100,000 cubic yards appeared to exist. The consultants had recommended that the entire switchback be changed to a tunnel design, an option that would cost in the tens of millions of dollars. We suggested that a study be performed to determine if the four slide areas were an indication of major overall instability. Four slope-indicator installations and piezometers were proposed. Stabilizing the separate slides would be much less expensive than constructing the long bypass tunnel.

A section of the Sha'ar Descent highway constructed with retaining walls, bridges, tunnels, rock sheds, shotcrete and extensive pattern bolting. Note the unusual reinforced concrete rock bolt baseplate design.

Many of the Dywidag rock bolts have failed due to anchor failure in the lower two rows and upper row. Following tensioning, the bolts should have been grouted full length.

Most of the rock anchor bolts that failed sailed across the highway and down the mountain slope. Two anchors in this photo became stuck part way out of the drill hole. This is an extremely dangerous condition.

The day following our return to Riyadh, I presented my seminar. I had prepared three Kodak carousel trays full of slides and gave the presentation using them. One initial problem occurred. The multimedia presentation system in the small theatre was very modern. The slide presentation unit was in the room adjacent to the theatre, and the screen in the wall between the two, calling for rear projection. All my slides had to be reversed in the carousels. Pierson helped me with the quick change.

I gave Elkaissi a brief résumé of my experience so that he could introduce me in Arabic. I had been advised previously that the Saudi engineers who were attending could speak and understand English and that I would not require translation. Following the introduction, I expanded on my experience and began the seminar with several dramatic case examples of rockfall accidents. This got their attention. I then spent about fifteen minutes outlining the subjects I proposed to cover.

I preferred to have the seminar very informal and said, "If you have any questions, raise your hand and we'll discuss the question at that time." They did ask a number of questions, and I appreciated this, since the type of questions they asked acknowledged to me that they understood what I was saying.

For two hours the presentation appeared to be going well. Then, at about 10:30 everyone abruptly stood up and left the room! I was stunned! No one had said a thing, and I hadn't spoken about a midmorning break. But there I was — alone in the room except for Elkaissi and Pierson. I looked at Elkaissi, turned my hands up and said, "What's up?"

He smiled and explained, "I should have told you. They go for prayers every morning and afternoon!"

I felt relieved. "I thought that I had said something wrong, something that antagonized them."

Elkaissi said, "No, you said nothing wrong. Prayers are a Muslim tradition. They will return in about thirty minutes."

They did and I resumed the seminar. However, I had to adjust to new time constraints. Prayer time reduced the seminar by about one hour.

The rest of the seminar went well, including the afternoon prayer time. The following day Pierson made his presentation.

The Finnish consultant, Saari, attended my presentation, and that evening he took us to a Saudi spa and "bar." The complex had a large swimming pool with small hot pools. The bar consisted of breathing the vapour from a hot water canister. I have to admit I still prefer a glass of good red wine, or a scotch and soda. Had any newcomer arrived, it would have looked like everyone was taking oxygen. However, it probably cleansed our lungs.

On the morning of Pierson's presentation, Dr. McCann had arranged for me to meet with engineer Abdullah Al Mogbel, Assistant Deputy Minister of Communication. I had developed my photographs the previous day and used them during the presentation to describe the problems that we had observed. The assistant deputy minister had a degree from Harvard University in civil engineering, and was most interested. Potential solutions for each of the problems were described. Several further studies were recommended. He asked McCann if copies of our report to JECOR could be provided to him. McCann said, "Yes, and a contract for the studies could also be arranged." Pierson and I prepared a report and proposal which was sent to JECOR.

Several months later we received word that the study would not proceed. Saudi Arabia was broke as a result of the Gulf War and the payment they had agreed to make to the United States for their involvement. Nevertheless, this was one of the most interesting projects I have been involved in, involving a review of faulty rock-bolt technology, and an introduction to the Islamic world. Both were surprises.

13

Reward

The Canadian Geotechnical Society

La Société canadienne de géotechnique

July 27, 1999

Mr. Chuck Brawner
1386 Sowden Street
North Vancouver (BC) V7P 1L8

Dear Chuck:

The Legget Award is the most prestigious award of the Canadian Geotechnical Society and it is with great pleasure that I am writing to you today. The five members of the selection committee have unanimously chosen you as the 1999 Legget Award winner among very valuable candidates.

The Canadian Geotechnical Society is particularly proud to have members like you, who contributed so much to the development and practice of Geotechnical Engineering.

Let me personally and as President of the Canadian Geotechnical Society congratulate you for this well-deserved honour. I look forward to meet you at our annual conference in Regina where the award will be presented to you.

With best regards,

Guy Lefebvre
President

Introduction of C. O. Brawner at the Legget Award ceremony of the Canadian Geotechnical Society by Mr. Fred Matich, MAJM Corporation, Toronto (1999):

"I have the honour and great pleasure of introducing this year's recipient of the coveted and prestigious R. F. Legget Award, Mr. C. O. (Chuck) Brawner, a dedicated Canadian consulting engineer and educator who is also foremost among Canada's internationally known geotechnical experts.

"Most of you will know Mr. Brawner by professional reputation and some will have had the privilege (as I have) of associating with him on major projects, and enjoying a long-lasting friendship. However, comparatively few will probably realize the great depth and diversity of his experience and accomplishments, as well as his significant pioneering role and other technological contributions to the geotechnical and mining communities in Canada and around the world. Unfortunately, there is time today to touch on only some of the highlights of his long and distinguished career.

"I will be referring in more detail later to Mr. Brawner's contributions to the well being of the geotechnical profession. It is opportune to mention at the outset, however, that this has included service on a number of National Research Council of Canada technical committees under the leadership of the late Dr. R. F. Legget. Those of you who were also privileged to so serve at the time will, I am sure, agree that Dr. Legget held Mr. Brawner in high esteem.

"Mr. Brawner received his Bachelor of Science degree from the University of Manitoba in 1953. His career began with the B. C. Department of Highways where he held the positions of Soils Engineer and Senior Materials Engineer in the period 1953 to 1963, except for time to complete a Master of Engineering degree in civil engineering at Nova Scotia Technical University in 1958.

"In 1963, he decided to enter private practice with the late Dr. H. Q. Golder and other colleagues, forming the consulting firm of Golder, Brawner and Associates, of which he became Principal and President. The firm grew in size and reputation both nationally and internationally in all aspects of applied geotechnical engineering, and became particularly well known for its expertise in specialty areas such as soil and rock mechanics; muskeg engineering; landslide evaluation; earth and tailings dams; and stability in mining. Early in his Consulting career, he saw the potential demand for rock slope engineering. His vision and leadership skills can be gauged from the fact that he took a month off to meet and discuss rock mechanics and slope stability with eminent people in this field in Canada, England, Germany, Portugal and the United States. Assignments on other major mining projects worldwide followed, and Mr. Brawner became identified as an authority on rock slope engineering. He has been instrumental in training and developing many geotechnical engineers who have gone on to become well known in the profession in their own right.

"Mr. Brawner made a major career change in 1978 when he joined the University of British Columbia, where he was appointed Professor of Geomechanics in the Departments of Mining, Geological and Civil Engineering. He became a popular and successful educator, not only at UBC but also at many venues where he was invited to present lectures or courses, including Universities and Institutions in such countries as the U.S.A., China and Russia. He retired from the University in 1994 and is currently Professor

Emeritus. He continued his specialist consulting engineering while at the University.

"While contributing to the development of a successful consulting company, Mr. Brawner provided bold and innovative advice in solving numerous practical issues in applied geotechnical engineering. He has continued to do so through his practice as a specialist Geotechnical Engineer, and has for years been much sought after on major engineering and mining projects, serving either as an individual expert or as a member of a board of consultants. The diversity of projects he has been involved with had their geographical distribution in about 30 different countries, is impressive and too great to describe in detail at this time. They are attested to, however, by the fact that the list includes (*i*) expert review for two of the world's largest mining operations, namely Syncrude Canada Limited's Tar Sand Mine, and Codelco Chile's Chuquicamata open pit copper mine, and (*ii*) advice on correction of distress experienced recently by tailings disposal systems at the OMAI Mine in Guyana, and the Marcopper Mine in the Philippines, respectively. Mr. Brawner also has a commendable record of original research in a variety of geotechnical fields, and has pioneered developments (such as special geotechnical processes and monitoring instrumentation) particularly for the mining industry.

"Mr. Brawner has shared his knowledge and experience through publication of numerous technical papers. He has convened a number of conferences and edited or co-edited many conference texts or proceedings, and given freely of his time on a number of committees and task forces. These include NRC's Sub-Committee on Muskeg and the Canadian Advisory Committee on Rock Mechanics, to which Mr. Brawner made vital contributions. He has been an invited keynote speaker at National and International conferences, including the 40th Canadian Geotechnical Conference. He was one of the first to recognize the importance of resolving disputes in engineering and construction matters, and become a chartered Arbitrator.

"I hope that this serves to give you at least some insight into the notable career and accomplishments of our new Award Recipient. I should like to also quote the following comments by distinguished Canadian Geotechnical Engineer and mutual friend, Dr. N. R. Morgenstern, which are most apt:

'I believe that the most substantial contribution made by C. O. Brawner was his early recognition of the role of Geotechnical Engineering in the Mining Industry and his dedication, over a lifetime, to fostering this role.

One of the characteristics of the Canadian Geotechnical Community that distinguishes it from others is the capacity for solving an enormous range of difficult problems encountered by the Mining Industry, particularly related to tailings management and slope stability. As the Canadian Mining Industry has become more international, so has the Canadian Geotechnical community, building on this special relationship.

C. O. Brawner was one of a few people who understood the challenges, gathered human resources to deal with them, developed the technology and created a professional environment for growth in all senses that has been of enormous benefit to the whole Canadian geotechnical community.'

"In reviewing Mr. Brawner's successful career path with its significant diversity and commitment, and contributions to the geotechnical profession, much credit is clearly due to his wife Patricia for her constant dedication and support. This has included such highly interesting but challenging activities as travel to a variety of countries abroad, including the Peoples' Republic of China, accompanying her husband on important assignments. Chuck and Patricia are blessed with four children and seven grandchildren. The children have a diverse range of careers. Cathy, who had been a school teacher, is a homemaker; Randy is a senior executive with a U.S. stock brokerage firm; Suzanne a professional nurse; while Gregory is a pilot with Continental Airlines. Chuck has found time to travel recreationally with Patricia and enjoy outings with family members to play golf and go fishing, including big-game fishing in New Zealand.

"Mr. Brawner is a Fellow of the Canadian Academy of Engineering, The Engineering Institute of Canada, and The Canadian Institute of Mining and Metallurgy. He is the recipient of many other awards, including the 1986 Daniel C. Jackling Award — the highest award of the American Society of Mining Engineers. I am sure you will agree that he is deserving in all respects of his appointment to the honour roll of Canada's geotechnical engineers, and will join me in extending congratulations to Mr. Brawner on winning the 1999 R. F. Legget Award, the highest honour bestowed by the CGS."

Reply by C. O. Brawner

I am greatly honoured to receive this major award from the Canadian Geotechnical Society.

First, I would like to recognize the outstanding contribution that Dr. Legget made to the Canadian geotechnical field during his time with the National Research Council in Ottawa. For some years I had the opportunity to be a member of the Muskeg and Geotechnical Research committees while he was at NRC and greatly benefited from my association with him.

In addition, I wish to acknowledge the professional association with many great geotechnical engineers, in particular, Dr. G. Meyerhof, Dr. N. R. Morgenstern, Mr. Fred Matich, Dr. R. W. Hardy, Dr. Hugh Golder, Dr. Victor Milligan, Mr. L. Soderman and Mr. John Seychuck.

I also wish to acknowledge my former Associate Partners in Golder Brawner and Associates for their excellent engineering assistance — Robert Wilson, Norman McCammon, David Campbell, David Pentz, Dr. John Sharp, Dr. Brian Stimpson, Glen Gilchrist, Herbert Hawson, Peter Stacey, Duncan Wyllie, Ross Hammet, Richard Butler, the late Bruce Kennedy, Norman Norrish, Adrian Brown and Richard Talbot.

In responding to this award I would like to comment on a number of issues that have been significant in my career which may be of benefit to other engineers, particularly consultants.

1. Business Development

Few engineers receive any formal business training. It is generally learning by trial and error. Competition in consulting engineering is substantial. Many clients shop for the lowest price rather than for experience or special expertise. Larry Soderman taught me that you must emphasize your experience in an effective manner and show that you have the staff and capability to perform the project effectively, efficiently and economically. In my experience, this approach led to the development of corporate and specialty brochures, newsletters, interviews with the media, development and presentation of papers, attendance and development of display booths at conferences, and presentation of specialty in-house seminars to clients.

Several important rules should be followed:

- Meet your deadlines.
- Never submit an invoice in excess of the project estimate, without advance notice and detailed justification.
- Look for new projects when you are still busy.
- Maintain your existing clients.

2. International Engineering

Working in the international field provides a great range of personal and professional challenges. In geotechnical engineering, this may mean different geological, soil, rock and hydrological conditions. Along with that go different cultures, customs, financial, environmental, regulatory and work ethic considerations. With proper recognition and preparation, these can be interesting, challenging and often fun. If you don't prepare, you must expect problems.[1]

Several recommendations are provided.

- The first time you go to a foreign country arrange to contact the Dept. of Foreign Affairs in Ottawa for information. Also arrange to meet with the Senior Commercial officer in the Canadian Embassy or Consulate in the country you are going to, on the day of your arrival. Request information and advice regarding customs, moral code, dress code, hospitality code, method of salutation, money exchange, tipping, hand signals, the internal wage scale, *etc.*
- In Third-World countries never drive a car, hire a driver. If he kills a chicken, he goes to jail, not you.
- Beware of corruption. If you pay once, you'll pay forever.
- Don't take anything you wouldn't declare through customs.
- Beware and budget for the mañana (tomorrow) syndrome.
- You are a foreigner in their country. Respect their culture.

3. Environment

For any project you work on, respect the environment. First of all become familiar with the local regulations. In some Third-World countries, regulations do not exist. In this case recommend that the relevant regulations in Canada be used. As an example, Cambior Inc., a gold mining company from Montreal and Val d'Or, Quebec, developed the OMAI Gold Mine in Guyana, South America, which had no mining regulations. They recommended to the Prime Minister of Guyana that environmental regulations in Quebec be selected to be effective in Guyana. This recommendation was accepted — and followed.

In 1966, B.C. Hydro retained Golder Brawner and Associates to develop slope protection for the village of Nakusp waterfront. This was required because of raising of the Arrow Lakes about 90 feet, via construction of the Keenleyside Dam near Castlegar. Hydro wanted to place rip rap along the slope in front of the town for some 3500 feet. This was the standard erosion control procedure at that time. However, it would have destroyed the use of and access down the slope. The alternative of a soil cement slope (now called roller compacted concrete) was proposed, the first of its kind in Canada.[2] The finished slope is attractive, allows ready access to the beach and recreation in the winter after snowfall. After 33 years this slope is still in excellent condition.

4. Teaching and Case Examples

We must publish case histories, not only of successful and unique projects, but also to provide details of failures, forensic investigations, subsequent stabilization, and lessons learned. Many clients hesitate to allow publication of problems, partly due to liability reasons. However, we learn more from failures than we do from successes.

One of the areas that I have been involved in is mine waste disposal; the design and construction of tailings dams and waste dumps. Both of these structures represent a cost. They generate no profit. Hence, there has been a tendency by some mining companies to limit the extent of geotechnical investigations. History reveals this to be a very serious omission. We have learned much about earthquakes and liquefaction from tailings dam failures caused by the El Cobre earthquake in Chile, the necessity of proper filter design and construction at the OMAI gold mine in Guyana, the implication of hydraulic fracture at the Marcopper project in the Philippines, the impact of a thin strain sensitive layer in the foundation soils at Los Frailes in Spain, and the consequences of overloading loose tailings at Pinto Valley in Arizona. We must press for technical publication of failures and case examples.

5. Concern Over Project Size

As projects become larger, either for the economy of scale or for the need to move more people or product, the scale of geotechnical engineering has increased significantly. We have gone from two-lane highways to ten-lane freeways, to tailings dams up to 200 meters high, to waste dumps up to 400 meters high, and to open pit mines over 1000 meters deep. Younger engineers have a tendency to fear this great increase in scale.

Major projects are one of the engineering challenges of today. Provided the fundamental concepts of geotechnical engineering are understood and adhered to, large projects should not be feared. There may be financial concerns, but proper application of engineering principles should result in safe and practical solutions. Dr. Karl Terzaghi has proved that on many projects.[3] With the assistance of computers to evaluate multiple options, the availability of extensive technical literature, and the use of the observational approach, sound geotechnical engineering solutions are more readily attainable.

6. Advances in Surface Rock Mechanics

I first became involved in Rock Mechanics in Surface Mining in 1966 on Kennecott Copper's B.C. Molybdenum mine near Alice Arm, B.C. A major rockfall occurred from an 80 meter high excavated slope and damaged construction in progress at the main mineral processing building. The slope had been excavated at 75°. The rock was a hard limestone that had bedding which dipped out of the slope at about 45°. The friction angle along the joints was less than that. Failure had to occur. A modified design was developed to excavate the rock to the dip angle or hold it up with rock bolts. This stabilization was successful.

Some months later Kennecott called again. This time they had three projects for me to review, including the Bingham Canyon mine near Salt Lake City. This was the deepest open pit mine in the world (800 m deep). I concluded I was not ready for this assignment. I delayed my involvement with Kennecott for two months.

In the meantime, I performed a literature search at U.B.C., Vancouver. I only found eleven relevant, significant articles on rock slope stability. Based on this review, I spent several days each with Dr. Evert Hoek in England, Drs. Leopold Muller and Klaus John in Germany, Dr. Manual Rocha in Portugal, and the U.S. Bureau of Mines in Denver. The scale of rock mechanics problems is much greater in mining engineering than in transportation engineering, with which I was more familiar. I concluded that four major errors were being made in rock slope engineering which the mining industry was not recognizing.

1. Water is not a lubricant, as claimed by many geologists and miners.
2. Excess blasting was damaging final pit slopes.
3. Drainage of rock slopes would increase slope stability and allow steeper slopes.
4. Stable slope angles less than the typical angle of repose of broken rock (37°) may be required. For example, in clayey rocks.

I prepared an article describing these misconceptions and published it in the *Western Miner* magazine.[4] This led to Surface Rock Mechanics becoming a major speciality of Golder Brawner and the author. Since then, many advances have been made in this field.

7. Development of New Ideas

One of the greatest challenges in engineering is to develop or apply new geotechnical concepts or procedures. At Golder Brawner we had a number of successes. These included:

- Application of pre-consolidation of peat or muskeg to develop adequate foundations for major highways (the Burnaby freeway, Vancouver).
- Development of dredged sand islands as drill platforms to allow oil well drilling in the Arctic Ocean (Humble Oil).
- Design and construction of major slope erosion protection using soil cement (B.C. Hydro – Nakusp, B.C.)
- Development of vacuum horizontal drainage to stabilize major rock slides (Malibu, California and Bonneville Dam, Oregon).
- Application of tripod and electronic distance measurement to monitor mining instability (Fording Coal, B.C., Canadian Johns Manville, Quebec).
- Application of drainage adits to stabilize major mining slopes (Twin Buttes Mine, Arizona and Atlas Mine, Philippines).
- Application of controlled blasting to improve stability and steepen open pit mine slopes (Cleveland Cliffs Mine, Michigan).
- Development of recognition of potential block type failure in tar sand mining (Syncrude, Alberta).
- Development of the drop raise procedure to fill major underground openings in advance of open pit mining (Mascot Gold, Hedley, B.C.).

- Development of a rock slope stability priority rating system to prioritize slope stabilization (C.P. Rail).

The successful development and application of new procedures is one of the great pleasures in engineering. Don't hesitate to explore and test new ideas.

8. Review Boards

As geotechnical projects became larger, more complicated and more expensive, the concept of "two heads are better than one" becomes a significant reality. The implications of failure, financial and human loss, and the increasing potential for litigation require a second look at engineering investigation, evaluation, design and construction programs on major projects. Many projects are sufficiently complex that multiple specialties are involved.

In recent years an increasing number of organizations have required periodic reviews by external specialists. The responsibility of this review is to assess recent experience, review proposed new design and construction, and recommend new approaches and financially attractive alternatives. The Review Board at Syncrude Canada is an example. The Board consists of Dr. Norbert Morgenstern, Fred Matich, Donald Pollock and the author (1999). The group meets twice a year for one week. Syncrude staff present a summary of ongoing progress, recent problems, and programs for the near future for the Board's review. Typically eight to ten formal questions are directed to the Board for comment. Generally one day is provided for the Board to prepare a written report. This report is presented to Syncrude during a final meeting with management. The Board completes its report before leaving the site.

The Review Board has saved Syncrude tens of millions of dollars and no critical geotechnical failures have occurred during the 26 year tenure of the Board. Such reviews are strongly recommended on major engineering projects.

In summary, a number of experiences related to geotechnical and consulting engineering have been outlined. This experience has been challenging, educational, enjoyable and gratifying. I can honestly recommend the geotechnical field to any aspiring engineer.

Thank you for this great honour.

References

1. Brawner, C. O., 1978, "World Engineering," *B.C. Professional Engineer*, January.
2. Staff reporter, 1967, "Soil Cement Stabilizes Lakefront Slope for B. C. Town," *Heavy Construction News*, September 4.
3. Goodman, R. E., 1998, *Karl Terzaghi, The Engineer as Artist*, ASCE Press, Reston, Virginia.
4. Brawner, C. O., 1966, "Slope Stability in Open Pit Mines," *Western Miner*, October, Vancouver.